Nuclear Principles in Engineering

Second edition

T0180120

Tatjana Jevremovic

Nuclear Principles
in Engineering

Second edition

Tatjana Jevremovic
Purdue University
School of Nuclear Engineering
West Lafayette IN 47907-1290
1290, Nuclear Engineering Building
USA
tatjanaj@purdue.edu

ISBN 978-1-4419-4671-3   e-ISBN 978-0-387-85608-7
DOI 10.1007/978-0-387-85608-7

© Springer Science+Business Media, LLC 2010
All rights reserved. This work may not be translated or copied in whole or in part without the written permission of the publisher (Springer Science+Business Media, LLC, 233 Spring Street, New York, NY 10013, USA), except for brief excerpts in connection with reviews or scholarly analysis. Use in connection with any form of information storage and retrieval, electronic adaptation, computer software, or by similar or dissimilar methodology now known or hereafter developed is forbidden.
The use in this publication of trade names, trademarks, service marks, and similar terms, even if they are not identified as such, is not to be taken as an expression of opinion as to whether or not they are subject to proprietary rights.

Printed on acid-free paper

springer.com

This book is dedicated to my mother

Foreword

viii

Nuclear Principles in Engineering is an appropriate starting point for the new series *Smart Energy Systems: Nanowatts to Terawatts*. Not only because the nuclear universe stands at the boundary of human knowledge with respect to scale, but also, and most importantly, because nuclear ideas have a largely untapped potential for new sources of energy. When viewed in this light, nuclear principles offer renewed hope for energy innovation much needed by a global community confronting the inescapable environmental and geological limitations of fossil fuels.

The realm of nuclear processes occupies tiny microscopic dimensions, in the range of 10^{-15} meter or *femtometer*. It is a realm inaccessible by our senses, yet intelligible through the power of modern physics. The book brings the nuclear universe into clear view for the benefit of technical pedagogy and technological development. A plethora of existing technologies can be traced to the fruitful application of nuclear principles, including, but not limited to, weaponry, atomic and nuclear energy, medicine and instrumentation. The number is likely to grow as innovations are needed in smart materials, nanostructures, space, homeland security and biomedical engineering.

In recent years few books have appeared articulating nuclear principles for engineers. The enthusiasm of the 1950s and 60s (the atomic age) gave way to a much impeded if not diminished interest. But nuclear principles are far from fading hues of past scientific theories. Witnessing a renaissance in applications of nuclear technology, the book is aimed at engineering students who need material in a compact and easily digestible form. Professionals and students of science may benefit as well.

With nuclear principles, energy shares the view that much is yet to be gained from converting tiny specks of matter into useful work. This book appears on the centennial of Einstein's famous formula $E = mc^2$. A century of modern physics and half a century of accrued technical experience with nuclear power strongly support a renewed optimism on the technological potential of nuclear ideas. Professor Jevremovic's book presents principles that have stood the test of time and open new vistas for future energy.

Lefteri H. Tsoukalas, Ph.D.
School of Nuclear Engineering
Purdue University
West Lafayette
February 2005

Preface

In the second edition I have updated some of the materials, improved description of some of the fundamental principles and provided more examples with visualizations for easier understanding of the abstract phenomena. For example, an extensive graphical interpretation of neutron transport angular flux, scalar flux or current is provided and discussed. Some of the chapters now end with brief reviews of the application of the chapters concepts in the real world. For example, description of nuclear resonance fluorescence that gains growing attention in homeland security. There are also some more problems listed at the end of each chapter. Overall there is more material in the second than in the first edition covering a broad range of basic principles in nuclear physics, neutron transport and neutron physics of interest in the undergraduate or graduate education in engineering, physics, health sciences or similar.

I am indebted to students and colleagues who pointed out errors in the first edition. I have followed the comments received and tried to incorporate those I have found suitable for the presented scope of this book. In completing the second edition I would like to thank my students who helped with some of the graphics found in Chapters 2, 5, 6, 7 and 8. They are Shanjie Xiao, Nader Satvat, Yang Xue, Kevin Mueller, John Perry and Manuel Sztejnberg.

Tatjana Jevremovic, Ph.D.
School of Nuclear Engineering
Purdue University
West Lafayette
July 2008

Preface to First Edition

This book is an introduction to nuclear principles with special emphasis on engineering applications. Topics such as neutron physics, nuclear structure and radiation interactions are illustrated through numerous examples that include detailed solutions and links to theory. The reader will find plenty of descriptive easy-to-grasp models and analogies with rather simplified mathematics. A mathematical formula says little unless we understand the physical context. Hence, priority is given to developing physical intuition rather than mathematical formalism.

Nuclear engineering is a broad discipline that requires knowledge (of reasonable depth) in physics, mathematics and computation. The discipline is grounded in the scientific understanding of the subatomic realm and energy–matter processes that are taking place at the *femtometer* range (10^{-15} meter). Several areas of application are driving a renaissance in nuclear engineering including, but not limited to, new safe nuclear reactor development, a revolution in nuclear medicine, nuclear space propulsion and homeland security.

This book offers background and a basis for technology development in inherently safe reactors, medical imaging and integrated cancer therapies, food technology, radiation shielding, and nuclear space applications. It is intended to be a resource for practicing engineers and a text for university students in science and engineering.

Tatjana Jevremovic, Ph.D.
School of Nuclear Engineering
Purdue University
West Lafayette
February 2005

Acknowledgments for the First Edition

The author wishes to thank

Daniel Mundy and Shaun Clarke for innumerable hours spent editing and preparing all aspects of this book and for their invaluable technical help.

Godfree Gert for his help in editing and creating various technical figures.

James Kallimani for creative illustrations contained herein.

Dr. Xing Wang for his help in the creation of various technical figures.

Kevin Retzke for his AutoCAD expertise.

Mathieu Hursin for generating valuable examples using the AGENT code.

Josh Walter for contributions of his hard-earned data.

Frank Clikeman for sharing his notes on neutron physics and radiation interactions and allowing me to use some of his example problems.

Chris N. Booth for allowing me to adapt some of his example problems for use in this book.

Wael Harb, M.D. for his priceless friendship and collaboration in research.

To my colleagues in the School of Nuclear Engineering at Purdue University for their support: Professors Chan Choi, Thomas Downar, Mamoru Ishii, Martin Lopez de Bertodano, Paul Lykoudis, Sean McDeavitt, Karl Ott, Shripad Revankar, Alvin Solomon, Rusi Taleyarkhan, Lefteri H. Tsoukalas, Karen Vierow.

Contents

Chapter 1
NUCLEAR CONCEPTS
From nano-watts to tera-watts

The dreams of ancient and modern man are written in the same language as the myths whose authors lived in the dawn of history. Symbolic language is a language in which inner experience, feelings and thoughts are expressed as if they were sensory experiences, events in the outer world. It is a language which has a different logic from the conventional one we speak in the daytime, a logic in which time and space are not the ruling categories but intensity and association. *Erich Fromm* (The Forgotten Language, 1937)

1. INTRODUCTION

Early 20th century marked tremendous and fascinating discoveries in physics and chemistry. For the first time in human history hard evidence was produced supporting the existence of atoms. In his book *Imagined Worlds*, the eminent astrophysicist Freeman Dyson calls the changes in physics that occurred in the 1920s, a *concept-driven revolution*; theory had primacy over experiment. Quantum mechanics and the theory of relativity explained atomic and nuclear structures. Yet, the technology for accessing the atomic and subatomic levels remained rather primitive. It was not until decades later that serious technological applications appeared.

The middle of the 20th century marked the advent of nuclear technology. First weaponry, which left a trace of fear and apprehension in the meaning of the word "nuclear". After all, the press release for the new technology became Hiroshima and Nagasaki. Nuclear power for naval and terrestrial applications, nuclear medicine, radiochemistry, imaging and space exploration came later.

T. Jevremovic, *Nuclear Principles in Engineering*,
DOI 10.1007/978-0-387-85608-7_1, © Springer Science+Business Media, LLC 2009

Significant institutional development took place concomitantly to regulate and protect the public and the environment from the deleterious effects of radiation. National authorities such as the Nuclear Regulatory Commission in the United States and international bodies such as the UN International Atomic Energy Agency as well as trade and professional organizations were formed. Nuclear technology cannot be developed and deployed without serious technical and institutional safeguards.

2.   TERRESTRIAL NUCLEAR ENERGY

A major requirement for sustaining human progress is to adequately provide, generate and distribute energy. In the last 50 years we have seen nuclear energy grow to become an important source of carbon-free electricity. Concerns about global climate change and energy supply/demand imbalances bring renewed attention to nuclear energy. The unparalleled safety record of light water reactors (LWR) and the high capacity factors achieved by nuclear generators give plenty of motivation for new nuclear power expansion. Whereas in the 1990s nuclear power plants were considered expensive dinosaurs, there is a growing worldwide interest in new generation with US utilities clamoring for permission to build new plants. There is every indication that the successes of LWRs, global warming and growing worldwide energy challenges generate a unique confluence of reasons for a serious reexamination of the nuclear option.

Nuclear power comes mainly from the fission of uranium, plutonium or thorium. The fission of an atom of uranium produces several million times more energy than the energy produced by the combustion of an atom of carbon in fossil fuels, giving nuclear power an extraordinary advantage in power density. Energy released in fission is converted into electric energy (this type of electricity represents 80% of the electricity generated in France and over 22% in the United States). More than 400 nuclear power plants produce over 15% of the world's electricity. Having accumulated over 12,000 years of operational experience with civilian nuclear power, mankind is becoming more confident about the economic, safety and environmental benefits of nuclear power generation.

For the vast majority of nuclear reactors the fuel is slightly enriched uranium, material which is relatively abundant and ubiquitous. Nuclear power plants typically use enriched uranium in which the concentration of ^{235}U is increased from 0.7% (as found in nature) to about 4–5%. At present, global reserves of uranium are deemed sufficient for at least 100 years. In the very long term, however, breeder reactors are expected to be used to breed new fuel. Breeder reactors can generate nearly 100 times as much fuel as they consume.

One of the main issues with nuclear power is the problem of nuclear waste. Significant technical progress has been made in this area and a number of countries, including the United States, move toward addressing the political aspects of the problem. It is important to note that nuclear power takes full responsibility for its waste. Radioactive waste coming from nuclear power reactors is small in quantity and could be turned into useful nuclear fuel with known chemical processes.

The nuclear industry is developing and upgrading reactor technologies for nearly 50 years. Future reactor designs focus on safety, economics and proliferation-resistant fuel cycles. Great attention is paid to fuel improvements targeting, for example, the capability of light water reactors to burn plutonium, hence reducing the amount of radioactive waste.

3. SPACE EXPLORATION AND NUCLEAR POWER

Radioisotope generators in space have been providing electrical power for a variety of spacecrafts. For example, Cassini, the first craft ever to orbit Saturn, is powered by a radioisotope thermoelectric generator (RTG). After 6 years of travel to the Saturn rings, Cassini reached its destination in 2004 and is scheduled to remain in orbit until 2008. RTGs are a proven technology for missions to distant space destinations. They consist of a radioisotope (for example, ^{238}Pu, a non-weapon-grade material, because of its long half-life ~ 87 years) and a thermoelectric conversion system. Heat produced from the radioisotope is converted directly to electricity using thermocouples. For example, Cassini is powered by three RTGs (with nearly 33 kg of plutonium) that produce 750 W of power. The power generated diminishes somewhat with time due to the exponential decline of radioactivity. At the end of the 11th year of operation the Cassini system will produce close to 630 W of power. The development of such systems by the US Department of Energy generated astonishing success for missions to Moon, Neptune and even beyond the Solar System. Famous spacecrafts such as Pioneer 10 and 11, Apollo, Galileo and Voyager were powered by RTGs. Thus far, 44 RTGs have powered 24 US space vehicles. Russia has also developed RTGs using ^{210}Po. There are currently two Russian generators in orbit powering satellites. RTGs using short-lived radionuclides can power small devices deployed in remote areas on Earth or other planets. Such systems could stay intact and power instrumentation for collecting data that include climate variables, chemical composition of air or soil, salinity, ozone and temperature. After a gap of several decades, there is new interest in the

United States and Russia for nuclear power in space missions. In 2002, NASA announced the Nuclear Systems Initiative for space code-named *Project Prometheus*. It focuses at space mission design enabling nuclear-powered manned missions to Mars and distant planet exploration.

The Jupiter Icy Moon Orbiter (JIMO) is a spacecraft currently in development, powered by a nuclear reactor to explore Jupiter's dark and cold satellites. A major limiting factor for long-term space travel or manned mission to distant planets is radiation protection for the crew and the electronics. Nuclear principles will be used for the design of light but effective radiation shield.

4.   MEDICINE AND NUCLEAR PRINCIPLES

Soon after the German physicist Wilhelm Conrad Roentgen discovered them, X-rays revolutionized medicine. A century later advanced three-dimensional imaging, computerized treatment planning and high-energy X-ray machines have revolutionized the diagnostics and treatment of heart disease, cancer and surgery. A remarkable application of nuclear principles has been the use of gamma ray narrow beams to irradiate small tumors with high precision, an instrument called the *gamma knife*.

In 1932 Chadwick discovered an electrically neutral constituent of the nucleus which he called the "neutron". Few years later it was recognized that neutron interactions producing short-range highly ionizing particles could be used to treat cancer. In the early 1940s, neutron capture therapy (NCT) was proposed. This is a bimodal radiotherapy utilizing directed uptake of neutron-absorbing isotopes in tumor tissue and subsequent neutron irradiation. Neutron-interaction products deposit most of the energy from highly exothermic capture reactions in relatively small space. This is in the order of cellular dimensions thus delivering to tumor cells a far greater dose than what is incurred in surrounding healthy tissue. NCT has a great advantage particularly if the tumor is not imagable or difficult to spatially define. It has been applied clinically as a post-operative sterilization of potentially remnant brain tumors. The most prominent element used in NCT is ^{10}B, which undergoes a neutron interaction producing alpha particle and ^{7}Li. The potential for other elements has been also studied. Gadolinium, lithium and uranium can strongly absorb thermal neutrons and hence they are considered for NCT. The products of neutron capture in ^{157}Gd, for example, are quite different than neutron capture with other isotopes creating a mixture of prompt and cascade-induced photons and electrons. A novel application of the nuclear principles upon which NCT is based is application to breast and lung cancer. Recent literature points to novel ways of combining the NCT principles with the identified genomic signature of specific cancers. For example, it has been shown that 30% of breast cancers overexpress certain proteins, a fact that can be exploited for custom-made treatments. Monoclonal antibodies (MABs) are currently used as part of chemotherapy for metastatic and late-stage breast cancer. A recent study explores the possibility and effectiveness of using the MABs as a targeting vehicle for boron to breast cancer cells. This approach is called targeted (radiation) therapy. In such therapies the radiation or drug agent is brought directly to the cancer cells. This radioimmunotherapy combines radionuclides with MABs to deliver radiation to designated areas where it produces high irradiation effects.

A startling new picture of how cells respond to radiation is beginning to emerge from microbeam studies in which individual cells are targeted with a precise dose of radiation. Cells damaged by radiation communicate with neighboring cells using messenger molecules (cytokines) that can be transmitted between the cells. As a result, cells not hit by radiation, called *bystander cells*, generate molecular and cellular responses similar to cells that are irradiated. Study of bystander cell effects will have profound implications in planning for radiation therapy and also for the assessment of health risks of low radiation doses. On the other hand, the precise and non-invasive nature of microbeams is useful in radiobiology, cell and biomolecular diagnostics and intracellular micromanipulations. For example, biological tissues are mostly transparent to photon radiation giving the unique possibility to act on cell structure without changing the features or disturbing the vital functions. Recent advances in tissue and molecular engineering call for new technologies to analyze and possibly modify cell and tissue behavior while minimizing undesirable signaling (contamination) in the broader cellular environment.

Neutrons offer powerful tools for the investigation of macromolecular structures, such as the structure of proteins, membranes, polymers and other complex biological materials. The use of cold neutrons rather than thermal neutrons improves the detection limits of miniscule amounts of light elements such as hydrogen. They are widely used as a *microscopic probe* in fields ranging from elementary physics to biological science. Cold neutrons are finding a fabulous application in depth profiling of light element spatial deposition; for example, mapping the spatial distribution of boron atoms in a tumor region (thus providing information that may profoundly advance the BNCT). Cold neutrons are of great interest since they are non-invasive and a sample can be reused for other profiling tests by different techniques.

5.   NUCLEAR PRINCIPLES FOR HOMELAND SECURITY

"The National Strategy for Homeland Security and the Homeland Security Act of 2002 served to mobilize and organize our nation to secure the homeland from terrorist attacks. This exceedingly complex mission requires a focused effort from our entire society if we are to be successful. To this end, one primary reason for the establishment of the Department of Homeland Security was to provide the unifying core for the vast national network of organizations and institutions involved in efforts to secure our nation." [from the Homeland Security strategic plan as of 2008]. It requires the innovative and sturdy scientific and technological research and development on a variety of threats including but not limited to agricultural, chemical, biological, nuclear and radiological, explosive and cyber. One of the highly prioritized strategically identified treats is the potential of smuggling nuclear materials into the country through the cargo shipping. Many novel approaches have been developed and are emerging in recent years to find the accurate and fast technique for detection of hidden nuclear materials.

The necessity of detecting nuclear materials convertible to weapons (such as U^{233}, U^{235} and Pu^{239}) was recognized as early as the Manhattan Project. When Dr. J. Robert Oppenheimer was asked by a Congressional Committee how to detect nuclear materials hidden in baggage, he had a simple answer *"use a screwdriver"*. Oppenheimer referred to the fact that well-shielded fissionable materials are almost impossible to detect due to their weak nuclear signatures. More recently, neutron and gamma interrogation techniques with nuclear resonance fluorescence have provided very exciting opportunities for reliable, non-invasive detection. In neutron interrogation technique an external neutron source prompts the fission reaction in the nuclear material if present in the inspected cargo. Either delayed neutrons or gamma rays are monitored to indicate the presence of a fissile material. Due to low yield as well as penetration characteristics of the emitted delayed neutrons (easy scattering and absorption of low-energy neutrons), the high-energy gamma rays are generally monitored as the primary signatures. However, the produced signals are too complex to easily identify the isotope-specific signature. Gamma interrogation methods are based on the well-established physics of nuclear resonance fluorescence (NRF) [basic principles are described in Chapter 3]. Nuclei, like atoms, have discrete and characteristic quantum energy states. In interaction with the high-energy photons (>2 MeV), a nucleus is excited to a higher energy state due to the absorption of the energy (photon). After a very short period of time the excited nucleus rearranges by releasing a gamma ray and thus returning to a ground state. The emitted gamma ray represents a specific signature for that nucleus and is used to identify its presence. The NRF gamma interrogation imposes great challenges on gamma sources and requires high-resolution detectors. Recently the muon radiography has attracted great attention in detecting the nuclear materials. Muons are the elementary particles produced by the cosmic radiation (mostly made of energetic protons) striking the Earth's atmosphere; there are around 10,000 muons hitting every square meter of the Earth's surface. Muons are largely deflected by the high-Z materials (such as uranium or plutonium, but also the shielding materials such as lead, steel or aluminum) and slightly deflected by the low-Z materials (such as plastics or glass). In addition, the muons can penetrate very large thicknesses of the high-Z materials. The detection system envisions two sets of the detectors, one placed above the cargo or the vehicle and another set below the cargo or the vehicle. The deflection of the incoming muons is registered indicating the presence or the absence of high-Z and nuclear materials.

6.   BOOK CONTENT

This book offers an overview of basic nuclear principles in engineering including but not limited to physical processes in radiation interaction with matter, neutron transport and reactor physics, nuclear and atomic structure and radioactive decay. The understanding of principles is essential in developing engineering applications. For example, prediction of nuclear parameters in reactors and accurate radiation treatment in medicine are both based on principles of radiation interactions with matter. They share the same tools for predicting energy deposition along different pathways.

The book material is organized as follows:
− Atomic structure principles are described in Chapter 2. This knowledge is important in analyzing the probabilities of interaction leading to ionization of a medium (of extreme importance in biological tissues) and in understanding the energy levels and electronic configuration of atoms.
− The majority of nuclear interactions involve electron clouds or nuclei of a medium. For example, in a reactor we find interactions of neutron with nuclear fuel and structure materials. Understanding these interactions is of great importance in predicting reactor power, achieving reactor control and selecting fuel characteristics. In order to predict such parameters with high accuracy, knowledge of the nature of particle interactions and the structure of nucleus is of great importance and is described in Chapter 3.

– A brief overview of quantum mechanics starting from the concept of Planck's quanta and the de Broglie wavelength through Heisenberg principle and Schrödinger equation is provided in Chapter 4.
– Radioactivity, a phenomenon discovered at the end of 19th century, has found applications in many scientific and engineering approaches (radioactive dating, radioisotope generators, nuclear medicine) and is described in detail in Chapter 5.
– The interaction of various particles with matter is described in Chapter 6. The concept of stopping power, range of interactions and the attenuation of radiation beam are essential aspects in particle transport and applications of radiation effects.
– Chapter 7 focuses on description and analysis of the cross sections for neutron interactions; the nature of neutron interactions; and basic principles of the fission process.
– Reactor steady-state and kinetic physics is described in Chapter 8. The basic principles of neutron diffusion theory, reactor power, fission chain reaction, critical mass, spatial distribution of neutron flux and reaction rates are described with details needed to pursue analysis of reactor behavior giving a solid background for understanding time-dependent physics parameters of thermal reactors.
– Aspects of reactor control, the effects of neutron poisoning as well as temperature coefficients of reactivity are summarized in Chapter 9.

Chapter 2

ATOMIC THEORY
Basic Principles, Evidence and Examples

> Among all physical constants there are two which will be universally admitted to be
> of predominant importance; the one is the velocity of light, which now appears in
> many of the fundamental equations of theoretical physics, and the other is the
> ultimate, or elementary, electrical charge, *Robert Millikan* (1868–1953)

1.   INTRODUCTION

Around the 5th century BC, Greek philosopher Democritus invented the concept of the *atom* (from Greek meaning "indivisible"). The atom, *eternal, constant, invisible and indivisible*, represented the smallest unit and the building block of all matter. Democritus suggested that the varieties of matter and changes in the universe arise from different relations between these most basic constituents. He illustrated the concept of atom by arguing that every piece of matter could be cut to an end until the last constituent is reached. Today the word *atom* is used to identify the basic component of molecules that create all matter, but it is known that the atom itself is made of particles even more fundamental, some of which are elementary. The first theoretical and experimental models of the structure of matter came as late as the 19th century, which is the time marked as the beginning of modern science. At that time a more empirical approach, mainly in chemistry, opened a new era of scientific investigations.

The work of Democritus remained known through the ages in writings of other philosophers, mainly Aristotle. Modern Greece has honored Democritus as a philosopher and the originator of the concept of the atoms

T. Jevremovic, *Nuclear Principles in Engineering*,
DOI 10.1007/978-0-387-85608-7_2, © Springer Science+Business Media, LLC 2009

through their currency; the 10-drachma coin, before Greek currency was replaced with the euro, depicted the face of Democritus on one side, and the schematic of a lithium atom on the other.

This chapter introduces the structure of atoms and describes atomic models that show the evidence for the existence of atoms and electrons.

2.   ATOMIC MODELS

2.1   The Cannonball Atomic Model

All matter on Earth is made from a combination of 90 naturally occurring different atoms. Early in the 19th century, scientists began to study the decomposition of materials and noted that some substances could not be broken down past a certain point (for instance, once separated into oxygen and hydrogen, water cannot be broken down any further). These primary substances are called *chemical elements*. By the end of the 19th century it was implicit that matter can exist in the form of a pure element, chemical compound of two or more elements or as a mixture of such compounds. Almost 80 elements were known at that time and a series of experiments provided confirmation that these elements were composed of atoms. This led to a discovery of the *law of definite proportions*: two elements, when combined to create a pure chemical compound, always combine in fixed ratios by weight. For example, if element A combines with element B, the unification creates a compound AB. Since the weight of A is constant and the weight of B is constant, the weight ratio of these two will always be the same. This also implies that two elements will only combine in the defined proportion; adding an extra quantity of one of the elements will not produce more of the compound.

Example 2.1: The law of definite proportion
Carbon (C) forms two compounds when reacting with oxygen (O): carbon monoxide (CO) and carbon dioxide (CO_2).

1g of C + 4/3 g of O $\rightarrow$ 2 1/3 g of CO; 1 g of C + 8/3 g of O $\rightarrow$ 3 2/3 g of CO_2

The two compounds are formed by the combination of a definite number of carbon atoms with a definite number of oxygen atoms. The ratio of these two elements is constant for each of the compounds (molecules): C:O = 3:4 for CO and C:O = 3:8 for CO_2.

The first atomic theory with empirical proofs for the law of definite proportion was developed in 1803 by the English chemist John Dalton (1766–1844). Dalton conducted a number of experiments on gases and liquids and concluded that, in chemical reactions, the amount of the elements combining to form a compound is always in the same proportion. He showed that matter is composed of atoms and that atoms have their own distinct weight. Although some explanations in Dalton's original atomic theory are incorrect, his concept that chemical reactions can be explained by the union and separation of atoms (which have characteristic properties) represents the foundations of modern atomic physics. In his two-volume book, *New System of Chemical Philosophy,* Dalton suggested a way to explain the new experimental chemistry. His atomic model described how all elements were composed of indivisible particles which he called atoms (he depicted atoms like cannonballs, Fig.2-1) and that all atoms of a given element were exactly alike. This explained the law of definite proportions. Dalton further explained that different elements have different atoms and that compounds were formed by joining the atoms of two or more elements.

Figure 2-1 Cannonball atomic model (John Dalton, 1803)

In 1811, Amadeo Avogadro, conte di Quaregna e Ceretto (1776–1856), postulated that equal volumes of gases at the same temperature and pressure contain the same number of molecules. Sadly, his hypothesis was not proven until 2 years after his death at the first international conference on chemistry held in Germany in 1860 where his colleague, Stanislao Cannizzaro, showed the system of atomic and molecular weights based on Avogadro's postulates.

Example 2.2: Avogadro's law

As shown in Example 2.1, the ratio of carbon and oxygen in forming CO_2 is 3:8. Here is the explanation of this ratio: since a single atom of carbon has the same mass as 12 hydrogen atoms, and two oxygen atoms have the same mass as 32 hydrogen atoms, the ratio of the masses is 12:32 = 3:8. This shows that the description of the reaction is independent of the units used since it is the ratio of the masses that determines the outcome of a chemical reaction. Thus, whenever you see wood burning in a fire, you should know that for every atom of carbon from the wood, two oxygen atoms from the air are combined to form CO_2; the ratio of masses is always 12:32.

It follows that there must be as many carbon atoms in 12 g of carbon as there are oxygen atoms in 16 g of oxygen. This measure of the number of atoms is called a *mole*. The mole is used as a convenient measure of an amount of matter, similarly as "a dozen" is a convenient measure of 12 objects of any kind. Thus, the number of atoms (or molecules) in a mole of any substance is the same. This number is called Avogadro's number (N_A) and its value was accurately measured in the 20th century as 6.02×10^{23} atoms or molecules per mole.

For example, the number of moles of hydrogen atoms in a sample that contains 3.02×10^{21} hydrogen atoms is

$$\text{Moles of H atoms} = \frac{3.02 \times 10^{21} \text{ atoms H}}{6.02 \times 10^{23} \text{ atoms/mole}} = 5.01 \times 10^{-3} \text{ moles H}$$

2.2   The Plum Pudding Atomic Model

Shortly before the end of the 19th century, a series of new experiments and discoveries opened the way for new developments in atomic and subatomic (nuclear) physics. In November 1895, Wilhelm Roentgen (1845–1923) discovered a new type of radiation called *X-rays*, and their ability to penetrate highly dense materials. Soon after the discovery of X-rays, Henri Becquerel (1852–1908) showed that certain materials emit similar rays independent of any external force. Such emission of radiation became known as *radioactivity* (described in all details in Chapter 5).

During this same time period, scientists were extensively studying a phenomenon called *cathode rays*. Cathode rays are produced between two plates (a cathode and an anode) in a glass tube filled with the very low-density gas when an electrical current is passed from the cathode to the high-voltage anode. Because the glowing discharge forms around the cathode and then extends toward the anode, it was thought that the rays were coming out of the cathode. The real nature of cathode rays was not understood until 1897 when Sir Joseph John Thomson (1856–1940) performed experiments that led to the discovery of the first subatomic particle, the *electron*. The most important aspect of his discovery is that the cathode *rays* are the *stream of particles*. Here is the explanation of his postulate: from the experiment he observed that cathode rays were always deflected by an electric field from the negatively charged plate inside the cathode ray tube, which led him to conclude that the rays carried a *negative electric charge*. He was able to determine the speed of these particles and obtain a value that was a fraction of the speed of light (one tenth the speed of light, or roughly 30,000 km/s or 18,000 mi/s). He postulated that anything that carries a charge must be of material origin and composed of particles. In his experiment, Thomson was able to measure the charge-to-mass ratio, *e/m*, of the cathode rays; a property that was found to be constant regardless of the materials used. This ratio was also known for atoms from electrochemical analysis, and by comparing the values obtained for the electrons he could conclude that the electron was a very small particle, approximately 1,000 times smaller than the smallest atom (hydrogen). The electron was the first subatomic particle identified and the fastest small piece of matter known at that time.

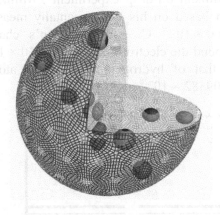

Figure 2-2. Plum pudding atomic model (J. J. Thomson, 1904)

In 1904, Thomson developed an atomic model to explain how the negative charge (electrons) and positive charge (speculated to exist since it was known that the atoms were electrically neutral) were distributed in an atom. He concluded that the atom was a sphere of positively charged material with electrons spread equally throughout like raisins in a plum pudding. Hence, his model is referred to as the *plum pudding model*, or *raisin bun atom* as depicted in Fig. 2-2. This model could explain

- The neutrality of atoms
- The origin of electrons
- The origin of the chemical properties of elements
  However, his model could not answer questions regarding
- Spectral lines (according to this model, radiation emitted should be monochromatic; however, experiments with hydrogen shows a series of lines falling into different parts of the electromagnetic spectrum)
- Radioactivity (nature of emitted rays and their origin in the atom)
- Scattering of charged particles by atoms

Thomson won the Nobel Prize in 1906 for his discovery of the electron. He worked in the famous Cavendish Laboratory in Cambridge and was one of the most influential scientists of his time. Seven of his students and collaborators won Nobel Prizes, among them his son who, interestingly, won the Nobel Prize for proving the electron is a wave (Chapter 4).

2.3   Millikan's Experiment

In 1909 Robert Millikan (1868–1953) developed an experiment at the University of Chicago to measure the charge of the electron. The experiment is known as the "Millikan oil-drop experiment". Millikan determined the mass of the electron based on his experimentally measured value of the electron charge, 1.60×10^{-19} C, and Thomson's charge-to-mass ratio, 1.76×10^{8} C/g. He found the electron mass to be 9.10×10^{-28} g (about 2000 times smaller than that of hydrogen, the lightest atom); the presently accepted value is 9.109382×10^{-28} g.

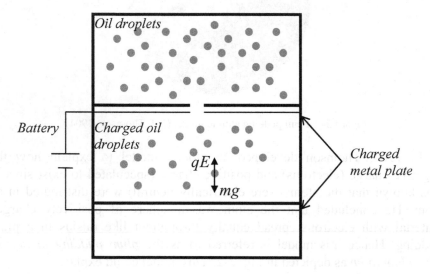

Figure 2-3. Schematics of Millikan's oil-drop experiment (1909)

How was the charge of an electron measured from the oil drops? Millikan's experimental apparatus consisted of a chamber with two metal plates connected to a voltage source where the oil droplets were allowed to fall between (Fig.2-3). In the absence of a voltage (electrical field, E, equal to zero) droplets were allowed to fall until they reached their terminal velocity (when the downward force of gravity, mg, is balanced with the upward force of air resistance). By measuring the terminal velocity he was able to determine the mass of the oil droplets. By introducing an electrical field, the forces (gravitational and electrical) could be balanced and the drops would be suspended in mid-air. The resulting force is zero, because the gravitational force is equal to the electrical force:

$$mg = qE \tag{2-1}$$

where m is the mass of the oil drop, g is the acceleration of gravity and the total charge of the oil droplet, $q = N{\cdot}e$, is an integer times the charge of one electron (because the electron cannot be divided to produce a fractional charge). By changing the electric charge of oil droplets (irradiating them with X-rays known at that time to ionize the molecules), Millikan found that the charge was always a multiple of the same number. In 1913 he published the value of -1.5924×10^{-19} C. The accepted value today is $-1.60217653 \times 10^{-19}$ C (see also Appendix 2). Robert Millikan was awarded the Nobel Prize in 1923 for this work.

2.4 The Planetary Atomic Model

2.4.1 Disproof of Thomson's Plum Pudding Atomic Model

Thomson's atomic model described the atom as a relatively large, positively charged, amorphous mass of a spherical shape with negatively charged electrons homogeneously distributed throughout the volume of a sphere, the sizes of which were known to be on the order of an Ångström ($1 \text{ Å} = 10^{-8}$ cm $= 10^{-10}$ m). In 1911 Geiger and Marsden carried out a number of experiments under the direction of Ernest Rutherford (1871–1937) who received the Nobel Prize in chemistry in 1908 for investigating and classifying radioactivity. He actually did his most important work after he received the Nobel Prize and the 1911 experiment unlocked the hidden nature of the atom structure.

Rutherford placed a naturally radioactive source (such as radium) inside a lead block as shown in Fig. 2-4. The source produced α particles which were collimated into a beam and directed toward a thin gold foil. Rutherford hypothesized that if Thomson's model was correct then the stream of α particles would pass straight through the foil with only a few being slightly deflected as illustrated in Fig. 2-5. The "pass through" the atom volume was expected because the Thomson model postulated a rather uniform distribution of positive and negative charges throughout the atom. The *deflections* would occur when the positively charged α particles came very close to the individual electrons or the regions of positive charges. As expected, most of the α particles went through the gold foil with almost no deflection. However, some of them rebounded almost directly backward – a phenomenon that was not expected (Fig. 2-6). The main challenge was to explain what caused such a large deflection angle and what caused other particles to go through the atom without noticeable scattering.

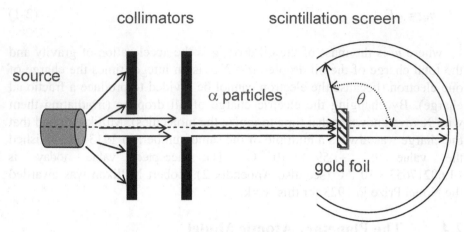

Figure 2-4. Schematics of Rutherford's experiment (1911)

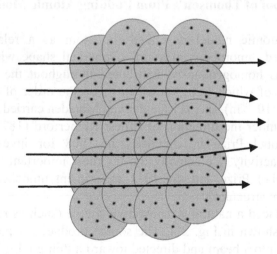

Figure 2-5. Expected scattering of α particles in Rutherford's experiment

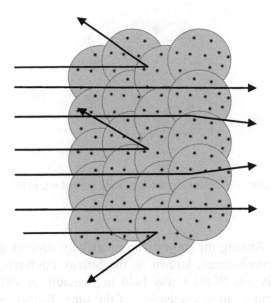

Figure 2-6. Actual scattering of α particles in Rutherford's experiment

Rutherford explained that most of the α particles pass through the gold foil with little or no divergence not because the atom is a uniform mixture of the positive and negative charges but because the atom is largely empty space and there is nothing to interact with the α particles. He explained the large scattering angle by suggesting that some of the particles occasionally collide with, or come very close to, the "massive" *positively charged nucleus* that is located at the center of an atom. It was known at the time that the gold nucleus had a positive charge of 79 units and a mass of about 197 units while the α particle had a positive charge of 2 units and a mass of 4 units. The repulsive force between the α particle and the gold nucleus is proportional to the product of their charges and inversely proportional to the square of the distance between them. In a direct collision, the massive gold nucleus would hardly be moved by the α particle. The diameter of the nucleus was shown to be ~1/105 the size of the atom itself, or ~10^{-13} m. Clearly these ideas defined an atom very different from the Thomson's model.

Ernest Solvay (1838–1922), a Belgian industrial chemist, who made a fortune from the development of a new process to make washing soda (1863), was known for his generous financial support to science, especially

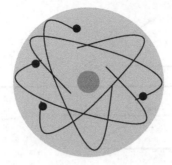

Figure 2-7. Planetary atomic model (Rutherford, 1911)

physics research. Among the projects he financially supported, was a series of international conferences, known as the Solvay conferences. The *First Solvay Conference on Physics* was held in Brussels in 1911 and it was attended by the most famous scientists of the time. Rutherford was one of them; he presented the discovery of the atomic nucleus and explained the structure of the atom. According to his explanation, the electrons revolve around the nucleus at relatively large distances. Since each electron carries one elementary charge of negative electricity, the number of electrons must equal the number of elementary charges of positive electricity carried by the nucleus for the atom to be electrically neutral. The visual model is similar to the solar planetary system and is illustrated in Fig. 2-7.

2.4.2   Idea of a Nucleus in the Center of an Atom

Rutherford's scattering experiment showed that a positive charge distributed throughout the volume of Thomson's atom could not deflect the α particles by more than a small fraction of a degree. A central assumption of Thomson's atomic model was that both the positive charge and the mass of the atom were distributed nearly uniformly over its volume. The electric field from this charge distribution is the field that must scatter the α particles, since the light-weight electrons would have a negligible impact. The expected deflection of an α particle from the gold nucleus according to Thomson's atomic model is shown in Fig. 2-8. The thickness of the gold foil used by Rutherford was about 400 atoms (or $\sim 5 \times 10^{-7}$ m). The gold atom has a positive charge of $79e$ (balanced by 79 electrons in its neutral state).

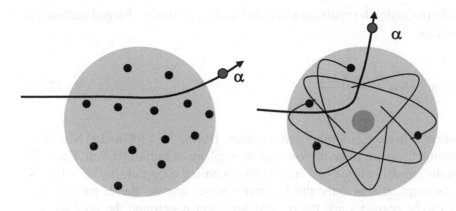

Plum pudding atomic model   Planetary atomic model

Figure 2-8. Trajectory of the α particle in the electric field of an atom in Rutherford's experiment according to the plum pudding and planetary atomic model

Neglecting the electrons, the maximum electric force the α particle would encounter is that at the surface of the positively charged sphere.

Johannes Kepler was the first to mathematically formulate Tycho Brahe's precise measurements of the motion of planets, showing that the orbit of the planets around the Sun is elliptical. Newton later proved that these elliptical orbits are a consequence of the attractive gravitational force (GmM/r^2). He also established that the motion of heavenly bodies in the field of a central attractive force with a $\sim 1/r^2$ dependence (such as the gravitational field of the Sun) is always a conical section, depending on the initial conditions: a *hyperbola* (body has sufficient kinetic energy to avoid capture by the gravitational field), an *ellipse* (the body is captured) and a *parabola* (a limiting case between these two). The scattering of particles in the electric field follows the same law that describes the motion of bodies in a gravitational field, except that the force can be both attractive and repulsive (the latter being the case for α particles and a positively charged nucleus). These two forces, electric and gravitational, are generated according to modern quantum physics by the exchange of a massless particle (or field quantum). In the case of the electric force the field quantum is a photon and in the case of the gravitational force the field quantum is called a graviton.

If the mass of an α particle is m with charge $q = 2e$, and the charge of the gold foil nucleus is $Q = Ze = 79e$, then the electric force acting on the α particle (a Coulomb repulsion force due to the positively charged nucleus) is written as

$$F = \frac{kQq}{r^2} = \frac{k(79e)(2e)}{r^2} \tag{2-2}$$

where k is the Coulomb force constant, $1/(4\pi\varepsilon_0) = 8.9876\times10^9$ N m^2/C ^{2}. Assuming the atom to be represented by a sphere of radius 10^{-10} m, Eq. (2-2) gives the repulsive force that acts on the incoming α particle as 3.64×10^{-6} N. The assumption that only the Coulomb force acts on the α particle was shown to be correct since the α particles never penetrated the gold nucleus and Rutherford's theoretical explanation agreed with the experimental measurements for all cases. Due to the nature of the Coulomb force acting on the α particle (inverse square law), the α particle follows a hyperbolic trajectory (Fig. 2-8) that is characterized by the *impact parameter, b*. The impact parameter represents the distance from the nucleus perpendicular to the line of approach of the incident α particle. The angle of deflection, θ, of any α particle is related to the impact parameter through the following relation (see Example 2.4)

$$b = \frac{k(Ze)(ze)}{2E\tan(\theta/2)} = \frac{k(79e)(2e)}{2E\tan(\theta/2)} \equiv \frac{(79e)(2e)}{8\pi\varepsilon_0 E}\cot(\theta/2) \tag{2-3}$$

where E denotes the kinetic energy of the incident α particle $(= mv^2/2)$. It follows from Eq. (2-3) that the impact parameter is smaller for larger scattering angles and larger energy of the incident particle. From the distribution of the α particle's scattering angles, Rutherford has concluded that the structure of an atom most likely mimics the solar planetary system. The size of the nucleus at the center of the atom was estimated based on the kinetic energy of the incident α particle and its potential energy at the point of the closest approach, d. The closest approach occurs in the case of a head-on collision in which the α particle comes to a rest before it bounces back at an angle of 180° (Fig. 2-9). At that point the kinetic energy is zero, and the potential energy equals the initial kinetic energy

$$E = \frac{k(79e)(2e)}{d} \tag{2-4}$$

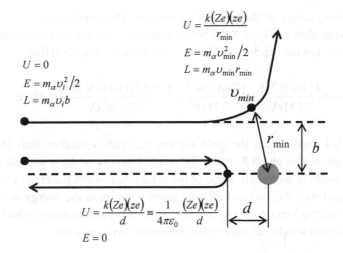

$$U = \frac{k(Ze)(ze)}{r_{min}}$$

$$E = m_\alpha v_{min}^2/2$$

$$L = m_\alpha v_{min} r_{min}$$

$U = 0$

$E = m_\alpha v_i^2/2$

$L = m_\alpha v_i b$

v_{min}

r_{min}

b

$$U = \frac{k(Ze)(ze)}{d} \equiv \frac{1}{4\pi\varepsilon_0}\frac{(Ze)(ze)}{d}$$

d

$E = 0$

Closest approach to nucleus

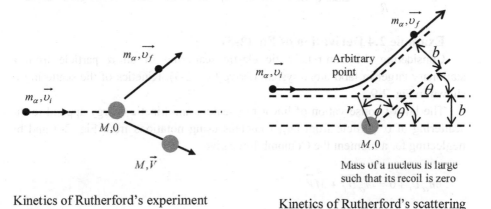

$m_\alpha, \vec{v}_f$

$m_\alpha, \vec{v}_i$

$M, 0$

$M, \vec{V}$

Kinetics of Rutherford's experiment

$m_\alpha, \vec{v}_f$

$m_\alpha, \vec{v}_i$

Arbitrary point

b

θ

r

ϕ

θ

b

$M, 0$

Mass of a nucleus is large such that its recoil is zero

Kinetics of Rutherford's scattering

Figure 2-9. Deflection of α particle (by angle θ) by the target nucleus (of radius R)

Knowing the kinetic energy of the incident α particle, its closest approach to any nucleus in the gold foil (on the order of 10^{-14} m) and the approximate size of the gold nucleus (on the order of 10^{-15} m) may be determined. The unit of 10^{-15} m is designated as a Fermi, fm. The small volume of the nucleus implies its high density and the need for a strong attractive force in the nucleus to overcome the Coulomb repulsive force. It was also understood that this attraction must be of a very short range.

Example 2.3 Size of the gold nucleus in Rutherford's experiment

In the Rutherford's experiment the kinetic energy of incident α particle was 7.7 MeV. Estimate the upper limit size of the gold nucleus and comment on the effect of increased energy of the incident α particles in the experiment.

According to Fig. 2-8 the point of closest approach will determine the size of the nucleus. For the head-on collision it follows from Eq. (2-4) that

$$d < \frac{k(79e)(2e)}{7.7 \text{ MeV}} = \frac{(79)(2)ke^2}{7.7 \text{ MeV}} = \frac{(79)(2)(1.44 \text{ eV nm})}{7.7 \times 10^6 \text{ eV}} \approx 30 \text{ fm}$$

This implies that the gold nucleus has radius smaller than 30 fm (the actual measurement is about 8 fm). If the incident energy of the α particles in Rutherford's experiment is increased, some of the α particles would penetrate the nucleus, first in the head-on collisions and then for smaller angles as the energy is further increased. The limiting kinetic energy for the incident α particle above which the Rutherford experiment would not agree with theoretical explanation is

$$E \approx \frac{(79)(2)ke^2}{R} \cong 28.5 \text{ MeV}$$ where R represents the radius of the gold nucleus.

Example 2.4 Derivation of Eq. (2-3)

Considering the non-relativistic elastic scattering of an α particle from a stationary target in laboratory system derive Eq. (2-3). Kinetics of the scattering is shown in Fig. 2-9.

The law of conservation of linear momentum and kinetic energy applied to the scattering of α particle from target nucleus using notations from Fig. 2-9 and by neglecting for a moment the Coulomb force give

$$m_\alpha \vec{\upsilon}_i + 0 = m_\alpha \vec{\upsilon}_f + M \vec{V}$$

$$\frac{m_\alpha \upsilon_i^2}{2} + 0 = \frac{m_\alpha \upsilon_f^2}{2} + \frac{M V^2}{2}$$

where $\upsilon_i = |\vec{\upsilon}_i|$, $\upsilon_f = |\vec{\upsilon}_f|$ and $V = |\vec{V}|$. Squaring the first equation and using the second equation we obtain the following:

$$m_\alpha^2 \upsilon_i^2 = m_\alpha^2 \upsilon_f^2 + M^2 V^2 + 2m_\alpha M(\vec{\upsilon}_i \cdot \vec{V}) \quad \text{with} \quad m_\alpha^2 \upsilon_i^2 = m_\alpha^2 \upsilon_f^2 + m_\alpha M V^2$$

$$m_\alpha M V^2 = M^2 V^2 + 2m_\alpha M(\vec{\upsilon}_i \cdot \vec{V}) \quad \text{or} \quad V^2 \left(1 - \frac{M}{m_\alpha}\right) = 2(\vec{\upsilon}_i \cdot \vec{V})$$

We may discuss the last equation as follows:

- If the target is an electron in which case $M = m_e \ll m_\alpha$, there is no possibility for the large-angle scattering of the incoming α particle. The direction of scattering will be along or very close to the initial direction. [Because the thickness of the gold foil was very small, the multiple small-angle scattering to produce a large-angle scattering was not possible.]

- If the target is heavy nucleus, $M=m_{Au}>>m_{\alpha}$, the scattering angle may be large (because the left-hand side of the above equation becomes negative; the dot product of these two vectors is $\left|\vec{v}_i\right|\left|\vec{V}\right|\cos\theta$).

In deriving Eq. (2-3) we assume that the scattering of a non-relativistic α particle is from heavy nucleus that its recoil can be neglected (as indicated in Fig. 2-9). In the absence of any interaction the incoming particle approaching the target nucleus would travel in a straight line passing the nucleus at distance b (impact parameter). The incoming particle starting at the horizontal trajectory however slightly deflects from it due to the electrical repulsion making an angle of θ with the horizontal. This angle of deflection is related to the impact parameter. The impulse received by the α particle (due to Coulomb force, F) gives it a linear momentum. The change in that linear momentum in any direction over a given time interval is equal to the integrated force F in that direction over that time interval:

$$\Delta p = \int F dt \;\; \rightarrow \;\; \text{at arbitrary point along the particle trajectory}$$

Total time represents the time it takes for the α particle to travel through the atom. The arbitrary point as selected in Fig. 2-9 is the point of a closest approach. The line connecting that point to the target nucleus splits the angle $(\pi - \theta)$ in half. The component of the initial linear momentum is

$$-m_\alpha v_i \cos(\pi - \theta - \phi) = -m_\alpha v_i \cos\left(\pi - \theta - \left(\frac{\pi}{2} - \frac{\theta}{2}\right)\right)$$

$$= -m_\alpha v_i \cos\left(\frac{\pi}{2} - \frac{\theta}{2}\right) = -m_\alpha v_i \sin\frac{\theta}{2}$$

By symmetry, the terminal velocity will be same as initial, and its component of a linear momentum in the same direction as shown in Fig. 2-9 is $+m_\alpha v_i \sin(\theta/2)$. Thus, it follows that the net change in the linear momentum of the α particle is $2m_\alpha v_i \sin(\theta/2)$. The component of the central force in the indicated direction in Fig. 2-9 is $F\cos(\phi)$. Therefore,

$$2m_\alpha v_i \sin(\theta/2) = \frac{k(Ze)(ze)}{r^2} \int_{-\infty}^{+\infty} \cos\phi \, dt$$

The angular momentum of the α particle with respect to the target nucleus location (Fig. 2-9) is $m_\alpha r$ at the arbitrary point on its trajectory where the location of the α particle is described by angle ϕ and its distance r from the nucleus. At that point the α particle has a tangential velocity with respect to the location of the nucleus equal to $rd\phi/dt$ giving its angular momentum with respect to the nucleus of

$$L = m_\alpha r \upsilon_{\text{tangetial}} = m_\alpha r^2 \frac{d\phi}{dt}$$

The angular momentum is conserved; thus we have

$$m_\alpha r^2 \frac{d\phi}{dt} = m_\alpha \upsilon_i b$$

The variable of integration is conveniently changed to ϕ instead of time. From the conservation of the angular momentum it follows that

$$\frac{d\phi}{dt} = \frac{\upsilon_i b}{r^2} \quad \rightarrow \quad dt = \frac{r^2}{\upsilon_i b} d\phi \quad \text{and} \quad \begin{cases} t = -\infty & \mapsto \quad \phi = -(\pi - \theta)/2 \\ t = +\infty & \mapsto \quad \phi = +(\pi - \theta)/2 \end{cases}$$

giving

$$2m_\alpha \upsilon_i \sin(\theta/2) = \frac{k(Ze)(ze)}{\upsilon_i b} \int\limits_{-(\pi-\theta)/2}^{+(\pi-\theta)/2} \cos\phi \, d\phi$$

The integral is equal to $2\cos(\theta/2)$; thus we obtain Eq. (2-3):

$$2m_\alpha \upsilon_i \sin(\theta/2) = 2\frac{k(Ze)(ze)}{\upsilon_i b}\cos(\theta/2)$$

$$b = \frac{k(Ze)(ze)}{\upsilon_i^2 m_\alpha}\cot(\theta/2) = \frac{k(Ze)(ze)}{2E}\cot(\theta/2)$$

$$\equiv \frac{k(Ze)(ze)}{2E}\frac{1}{\tan(\theta/2)} \equiv \frac{(Ze)(ze)}{8\pi\varepsilon_0 E}\cot(\theta/2)$$

2.4.3   Rutherford's Scattering Formula

Rutherford's experiment eliminated Thomson's plum pudding atomic model on the basis of large-angle scattering. Relatively heavy α particles could not be turned around by much lighter electrons or by the combined mass of positive and negative charges if this mass were distributed uniformly over the whole volume (as demonstrated in Example 2.4). The electrostatic repulsion would only be strong enough to deflect incoming α particles through such large angles if the positive charge is concentrated (as he proposed in a central nucleus). This scattering of charged particles by the nuclear electrostatic field is called *Rutherford scattering*. The probability of large-angle scattering is very small due to the extremely small size of the nucleus relative to the whole atom (radius of 10^{-15} m versus 10^{-10} m); indeed, according to Rutherford's experiment, only 1 out of ~8000 events resulted in large-angle scattering.

Based on his planetary model of the atom, Rutherford was able to define the angular distribution of the scattered α particles. A particle with an impact parameter less than *b* will be scattered at an angle larger than θ (Fig. 2-9 and 2-10). Therefore, all particles hitting the gold foil through the area πb^2 (where *b* is the radius) will scatter at an angle θ or larger (Fig. 2-10). Assuming that the incident beam is made of N α particles and has a cross-sectional area A, the number of particles scattered by θ or larger is $\pi b^2/A$. Thus, the number of particles scattered through an angle of θ or larger by one gold atom in the foil is

$$\frac{N_{scatt}}{\text{atom}} = N\left(\frac{\pi b^2}{A}\right) \tag{2-5}$$

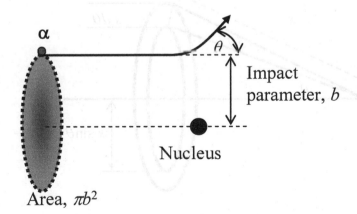

Figure 2-10. Correlation between the deflection angle of α particle and its impact parameter

The number of atoms encountered by the beam of particles in the gold is

$$N_{foil} = nAt \tag{2-6}$$

where *t* is the target (foil) thickness and *n* is the number of target atoms per unit volume. Therefore, it follows from Eqs. (2-5) and (2-6) that the total number of α particles scattered through an angle θ or larger by the gold foil in Rutherford's experiment is

$$N_{tot\text{-}scatt} = \frac{N_{scatt}}{atom} \cdot N_{foil} = \frac{N\pi b^2}{A} \cdot nAt = N\pi b^2 nt \tag{2-7}$$

or

$$N_{tot\text{-}scatt} = N\pi nt\left(\frac{Zzke^2}{2E\tan(\theta/2)}\right) \tag{2-8}$$

The number of particles that emerge between θ and θ + dθ is obtained by differentiating Eq. (2-8)

$$N_{tot\text{-}scatt,\ \theta\to d\theta} = Nnt\pi\left[\frac{Zzke^2}{E}\right]^2 \frac{\cos(\theta/2)}{\sin^3(\theta/2)}d\theta \tag{2-9}$$

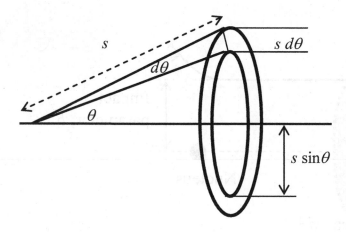

Figure 2-11. Detection of α particles after scattering through θ

At some distance *s* from the gold foil (where the detector is located) particles with a deflection angle between θ and θ + dθ pass through the annulus as shown in Fig. 2-11 and are uniformly distributed over the surface area

$$A_{ring} = (2\pi \sin\theta)(sd\theta) = 4\pi s^2 \sin(\theta/2)\cos(\theta/2) \tag{2-10}$$

The number of particles per unit area that pass through the annulus at distance s and at angle θ is

$$n(\theta) \equiv \frac{N_{tot\text{-}scatt,\theta \to d\theta}}{A_{ring}} = \frac{Nnt}{4s^2} \left[\frac{Zzke^2}{E} \right]^2 \frac{1}{\sin^4(\theta/2)} \tag{2-11}$$

This is called the *Rutherford scattering formula* or inverse square scattering formula. According to this formula, the number of particles scattered at a certain angle is proportional to the thickness of the foil and to the square of the nuclear charge of the foil and inversely proportional to the incident particle kinetic energy squared and to the fourth power of sin ($\theta/2$). This was confirmed in all of the experiments with gold foil. Rutherford derived Eq. (2-11) assuming that the only force acting between the nucleus and α particle is the Coulomb repulsive force, and since all of the experimental data agreed, this assumption was valid (see Example 2.3). However, some years later he repeated the experiment using the aluminum foil. The experimental results for small-angle scattering agreed with his formula, but large-angle scattering departed from it. Rutherford deduced that in the large-angle scattering that corresponded to a closer approach to the nucleus, the α particle was actually striking the nucleus. This meant that the size of the nucleus could be obtained by finding the maximum angle for which the Rutherford formula is valid and finding the incident particle's closest approach to the center of the nucleus.

2.4.4 Stability of the Planetary Atomic Model

The Rutherford planetary atomic model could not explain the following:
- How are the electrons (negatively charged bodies) held outside the nucleus (a positively charged body) despite the attractive electrostatic force? According to the planetary model electrons are revolving around the nucleus like planets around the Sun, though planets are electrically neutral and thus stay in their orbits. According to classical electromagnetic theory any charge placed in circular motion will radiate light (electromagnetic energy), which means that electrons orbiting around the nucleus would spiral inward and collapse into the nucleus due to the loss of kinetic energy. This would produce extremely unstable atoms.
- The radiated energy of photons from spiraling electrons would change in frequency during the deceleration process and produce a continuous spectrum; however, at that time, the spectra of some of the elements were known to show specific discrete lines.
- What does hold the positive charges in the nucleus together in spite of the repulsive electrostatic forces?

2.5   The Smallness of the Atom

Rutherford's gold foil experiment was the first indication and proof that the space occupied by an atom is huge compared to that occupied by its nucleus. In fact, the electrons orbiting the nucleus can be compared to a few flies in a cathedral. As a qualitative reference, a human is about two million times "taller" than the average *Escherichia coli* bacterium; Mount Everest is about 5,000 times taller than the average man; and a man is about ten billion times "taller" than the oxygen atom. If the atom were scaled up to a size of a golf ball, on that same scale a man would stretch from Earth to the Moon. Atoms are so small that direct visualization of their structure is impossible. Today's best optical or electron microscopes cannot reveal the interior of an atom.

The picture shown in Fig. 2-12 was taken with a scanning transmission electron microscope and shows a direct observation of cubes of magnesium oxide molecules, but details of the atoms cannot be seen.

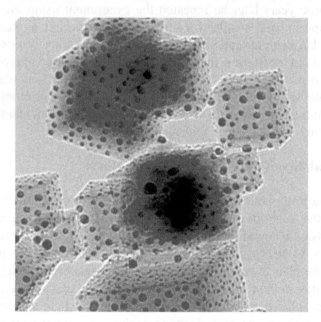

Figure 2-12. Magnesium oxide molecules as seen with scanning transmission electronic microscope produced at the Institute of Standards and Technology in the USA (Courtesy National Institute of Standards and Technology)

At the National Institute of Standards and Technology (NIST), however, the Nanoscale Physics Facility is used to manipulate and arrange atoms, one by one, into desired patterns. The image shown in Fig. 2-13 represents an 8 nm square structure with cobalt atoms arranged on a copper surface. Such arrangements of atoms are used to investigate the physics of ultra-tiny objects. The shown structure was observed with a scanning tunneling microscope at a temperature of 2.3 K (about −455° F): the larger peaks (upper left and lower right) are pairs of cobalt atoms, while the two smaller peaks are single cobalt atoms. The swirls on the copper surface illustrate how the cobalt and copper electrons interact with each other.

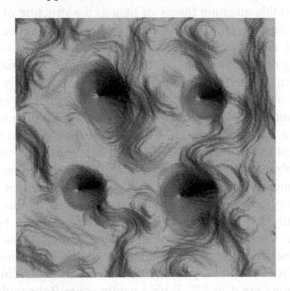

Figure 2-13. Nanoscale structure of cobalt and copper atoms produced at the Institute of Standards and Technology in the USA (Courtesy of J. Stroscio, R. Celotta, A. Fein, E. Hudson, and S. Blankenship, 2002)

2.6 The Quantum Atomic Model

2.6.1 Quantum Leap

In 1913, Niels Bohr (1885–1962) developed the atomic model that resolved Rutherford's atomic stability questions. His model was based on the work of Planck (energy quantization), Einstein (photon nature of light) and Rutherford (nucleus at the center of the atom).

In 1900, Max Planck (1858–1947) resolved the long-standing problem of black body radiation by showing that atoms emit light in bundles of radiation (called *photons* by Einstein in 1905 in his theory of the photoelectric effect). This led to formulation of Planck's *radiation law*: a light is emitted as well as absorbed in *discrete quanta* of energy. The magnitude of these discrete

energy quanta is proportional to the light's frequency (*f*, which represents the
color of light):

$$E = hf = \frac{hc}{\lambda}$$ (2-12)

where *h* is Planck's constant ($h = 6.63 \times 10^{-34}$ J s), *c* is the speed of light
and λ is the wavelength of the emitted or absorbed light.

Bohr applied this quantum theory of light to the structure of the electrons
by restricting them to exist only along certain orbits (called the *allowed*
orbits) and not allowing them to appear at arbitrary locations inside the
atom. The angular momentum of the electrons is quantized and thus
prohibits random trajectories around the nucleus. Consequently the electrons
cannot emit or absorb electromagnetic radiation in arbitrary amounts since
an arbitrary amount would lead to an energy that would force the electron to
move to an orbit that does not exist. Electrons are thus allowed to move from
one orbit to another. However, the electrons never actually cross the space
between the orbits. They simply appear or disappear within the allowed
states; a phenomenon referred to as a *quantum leap* or *quantum jump*.

For his theory of atoms that introduced the new discipline of quantum
mechanics in physics, Bohr received a Noble Prize in 1922. He was also a
founder of the Copenhagen school of quantum mechanics. One of his
students once noticed a horseshoe nailed above his cabin door and asked
him: "Surely, Professor Bohr, you don't believe in all that silliness about the
horseshoe bringing good luck?" With a gentle smile Bohr replied, "No, no,
of course not, but I understand that it works whether you believe it or not".

2.6.2   Absorption and Emission of Photons

In Bohr's atomic model, an electron jumps to a higher orbit when the
atom absorbs a photon and back to a lower orbit when the atom emits a
photon. In other words, a quantum leap to a higher orbit requires energy,
while a quantum leap to a lower orbit emits that energy (Fig. 2-14).

Bohr's atomic model resolved the problem of atomic instability (Section
2.4.4) by changing the classical mechanics into quantum mechanics. This
explains the existence of discontinuities in the absorption and emission of
energy which is determined by the allowable electronic states in atoms.
These allowed orbits are also called *stationary orbits* or *stationary states*.
Since the orbits are discrete and quantized, so are their energies. The
electrons in an atom can thus only have discrete energies. According to
Bohr's theory, in an electrically neutral atom, an electron is in its stationary
state and does not radiate energy as long as it is not disturbed. This

explained the stability of atoms but does not explain why electrons do not radiate energy while orbiting along their stationary trajectories. The theory also explained the reason for the discontinuities in the atomic spectra. When an electron jumps to a higher orbit a photon must be absorbed and its energy is equal to the energy difference of the two orbits. Conversely, a photon is emitted when an electron drops to a lower orbit and the photon energy is again equal to the energy difference of these two orbits (Fig. 2-14):

$$hf = E_n - E_m, \; n > m \tag{2-13}$$

Both, the emission and absorption of energy by an atom, correspond to the electron transition representing a movement of an electron from one level to another. The electrons of an electrically neutral atom are normally all in the lowest possible energy levels. The addition of energy excites the electrons and the resulting atom is in an excited state (absorption of energy by an atom). Generally the electrons remain in this excited state for a short time and soon return to a more stable, lower energy level by releasing the extra energy (emission of energy by an atom).

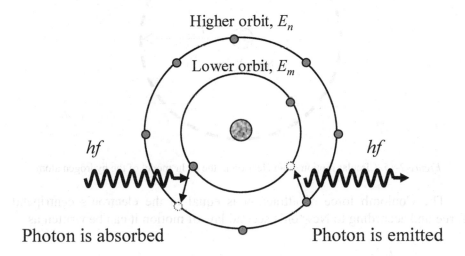

Figure 2-14. Schematic representation of a quantum leap of electrons according to the quantum atomic model (Niels Bohr, 1913)

2.6.3 The Bohr Model of the Hydrogen Atom

According to the Bohr atomic model, the hydrogen atom consists of an electron of mass *m* and charge *–e*, which orbits around a nucleus of charge

+*e* (Fig. 2-15). For simplicity, it is assumed that the electron orbits the nucleus in a circular motion and that the nucleus is fixed in its position (since the hydrogen nucleus consists of one proton that is much heavier than the electron, this assumption does not affect the final result). Thus, the only force that is acting on the electron is the attractive Coulomb force from the positively charged nucleus:

$$F = \frac{ke^2}{r^2}$$
(2-14)

where *k* is the Coulomb force constant, $k = 1/4\pi\varepsilon_0 = 8.99 \times 10^9$ N m^2/C^2 and $\varepsilon_0 = 8.8542 \times 10^{-12}$ C^2/N m^2 is the permittivity of free space.

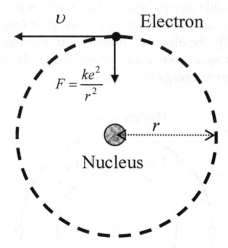

Figure 2-15. Circular motion of an electron in the Bohr model of the hydrogen atom

The Coulomb force of attraction is equal to the electron's centripetal force and according to Newton's second law of motion it can be written as

$$m\frac{v^2}{r} = \frac{ke^2}{r^2}$$
(2-15)

where v^2/r is the centripetal acceleration. Equation (2-15) can be rewritten as follows:

$$mv^2 = \frac{ke^2}{r}$$
(2-16)

and according to classical mechanics, this indicates possible values for electron velocity and its distance from the nucleus that range continuously from 0 to ∞. The electron's kinetic energy is $E = m\upsilon^2 / 2$ and its potential energy in the field of the proton is $U = -ke^2 / r$. By convention the potential energy is zero ($U = 0$) when the electron is far away from the nucleus ($r \to \infty$). For an electron in a circular orbit around a positively charged nucleus, kinetic and total energies are

$$E = -\frac{1}{2}U \tag{2-17}$$

$$E_{tot} = E + U = \frac{1}{2}U = -\frac{1}{2}\frac{ke^2}{r} \tag{2-18}$$

The negative value for the total energy indicates that the electron is bound to the nucleus and cannot escape to infinity. Since the distance from the nucleus ranges from 0 to ∞, it follows from Eq. (2-18) that the electron's total energy can have values between $-\infty$ and 0.

This analysis is based on classical mechanics and does not show that the energy of the electron is quantized. Bohr's hypothesis was that the electron's angular momentum ($L = m\upsilon r$) was quantized in multiples of Planck's constant (this is because Planck's constant has a unit of angular momentum) and for circular orbits (see also Chapter 4, Section 3.9)

$$L = m\upsilon r = n\frac{h}{2\pi} = n\hbar \quad (n = 1,2,3,...) \tag{2-19}$$

where $\hbar = h/2\pi = 1.055 \times 10^{-34}$ J s (read as "h bar").
Combining Eqs. (2-16) and (2-19) it follows that

$$m\left[\frac{n\hbar}{mr}\right]^2 = \frac{ke^2}{r} \quad \Rightarrow \quad r_n = \frac{n^2\hbar^2}{ke^2 m} = n^2 a_0 \quad (n = 1,2,3,...) \tag{2-20}$$

Eq. (2-20) gives quantized values for the radius of the electron's orbit. In addition, it defines the so-called *Bohr radius*, a_0:

$$a_0 = \frac{\hbar^2}{ke^2 m} = 0.0529 \text{ nm} \tag{2-21}$$

The possible energy levels are obtained from the possible electron orbit radii as follows:

$$E_{tot} = E + U = \frac{1}{2}U = -\frac{1}{2}\frac{ke^2}{r} \quad \text{and} \quad r_n = \frac{n^2\hbar^2}{ke^2m} = n^2 a_0 \text{ gives}$$

$$E_n = -\frac{ke^2}{2a_0}\frac{1}{n^2} \quad (n = 1,2,3,...) \tag{2-22}$$

The energy of the photons that are absorbed or emitted from the hydrogen atom during electronic transitions between orbits n and m ($n > m$, see Fig. 2-14) can be now determined:

$$E_\gamma = E_n - E_m = \frac{ke^2}{2a_0}\left[\frac{1}{m^2} - \frac{1}{n^2}\right] \tag{2-23}$$

In the chapters that follow this equation is explored further and connected to the work of Rydberg.

2.7   Atomic Spectra

A spectrum is defined as the distribution of light (electromagnetic radiation) as a function of its frequency or its wavelength. Newton performed the first light color spectrum experiment in 1666 by shining white light through a glass prism. The experiment produced a rainbow of colors and showed that what we observe as white light is a mixture of many different colors. In 1814 a German physicist, Joseph von Fraunhofer, noticed a multitude of dark lines, indicating that certain colors are missing in the solar light spectrum. These dark lines were caused by the absorption of some of the solar light's components by the gases in the Sun's outer atmosphere. A series of experiments followed and by the middle of the 19th century it was understood that gases absorb light (specific frequencies of light) that are characteristic of the gas constituents.

If white light is shone through a gas that consists of only one kind of atom, the gas will absorb light of frequency (energy) that is characteristic to that atom. If the light is then subsequently transmitted through a glass prism, the resulting spectrum will lack the colors corresponding to the absorbed frequencies. This spectrum is called the *absorption spectrum* and the dark lines correspond to the absorbed frequencies (see the hydrogen *absorption* spectrum in Fig. 2-16). By 1859 Robert Bunsen discovered that sufficiently heated gases also emit light and an *emission spectrum* is observed when the emitted light is transmitted through a glass prism (see the hydrogen *emission* spectrum in Fig. 2-16). The emission spectrum's bright lines correspond to the dark lines in the absorption spectrum. At the same time, his colleague, Gustav Kirchhoff, while analyzing the spectra of sunlight and heated sodium, realized that the dark lines in the solar spectrum represented the light frequencies that were absorbed by the sodium atoms in the solar gases.

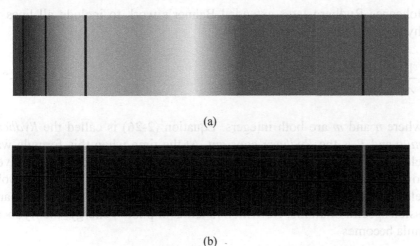

(a)

(b)

Figure 2-16. (a) Absorption and (b) emission spectra of the hydrogen atom

The emission and absorption spectra thus represent a "signature" of an atom. The Kirchhoff–Bunsen discovery was not fully understood until Bohr explained the transition of electrons between strictly defined orbits (energy levels), but it represents the beginning of the science of spectroscopy. By 1870 spectroscopy became a tool that was used to analyze the chemical compositions of the Sun and the stars.

2.7.1   The Balmer–Rydberg Formula

In 1885 a Swiss school teacher, Jakob Balmer (1825–1898), analyzed the hydrogen atomic spectral data and showed that the observed wavelengths correlate to the formula

$$\frac{1}{\lambda} = R\left[\frac{1}{4} - \frac{1}{n^2}\right] \qquad (2\text{-}24)$$

where R is a constant with a dimension of inverse length, according to Balmer equal to 0.0110 nm^{-1} for the hydrogen spectrum, and n is an integer with values of 3, 4, 5 and 6 that correspond to the four observed hydrogen spectral lines. Balmer correctly assumed that this dependence could not be a random coincidence and that other lines must exist (*n* can be greater than 6). The Balmer formula can be rewritten in the form

$$\frac{1}{\lambda} = R\left[\frac{1}{2^2} - \frac{1}{n^2}\right] \quad (n = 3,4,5,...) \tag{2-25}$$

Johannes Rydberg later extended Balmer's work to include all lines in the hydrogen atom emission spectrum:

$$\frac{1}{\lambda} = R\left[\frac{1}{m^2} - \frac{1}{n^2}\right] \quad (n > m) \tag{2-26}$$

where *n* and *m* are both integers. Equation (2-26) is called the *Rydberg formula* and *R* is the *Rydberg constant*. At the time when this formula was developed, it only represented empirical data and no explanation was given as to why the spectral lines obey such regularities. In 1913, Neils Bohr developed an atomic model that explained this nature of absorption and emission spectra of atoms. Rewritten in terms of photon energy, the Rydberg formula becomes

$$E_\gamma = hc/\lambda \text{ and } \frac{1}{\lambda} = R\left[\frac{1}{m^2} - \frac{1}{n^2}\right] \quad (n > m) \text{ gives}$$

$$E_\gamma = Rhc\left[\frac{1}{m^2} - \frac{1}{n^2}\right] \quad (n > m) \tag{2-27}$$

Recall from Section 2.6.3 (Eq. 2-23) that the energy of emitted or absorbed photons according to the Bohr atomic model is

$$E_\gamma = E_n - E_m = \frac{ke^2}{2a_0}\left[\frac{1}{m^2} - \frac{1}{n^2}\right]$$

From the last two relations it can be seen that Bohr's model predicts Rydberg formula and gives the value for the Rydberg constant

$$R = \frac{ke^2}{2a_0}\frac{1}{hc} = \frac{1.44 \text{ eV nm}}{2(0.0529 \text{ nm})(1240 \text{ eV nm})} = 0.0110 \text{ nm}^{-1} \tag{2-28}$$

which is in perfect agreement with the measured values. The term, hcR, is called the *Rydberg energy*, E_R:

$$E_R = hcR = \frac{ke^2}{2a_0} = \frac{m(ke^2)^2}{2\hbar^2} = 13.6 \,\text{eV} \tag{2-29}$$

Thus, the allowed energies of the electron in a hydrogen atom can be expressed in terms of the Rydberg energy

$$E_n = -\frac{E_R}{n^2} \tag{2-30}$$

and the energies of the photons emitted or absorbed by the hydrogen atom are given by

$$E_\gamma = E_n - E_m = E_R\left[\frac{1}{m^2} - \frac{1}{n^2}\right] \quad n > m \tag{2-31}$$

2.7.2 Properties of the Hydrogen Atom According to Bohr's Atomic Model

Bohr's model of the atom correctly predicts the following:
- Possible electron energies in a hydrogen atom are quantized and with values of $E_n = -E_R / n^2$ where $n - 1, 2, 3, \ldots$
- The lowest possible energy level corresponds to the *ground state* for which $n = 1$ and $E_1 = -E_R = -13.6 \,\text{eV}$.
- A minimum energy of $+13.6$ eV is needed to completely remove the electron from a hydrogen atom. This energy is called the binding energy of the hydrogen atom and it is in perfect agreement with the empirical value.
- The radius that corresponds to the ground state of a hydrogen atom is equal to the Bohr radius, $r_1 = a_0 = 0.0529$ nm which agrees well with measured values of the size of the hydrogen atom.
- The radius of the nth circular electron orbit is $r_n = n^2 a_0$.
- The orbits with radii greater than the ground state radius are called the *excited states* of an atom. There are infinitely many levels and all are between the ground state and the zero energy levels. For the hydrogen atom, the energies of excited states are $E_2 = -E_R / 4 = -3.4 \,\text{eV}$, $E_3 = -E_R / 9 = -1.5 \text{eV}$ and so on. These energy levels are generally plotted as illustrated in Fig. 2-17 in a format commonly referred to as *energy-level diagrams*. The transition from the ground state ($n = 1$) to the

$n = 2$ energy level is called the *first excitation level,* and the energy required to raise the hydrogen atom to that level is $E_1 - E_2 = 10.2\,\text{eV}$.

The spectral lines of the hydrogen atom as shown in Fig. 2-18 are given names based on the names of the scientists who discovered them:

1. *Lyman series*: transition to the ground state $m = 1$
2. *Balmer series*: transition to the level $m = 2$
3. *Paschen series*: transition to the level $m = 3$
4. *Bracket series*: transition to the level $m = 4$.

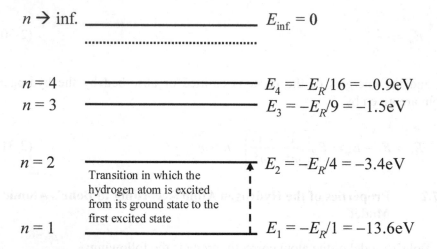

Figure 2-17. Energy-level diagram of the hydrogen atom

Example 2.5 Electron transitions in hydrogen atom

Calculate the wavelength and energy of the light emitted when the electron in a hydrogen atom falls from the first excited state to the ground level.

According to the Balmer formula

$$\frac{1}{\lambda} = R\left[\frac{1}{m^2} - \frac{1}{n^2}\right] \quad (n > m);$$

therefore, the wavelength of the emitted light is

$$\frac{1}{\lambda} = R\left[\frac{1}{1^2} - \frac{1}{2^2}\right] = \frac{3}{4}R \quad \Rightarrow \quad \lambda = \frac{4}{3R} = \frac{4}{(3)(0.0110\,\text{nm}^{-1})} = 121\,\text{nm}$$

and the required energy for this transition is

$$E_\gamma = E_R\left[\frac{1}{1^2} - \frac{1}{2^2}\right] = \frac{3}{4}E_R = \frac{3}{4}(13.6\text{ eV}) = 10.2\text{ eV}$$

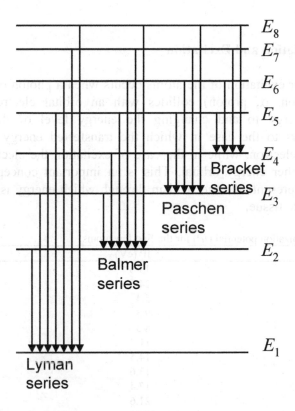

Figure 2-18. Spectral lines in hydrogen atom

Example 2.6 Orbiting velocity of the electron in hydrogen atom

Calculate the highest velocity, the smallest orbit radius and the time it takes for an electron to complete one revolution in a hydrogen atom.

The electron has its highest velocity and smallest orbit radius while in the ground state. The ground state radius in the hydrogen atom corresponds to the Bohr radius, $r_1 = a_0 = 0.0529$ nm. The highest velocity is thus

$$m\upsilon_1 r_1 = n\hbar \qquad r_1 = a_0 \qquad n = 1 \quad \Rightarrow$$

$$\upsilon_1 = \frac{\hbar}{ma_0} = \frac{1.05\times10^{-34}\text{ kg m}^2/\text{s}}{(9.31\times10^{-31}\text{kg})(0.0529\times10^{-9}\text{m})} = 2.1\times10^6\text{ m/s}$$

The time it takes for a ground state electron to complete one revolution around the nucleus is

$$t = \frac{2\pi a_0}{\upsilon_1} = 1.52 \times 10^{-16} \, s$$

2.7.3   Ionization and Excitation

Ionization or excitation of the atoms occurs when a photon or a charged particle (electron, α, proton) collides with an orbital electron, thereby transferring energy to and changing the energy level of the electron. Ionization refers to the case in which the transferred energy causes the ejection of an electron, while in the case of excitation the electron simply moves to a higher energy orbital. This is an important concept in health physics as it represents the mechanism through which energy is transferred from radiation to tissue.

Table 2-1. First ionization potential (*IP*) for the first few atoms

Atom	IP (eV)
Hydrogen	13.6
Helium	24.6
Lithium	5.4
Beryllium	9.3
Boron	8.3
Carbon	11.3
Nitrogen	14.5
Oxygen	13.6
Fluorine	17.4
Neon	21.6
Sodium	5.14

The ionization energy (also called the ionization potential, *IP*) of an atom is the amount of energy required to remove the least tightly bound electron from the atom. To remove a second electron requires remarkably more energy and the removal of each subsequent electron becomes increasingly more difficult. For most elements, the first ionization potential is on the order of several eV (Table 2-1). The first ionization potential of the hydrogen atom is described in Section 2.7.2. When a photon with energy greater than the ionization energy collides with a bound electron of an atom, the photon vanishes and the electron is ejected from the atom with a kinetic energy, E_{pe}, equal to the difference between the photon's initial energy and the ionization potential:

$$E_{pe} = hf - IP \tag{2-32}$$

This mechanism is called the *photoelectric effect* and is described in Chapter 6.

Example 2.7 Excitation of the hydrogen atom

Sketch the excitation of the hydrogen atom for the corresponding absorption and emission of light of energy 10.2 eV.

The absorption of a photon with energy 10.2 eV will move the electron from its ground state to orbit $n = 2$. Conversely, the jump back to ground state will emit a photon of energy 10.2 eV (Fig. 2-19).

$$E = hf = 10.2 \text{ eV}$$

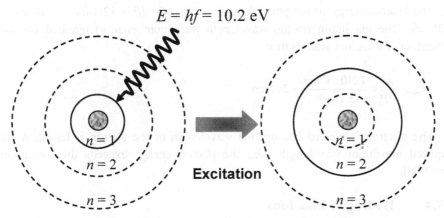

Energy is absorbed and the electron jumps from the ground state ($n = 1$) to its first excited state in a higher orbital ($n = 2$)

$$E = hf = 10.2 \text{ eV}$$

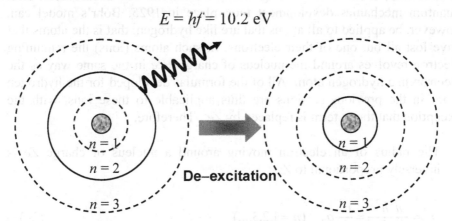

Energy is emitted and the electron falls from its excited state ($n = 2$) back to its ground orbital ($n = 1$)

Figure 2-19. Excitation and de-excitation of the hydrogen atom

Example 2.8 Ionization potential (*IP*)

For a photon of wavelength 10^{-7} m striking the outer orbital electron of a sodium atom, calculate the kinetic energy of the photoelectron (ejected electron). What is the maximum photon wavelength (minimum energy) required to ionize the sodium atom? The *IP* for sodium is given in Table 2-1.

The energy of the photon of wavelength 10^{-7} m is calculated by

$$E_\gamma = \frac{hc}{\lambda} = \frac{1240 \text{ eV nm}}{100 \text{ nm}} = 12.4 \text{ eV}$$

The kinetic energy of the photoelectron is $E_{pe} = hf - IP = 12.4$ eV $-$ 5.14 eV = 7.26 eV. The maximum photon wavelength (minimum energy) required for the ionization of a sodium atom is then

$$\lambda_{min} = \frac{hc}{IP} = \frac{1240 \text{ eV nm}}{5.14 \text{ eV}} = 241 \text{ nm}$$

The electron is ejected because the wavelength of the photon is less than the required maximum wavelength (i.e., the photon energy exceeds the ionization potential).

2.7.4   Hydrogen-Like Ions

The Bohr's atomic model was valid for the hydrogen atom. Any attempt to generalize it for atoms having multiple electrons was unsuccessful until quantum mechanics development took place in 1925. Bohr's model can, however, be applied to all atoms that are like hydrogen, that is the atoms that have lost all but one of their electrons. In such atoms (ions) the remaining electron revolves around the nucleus of charge $+Ze$ in the same way as the electron in a hydrogen atom. All of the formulas developed for the hydrogen atom in the previous sections are thus applicable to these ions, with the exception that the e^2 term is replaced by Ze^2. Therefore,

- The radius of an electron moving around a nucleus of charge Ze is inversely proportional to Z:

$$r_n = \frac{n^2 \hbar^2}{kZe^2 m} = \frac{n^2}{Z} a_0 \quad (n = 1,2,3,...)$$

(2-33)

- The potential energy of the electron in a hydrogen-like ion is

$$U = -\frac{kZe^2}{r} \tag{2-34}$$

- The total energy of the electron in a hydrogen-like ion is

$$E_{tot} = E + U = \frac{1}{2}U = -\frac{1}{2}\frac{kZe^2}{r} \tag{2-35}$$

- The allowed energies for the electron in a hydrogen-like ion are Z^2 times the corresponding energies in hydrogen atom:

$$E_n = -\frac{kZe^2}{2a_0}\frac{1}{n^2} = -Z^2\frac{E_R}{n^2} \quad (n = 1,2,3,...) \tag{2-36}$$

- The energies of the photons emitted and absorbed by the electron in hydrogen-like ions are

$$E_\gamma = E_n - E_m = Z^2 E_R \left[\frac{1}{m^2} - \frac{1}{n^2} \right] \quad n > m \tag{2-37}$$

Example 2.9 Helium ion and reduced mass correction

Calculate the ratio of the allowed energies in the helium ion to that in the hydrogen atom taking into account the effect of nuclear motion.

The assumption that the electron orbits around a fixed nucleus is not entirely correct. In reality, they both revolve around the common center of mass. Since the nucleus is much heavier than the electron, the center of mass is close to the nucleus, which is therefore almost stationary. In the equations for allowed energies as well as for the Rydberg energy, the electron mass, m, must be corrected for the motion of the nucleus (mass = M). This is done by replacing the electron mass with the so-called *reduced mass*, μ, which is defined as

$$\mu = \frac{m}{1 + m/M} \tag{2-38}$$

The reduced mass is always less than the actual mass of the electron. In a hydrogen atom, the nucleus consists of a single proton and $m/M \sim 1/1800$. The helium ion (He$^+$) nucleus is four times heavier than that of the hydrogen atom (H) and thus m/M is four times smaller:

$$E_n^{\text{He}^+} = 2^2 \frac{E_R}{n^2} = -4\frac{\mu^{\text{He}^+}\left(ke^2\right)^2}{2\hbar^2 n^2} \quad E_n^{\text{H}} = -\frac{\mu^{\text{H}}\left(ke^2\right)^2}{2\hbar^2 n^2}$$

$$\frac{E_n^{\text{He}^+}}{E_n^{\text{H}}} = \frac{4\mu^{\text{He}^+}}{\mu^{\text{H}}} = \frac{4\times\dfrac{1}{1+1/(4\times1800)}}{\dfrac{1}{1+1/(1800)}} = 4.0017$$

When nuclear motion is accounted for, the ratio of allowed energy levels in the helium ion to that of hydrogen increases from exactly 4 to 4.0017. This small difference is observed in the measurements of atomic and ionic spectra.

2.7.5   Empirical Evidence of Bohr's Theory

Although Bohr's theory was shown to be almost completely valid for the hydrogen atom, great success was also achieved when it was used to describe hydrogen-like ions (as discussed in Section 2.7.4). Bohr's theory also proved to be valid for calculating the allowed energy levels of the innermost electron in multi-electron atoms. The latter application approximates the charge of the outer electrons to be uniformly distributed in a sphere surrounding the innermost electron. It follows that, due to the spherical symmetry of the electric field, the innermost electron experiences no net force from the outer electrons. The only force acting on the innermost electron is the electrostatic force from the positively charged nucleus (Ze). The allowed energies for the innermost electron in multi-electron atoms are given by Eq. (2-36):

$$E_n = -\frac{kZe^2}{2a_0}\frac{1}{n^2} = -Z^2\frac{E_R}{n^2} \quad (n=1,2,3,...)$$

For example, the energy required to remove the innermost electron from its ground state orbit ($n = 1$) in an iron atom ($Z = 26$) is

$$E_1 = -Z^2 E_R = -(26)^2\,(13.6\,\text{eV}) \approx 9{,}194\,\text{eV}$$

For heavier atoms the energy needed to remove the innermost electrons is on the order of thousands of eV and thus photons emitted or absorbed in such transition are in the range of X-rays (see Chapter 3). Henry Moseley

(1887–1915), a British physicist (killed at the age of 27 in World War I), was measuring the wavelengths of X-rays emitted by various atoms when he discovered that the dependence on atomic number exactly followed Bohr's theory. His explanation of characteristic X-rays was that if an innermost electron ($n = 1$) is ejected, the vacancy created is filled by an outer electron.

The transition of the outer electron to the inner shell will produce the emission of a characteristic photon with energy that is equal to the difference in allowed energies of the levels involved in the electron jump. For example, in the transition of an electron from level $n = 2$ to level $n = 1$, traditionally called K_α, the energy of the emitted photon is given by Eq. (2-37):

$$E_\gamma = E_2 - E_1 = Z^2 E_R \left(1 - \frac{1}{4}\right) = \frac{3}{4} Z^2 E_R$$

Moseley measured the frequencies of emitted photons for about 20 different elements and found that frequency changes with the square of the atomic number Z. He then plotted the square root of the frequencies as a function of known values of Z and verified that it is a linear function. This helped in the identification of the atomic numbers of several elements that were not known at the time (one of which was technetium, $Z = 43$, which does not occur naturally and was produced artificially in 1937).

The plot shown in Fig. 2-20 indicates that the line does not start from the origin as the relation would suggest. After detailed examination of the plot, Moseley concluded that the line crosses the Z-axis at a point close to $Z = 1$, implying that $\sqrt{f} \propto (Z - 1)$ or $E_\gamma \propto (Z - 1)^2$. The prediction that characteristic X-rays are emitted with frequencies proportional to Z^2 was based on the assumption that inner electrons experience a force due to the positive charge of nucleus ($+Ze$) but are not affected by the charges of the other (outer) electrons in the atom.

In reality, however, the inner electrons do experience a force from the outer electrons in the form of a screening of the nuclear attraction force; in other words, the attractive force of the nucleus is somewhat diminished due to the presence of the outer electrons. This so-called *screening factor*, a, is usually close to unity and the energy of emitted (or absorbed) K_α X-rays is

$$E_\gamma = \frac{3}{4}(Z - a)^2 E_R \quad a \cong 1 \tag{2-39}$$

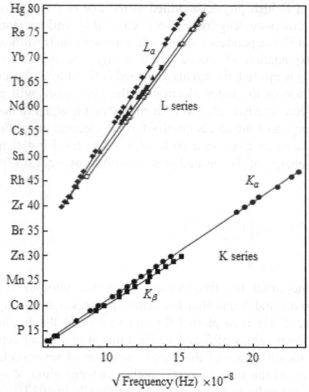

Figure 2-20. Plot of K_α X-ray characteristic lines known at the time of Moseley's experiments

Example 2.10 Characteristic K_α line

Estimate the wavelength of the characteristic K_α X-ray from niobium, which has the atomic number $Z = 41$. Assume that the screening factor is approximately equal to 1.

$$E_\gamma = \frac{3}{4}(Z-a)^2 E_R = \frac{3}{4}(41-1)^2(13.6\text{ eV}) = 16,320\text{ eV} \rightarrow \lambda = \frac{hc}{E_\gamma} = 0.076\text{ nm}$$

Example 2.11 Cascade of atomic vacancies

Calculate the wavelength and determine the spectral region for a krypton atom ($Z = 36$) when an electron from $n = 2$ fills a vacancy in the $n = 1$ level. What happened to the $n = 2$ level when the electron fell to the $n = 1$ level?

Allowed energies for these two levels, taking into account the screening effect, are according to Eq.(2-37):

$$E_1 = -(36-1)^2(13.6\text{ eV}) = -16,660\text{ eV}\, ; \, E_2 = -(36-1)^2\left(\frac{13.6\text{ eV}}{4}\right) = -4,165\text{ eV}$$

The energy of the emitted photon in this transition is 12,495 eV and the corresponding wavelength is 0.099 nm, which belongs to the X-ray region of the spectrum. After the $n = 2$ electron falls to the $n = 1$ level, an $n = 3$ or an $n = 4$ electron fills this orbital and emits another photon.

2.8   Atoms of Higher Z

2.8.1   Quantum Numbers

The light spectra of atoms with more than one electron are much more complex than that of the hydrogen atom (many more lines). The calculations of the spectra for these atoms with the Bohr atomic model are complicated by the screening effect of the other electrons (see Section 2.7.5). Examination of the hydrogen spectral lines with high-resolution spectroscopes shows these lines to have very fine structures, and the observed spectral lines are each actually made up of several lines that are very close together. This observation implied the existence of sublevels of energy within the principal energy level, which makes Bohr's theory inadequate even for the hydrogen atomic spectrum.

Bohr recognized that the electrons are most likely organized into orbital groups in which some are close and tightly bound to the nucleus, and others less tightly bound at larger orbits. He proposed a classification scheme that groups the electrons of multi-electron atoms into "shells" and each shell corresponds to a so-called quantum number n. These shells are given names that correspond to the values of the principal quantum numbers:

- $n = 1$ (K shell) can hold no more than 2 electrons
- $n = 2$ (L shell) can hold no more than 8 electrons
- $n = 3$ (M shell) can hold no more than 18 electrons, etc.

Moseley's work (described in Section 2.7.5) contributed to the understanding that the electrons in an atom existed in groups visualized as electron shells, and according to quantum mechanics, the electrons are distributed around the nucleus in *probability regions* also called the *atomic orbitals*. In order to completely describe an atom in three dimensions, Schrödinger introduced three quantum numbers in addition to the principal quantum number, n. There are thus a total of four quantum numbers that specify the behavior of electrons in an atom, namely

- principal quantum number, $n = 1, 2, 3, \ldots$
- azimuthal quantum number, $l = 0$ to $n - 1$
- magnetic quantum number, $m = -l$ to 0 to $+l$
- spin quantum number, $s = -1/2$ or $+1/2$.

The *principal quantum number* describes the shells in which the electrons orbit. The maximum number of electrons in a shell n is $2n^2$.

The sub-energy levels (*s*, *p*, *d*, etc.) are the reason for the very fine structure of the spectral lines and result from the electron's rotation around the nucleus along elliptical (not circular) orbits. The *azimuthal quantum number* describes the actual shape of the orbits. For example, $l = 0$ refers to a spherically shaped orbit, $l = 1$ refers to two obloid spheroids tangent to one another and $l = 2$ indicates a shape that is quadra-lobed (similar to a four leaf clover). For a given principal quantum number, *n*, the maximum number of electrons in an $l = 0$ orbital is 2, for an $l = 1$ orbital it is 6 and an $l = 2$ orbital can accommodate a maximum of 10 electrons.

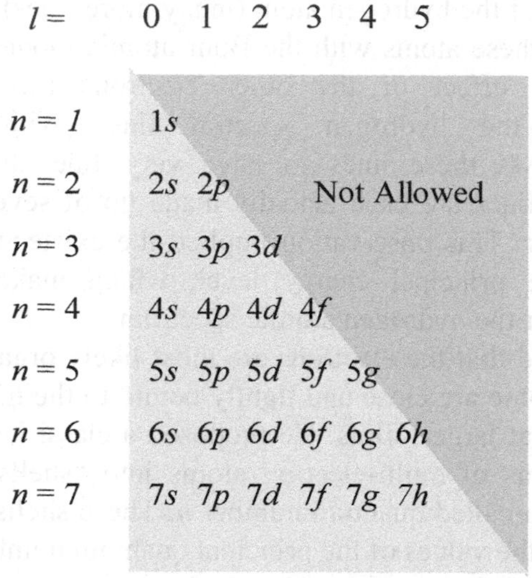

$l =$	0	1	2	3	4	5
$n = 1$	1s					
$n = 2$	2s	2p		Not Allowed		
$n = 3$	3s	3p	3d			
$n = 4$	4s	4p	4d	4f		
$n = 5$	5s	5p	5d	5f	5g	
$n = 6$	6s	6p	6d	6f	6g	6h
$n = 7$	7s	7p	7d	7f	7g	7h

Figure 2-21. Allowed combinations of quantum numbers

The *magnetic quantum number* is also referred to as the *orbital quantum number* and it physically represents the orbital's direction in space. For example when $l = 0$, *m* can only be zero. This single value for the magnetic quantum number suggests a single spatial direction for the orbital. A sphere is uni-directional and it extends equally in all directions, thus the reason for a single *m* value. If $l = 1$ then *m* can be assigned the values −1, 0 or +1. The three values for *m* suggest that the double-lobed orbital has three distinctly different directions in three-dimensional space into which it can extend. In the absence of any perturbing force (such could be an external magnetic field) the orbitals with the same *n* and *l* are equal in energy and are called *degenerate*. In the presence of a perturbing force caused by the magnetic field the orbitals would differ in energy, and thus this quantum number is called the *magnetic* quantum number.

The *spin quantum number* describes the spin of the electrons. The electrons spin around an imaginary axis (as Earth spins about the imaginary axis connecting the north and south poles) in a clockwise or counter-clockwise direction; for this reason there are two values, $-1/2$ or $+1/2$.

The allowed combination of quantum numbers is given in Fig. 2-21.

Example 2.12 Quantum numbers of the hydrogen atom

Write the quantum numbers of the ground and first-excited levels of the electron in a hydrogen atom. Comment on the values of angular momentum of the ground state atom using Bohr's atomic model. Use an energy-level diagram to indicate the quantum levels.

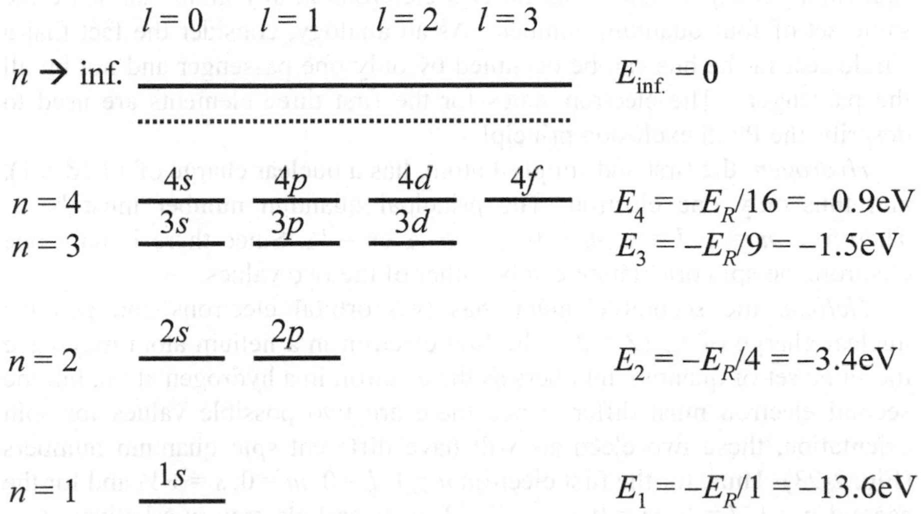

$l = 0$ $l = 1$ $l = 2$ $l = 3$

$n \rightarrow$ inf. _____ $E_{\text{inf.}} = 0$

$n = 4$ $\dfrac{\quad}{4s}$ $\dfrac{\quad}{4p}$ $\dfrac{\quad}{4d}$ $\dfrac{\quad}{4f}$ $E_4 = -E_R/16 = -0.9\text{eV}$

$n = 3$ $\dfrac{\quad}{3s}$ $\dfrac{\quad}{3p}$ $\dfrac{\quad}{3d}$ $E_3 = -E_R/9 = -1.5\text{eV}$

$n = 2$ $\dfrac{\quad}{2s}$ $\dfrac{\quad}{2p}$ $E_2 = -E_R/4 = -3.4\text{eV}$

$n = 1$ $\dfrac{\quad}{1s}$ $E_1 = -E_R/1 = -13.6\text{eV}$

Figure 2-22. Energy-level diagram for the hydrogen atom including the quantum numbers

From Fig. 2-21 it follows that for the ground level, $n = 1$. The only possible value for the azimuthal quantum number is then zero ($l = 0$), indicating that the ground state of a hydrogen atom has zero angular momentum. (In general the angular momentum is zero if the motion of the particle is directly toward or away from the origin, or if it is located at the origin.) This in turns gives only one value for the magnetic quantum number, $m = 0$. According to Bohr's atomic model, the ground state of a hydrogen atom has an angular momentum equal to $L = 1 \times \hbar$. However, the Schrödinger equation (see Chapter 4) predicts that $L = 0$. For the first excited level, $n = 2$, which gives two values for the azimuthal quantum number, namely $l = 0$ and $l = 1$. When $l = 0$, the only possible value for m is zero. However, when $l = 1$, m assumes three values, $m = 1$, 0 or -1 and this results in three possible orientations for the angular momentum. The energy-level diagram is shown in Fig. 2-22:

Ground state: $n = 1$, $l = 0$, $m = 0$
First excited level: $n = 2$, $l = 0$ or $l = 1$, $m = 0$ or $m = 1$, 0 or -1.

2.8.2   The Pauli Exclusion Principle

Quantum numbers describe the possible states that electrons can occupy in an atom. Additional rules are required to define how the electrons occupy these available states and thus explain the structure of multi-electron atoms and the periodic system of elements. An atom in its ground state has the minimum possible energy and electrons are distributed among the available and allowed states according to the principle formulated by the Austrian physicist Wolfgang Pauli (1900–1958). This principle, called the *Pauli exclusion principle,* states that no two electrons in any atom can share the same set of four quantum numbers. As an analogy, consider the fact that a single seat in the bus can be occupied by only one passenger and not by all the passengers. The electron states for the first three elements are used to describe the Pauli exclusion principle:

Hydrogen, the first and simplest atom, has a nuclear charge of $+1$ ($Z = 1$), and thus only one electron. The principal quantum number must be 1. Therefore, $n = 1$, $l = 0$, $m = 0$, $s = + \frac{1}{2}$ or $- \frac{1}{2}$. Since there is only one electron, the spin orientation can be either of the two values.

Helium, the second element, has two orbital electrons and positive nuclear charge of $+2$ ($Z = 2$). The first electron in a helium atom may have the same set of quantum numbers as the electron in a hydrogen atom, but the second electron must differ. Since there are two possible values for spin orientation, these two electrons will have different spin quantum numbers (Fig. 2-23). Thus, for the first electron $n = 1$, $l = 0$, $m = 0$, $s = + \frac{1}{2}$ and for the second $n = 1$, $l = 0$, $m = 0$, $s = - \frac{1}{2}$. The second electron in a helium atom exhausts all possibilities for $n = 1$. The anti-parallel orientation of the spins in the 1s state results in a zero magnetic moment, which is observed for the helium atom in its ground state, thus providing proof of the exclusion principle. If the spins of these two electrons were parallel (forbidden states), this would produce a non-zero magnetic moment, which has never been observed. In an excited helium atom as shown in Fig. 2-24, one electron can be in the 1s state and the other in 2s. In this case, according to Pauli exclusion principle, the spins of the two electrons can be parallel, which would give a non-zero magnetic moment, or anti-parallel, in which case the magnetic moment is zero. Both cases have been observed in reality and thus contribute evidence of the exclusion principle.

Lithium, which has three orbital electrons and atomic number $Z = 3$. The first two electrons occupy the 1s level with anti-parallel spins. The 1s level is thus filled and cannot accommodate any more electrons (all seats are taken!). Thus, the third electron, according to the exclusion principle, must occupy the next higher energy level and thus have a principal quantum number equal to 2. The lowest level in this state is the 2s level (Fig.2-25). This orbital may be circular or elliptical, i.e., the azimuthal quantum number may be either 0 or 1: (a) if $l = 0 \Rightarrow m = 0$ and (b) if $l = 1 \Rightarrow m = -1$, 0 or +1. Each of these states may contain two electrons, with each electron having a spin of $+ \frac{1}{2}$ or $-\frac{1}{2}$.

The Pauli exclusion principle also applies to any electron-like particle, *i.e.* a particle with a half-integer spin. For example, neutrons, like the electrons, have a half-integer spin and the arrangement of neutrons inside the nucleus is similar to that of the electrons in their orbits around the nucleus (see Chapter 4).

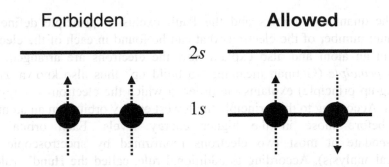

Figure 2-23. The ground state of a helium atom according to the Pauli exclusion principle

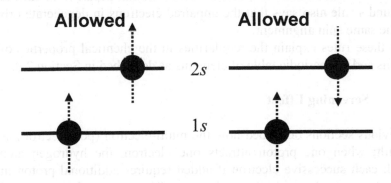

Figure 2-24. The lowest excited states of a helium atom

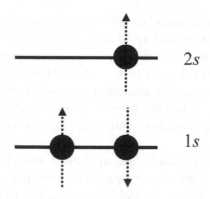

Figure 2-25. The ground state of a lithium atom

2.8.3 The Aufbau Principle

The quantum numbers and the Pauli exclusion principle define the maximum number of the electrons that can be found in each of the electron orbits in an atom and also explain how the electrons are arranged. The *aufbau principle* (German meaning "to build up" thus also known as the building-up principle) explains the order in which the electrons occupy the orbitals. According to this principle the lowest energy orbitals in an atom are filled before those in the higher energy levels. Each orbital can accommodate at most two electrons (confirmed by spectroscopic and chemical analysis). According to additional rule, called the Hund's rule, if two or more energetically equivalent orbitals are available (such as orbitals *p, d, f*) the electrons spread out before they start to pair. The reason for this is that because the electrons repel each other and because each orbital is directed toward a different section in space, the electrons can depart from each other. The Hund's rule also says that the unpaired electrons in degenerate orbitals have the same spin alignment.

All these rules explain the regularities in the chemical properties of the elements and the periodic table of elements as described in Section 2.9.

2.8.4 Screening Effect

Previous sections described how the multi-electron (polyelectron) atoms are built; when one proton attracts one electron, the hydrogen atom is created; each successive electron if added requires additional proton in the nucleus to assure the neutrality of the atom. The energy of attraction of one electron in one-electron atom (or ion) is proportional to Z^2/n^2. Thus, in an atom with one electron, that electron experiences the full charge of the positive nucleus and the effective nuclear charge can be calculated from Coulomb's law. In an atom with more than one electron, the outer electrons are simultaneously attracted to the positive nucleus but also repelled by other electrons. Because of this repulsion, each electron feels a nuclear charge that is smaller than the actual charge. The effective nuclear charge, Z_{eff}, on an electron is given with $(Z - a)$, as defined by Eq. (2-39). Parameter a as described in Section 2.7.5 represents the average number of electrons between the nucleus and the electron in question. It can be estimated using the Slater's rule or simply by deducing the number of electrons excluding the valence electrons from the total number of protons.

For a given electron, important screening (shielding) is only presented by electrons in the same or smaller shells; electrons in the larger shells do not importantly affect the Z_{eff}. The electrons in s orbit have the highest probability to be closest to the nucleus. The p or d or f or other higher orbits have regions in which electrons will never be found; therefore electrons in these orbits are never close to the nucleus as electrons in s orbit. Because of that the s orbits are called the penetrating orbits and these electrons have greater screening effect than electrons in higher orbits. For example, the electronic configuration of the carbon atom $1s^2 2s^2 2p^2$ indicates that the two electrons in $1s$ orbit shield better electrons in $2p$ than those in $2s$ orbit. The Z_{eff} is higher for the s electrons than it is for the p electrons (see Section 2.9.5).

2.9 The Periodic Table and Properties of the Elements

By the mid-19th century, several chemists had discovered that when the elements are arranged by atomic mass they demonstrate periodic behavior. In 1869, while writing a book on chemistry, Russian scientist Dmitri Mendeleev (1834–1907) realized this periodicity of the elements and he arranged them into a table that is today called the periodic table of elements. The table, as first published, was a simple observation of regularities in nature; the principles that defined this periodicity were not understood. Mendeleev's table contained gaps due to the fact that some of the elements were yet unknown. In addition, when he arranged the elements in the table he noticed that the weights of several elements were wrong.

In the modern periodic table, the elements are grouped in order of increasing atomic number and arranged in rows (Fig. 2-26). Elements with similar physical and chemical properties appear in the same columns. A new row starts whenever the last (outer) electron shell in each energy level (principal quantum number) is completely filled. Properties of an element are discussed in terms of their chemical or physical characteristics. Chemical properties are often observed through a chemical reaction, while physical properties are observed by examining a pure element.

The chemical properties of an element are determined by the distribution of electrons around the nucleus, particularly the outer, or valence, electrons. Since a chemical reaction does not affect the atomic nucleus, the atomic number remains unchanged. For example, Li, Na, K, Rb and Cs behave chemically similarly because each of these elements has only one electron in its outer orbit. The elements of the last column (He, Ne, Ar, Kr, Xe and Rn) have filled inner shells and all except helium have eight electrons in their outermost shells. Because their electron shells are completely filled, these elements cannot interact chemically and are therefore referred to as the inert, or noble, gases.

Group

Period	1	2	3	4	5	6	7	8	9	10	11	12	13	14	15	16	17	18
1	1 H																	2 He
2	3 Li	4 Be											5 B	6 C	7 N	8 O	9 F	10 Ne
3	11 Na	12 Mg											13 Al	14 Si	15 P	16 S	17 Cl	18 Ar
4	19 K	20 Ca	21 Sc	22 Ti	23 V	24 Cr	25 Mn	26 Fe	27 Co	28 Ni	29 Cu	30 Zn	31 Ga	32 Ge	33 As	34 Se	35 Br	36 Kr
5	37 Rb	38 Sr	39 Y	40 Zr	41 Nb	42 Mo	43 Tc	44 Ru	45 Rh	46 Pd	47 Ag	48 Cd	49 In	50 Sn	51 Sb	52 Te	53 I	54 Xe
6	55 Cs	56 Ba	71 Lu	72 Hf	73 Ta	74 W	75 Re	76 Os	77 Ir	78 Pt	79 Au	80 Hg	81 Tl	82 Pb	83 Bi	84 Po	85 At	86 Rn
7	87 Fr	88 Ra	103 Lr	104 Rf	105 Db	106 Sg	107 Bh	108 Hs	109 Mt	110 Ds	111 Rg	112 Uub	113 Uut	114 Uuq	115 Uup	116 Uuh	117 Uus	118 Uuo

6	57 La	58 Ce	59 Pr	60 Nd	61 Pm	62 Sm	63 Eu	64 Gd	65 Tb	66 Dy	67 Ho	68 Er	69 Tm	70 Yb		
7	89 Ac	90 Th	91 Pa	92 U	93 Np	94 Pu	95 Am	96 Cm	97 Bk	98 Cf	99 Es	100 Fm	101 Md	102 No		

Figure 2-26. The periodic table of elements

Each horizontal row in the periodic table of elements is called a period. The first period contains only two elements, hydrogen and helium. The second and third periods each contain eight elements, while the fourth and fifth periods contain 18 elements each. The sixth period contains 32 elements that are usually arranged such that elements from $Z = 58$ to 71 are detached from main table and placed below it. The seventh and last period is also divided into two rows, one of which, from $Z = 90$ to 103, is placed below the second set of elements from the sixth period. The vertical columns are called groups and are numbered from left to right. The first column, Group 1, contains elements that have a closed shell plus a single s electron in the next higher shell. The elements in Group 2 have a closed shell plus two s electrons in the next shell. Groups 3–18 are characterized by the elements that have filled, or almost filled, *p* levels. Group 18 is also called Group 0 and contains the noble gases. The columns in the interior of the periodic table contain the transition elements in which the electrons are present in the *d* energy level. These elements begin in the fourth period because the first *d* level (3*d*) is in the fourth shell. The sixth and the seventh shells contain 4*f* and 5*f* levels and are called lanthanides, or rare earth elements, and actinides, respectively.

The elements are also grouped according to their physical properties; for instance, they are grouped into metals, non-metals, and metalloids. Elements with very similar chemical properties are referred to as families; examples include the halogens, the inert gases, and the alkali metals. The following sections only focus on those atomic properties that are closely related to the principles of nuclear engineering.

2.9.1   Ground States of Atoms

The most common way to illustrate the electronic structure of the atoms in their ground states is to use the energy-level diagrams (like these shown in Fig. 2-23 and 2-25) or notations as shown in Table 2-2.

Table 2-2. Electron configuration of the first 18 elements

First shell	Second shell	Third shell
Hydrogen, H-1: $1s^1$	Lithium, Li-3: $1s^2 2s^1$	Sodium, Na-11: $1s^2 2s^2 2p^6 3s^1$
Helium, He-2: $1s^2$	Beryllium, Be-4: $1s^2 2s^2$	Magnesium, Mg-12: $1s^2 2s^2 2p^6 3s^2$
	Boron, B-5: $1s^2 2s^2 2p^1$	Aluminum, Al-13: $1s^2 2s^2 2p^6 3s^2 3p^1$
	Carbon, C-6: $1s^2 2s^2 2p^2$	Silicon, Si-14: $1s^2 2s^2 2p^6 3s^2 3p^2$
	Nitrogen, N-7: $1s^2 2s^2 2p^3$	Phosphor, P-15: $1s^2 2s^2 2p^6 3s^2 3p^3$
	Oxygen, O-8: $1s^2 2s^2 2p^4$	Sulfur, S-16: $1s^2 2s^2 2p^6 3s^2 3p^4$
	Fluorine, F-9: $1s^2 2s^2 2p^5$	Chlorine, Cl-17: $1s^2 2s^2 2p^6 3s^2 3p^5$
	Neon, Ne-10: $1s^2 2s^2 2p^6$	Argon, Ar-18: $1s^2 2s^2 2p^6 3s^2 3p^6$

How to determine the ground state of an atom? Simply as the data in Table 2-2 indicates, the number of electrons of that atom (*Z*) is to be assigned to the lowest energy levels taking into account the Pauli exclusion principle and the aufbau principle. The ground states of the first few elements are explained in Section 2.8.2 in the discussion of the quantum numbers and the electron spin orientation. Recall analyzing Table 2-2, for the first two elements, hydrogen and helium, H-1 and He-2, the 1*s* level is used to create the ground state. Starting from lithium, Li-3 through neon, Ne-10, 1*s* level is filled with the maximum allowed number of electrons, and the new shells are open, first the 2*s* level and then the 2*p* level. For sodium, Na-11, through argon, Ar-18, the levels, 1*s*, 2*s* and 2*p*, are already occupied; thus the electrons are filling the next shells, i.e., the energy levels 3*s* and then 3*p*. The atoms from Ar-18 to potassium, K-19, have electrons filling the orbits with the lower angular momentum and thus lower energies (these are the penetrating orbits as described in Sections 2.8.4 and 2.9.5), which would be $4s$ and not $3d$. Therefore, K-19 has the following electronic configuration: $1s^2 2s^2 2p^6 3s^2 3p^6 4s^1$.

Example 2.13 Electronic configuration

For Na and Li, write the electronic configurations in short notation based on the previous completed electron shell.

From Table 2-2 it follows that

Lithium, Li-3: $1s^2 2s^1 = $ [He] $2s^1$

Sodium, Na-11: $1s^2 2s^2 2p^6 3s^1 = $ [Ne] $3s^1$

Example 2.14 Electron energy shell diagram

Draw the energy diagram for the first six shells to illustrate the orders in which the electron energy levels (shells) are occupied taking into account all discussions presented in this and in the previous sections.

Shell	Energy levels				Max*	Z**
6	$6s$ $6p$	$5d$	$4f$		2+6+10 +14	86
5	$5s$ $5p$	$4d$			2+6+10	54
4	$4s$ $4p$	$3d$			2+6+10	36
3	$3s$ $3p$				2+6	18
2	$2s$ $2p$				2+6	10
1	$1s$				2	2

*Indicates the maximum number of electrons that can be found at that level
** Indicates the atomic number of the atom with the closed shells

2.9.2   Excited States of Atoms

For an atom to be in its excited state, one or more of its electrons is supposed to be moved from the ground (lowest energy) state to higher energy levels. Electron can be moved to any of the available levels and in many-electron atoms more than one electron can be moved to higher levels at the same time. Let us assume that the atom has its last and only one electron in its ground state located in the shell $3s$. From Table 2-2 it is clear that the levels $1s$, $2s$ and $2p$ are filled with the electrons indicating thus that the smallest energy needed to excite this atom would be to move the last electron to a higher level orbit. Since this last electron occupies shell $3s$ according to the schematics from Example 2.14, with the addition of energy, it can be moved to $3p$, or $4s$, or $4p$ and so on. The excitation energies of the most outer electrons (valence electrons) are usually in the order of few eV; these excitations are usually called the optical excitations (because the energy of few eV corresponds to a visible spectrum of light). The higher incoming energy of the photon is required to excite the atom by moving some of the inner electrons to higher energy levels. The highest energy needed is to remove the $1s$ electrons and is nearly equal to $-Z^2 E_R$ (Section 2.7.4). The $1s$ electron when excited can move only to those shells that Pauli exclusion principle allows; in other words these electrons cannot occupy already filled shells. Therefore, $1s$ electrons are moved to the valence shells or the higher levels that require much smaller energy. The difference (in energy required to remove $1s$ electron in the order of keV or higher to valence shell of energy in the order of few eV) is therefore emitted in the form of X-rays and this type of atom excitation is called the X-ray excitations. When $1s$ electron is removed, the formed vacancy is filled by any of the electrons from the higher energy levels. The difference in energy is again emitted in the form of X-rays. This is described by the Mosley experiment in Section 2.7.5.

2.9.3 Atomic Radius

The size of an atom, expressed as the atomic radius, represents the distance between the nucleus and the valence, or outermost, electrons. The boundary between the nucleus and the electrons is not easy to determine and the atomic radius is therefore approximated. For example, the distance between the two chlorine atoms of Cl_2 is known to be nearly 2Å. In order to obtain the atomic radius, the distance between the two nuclei is assumed to be the sum of the radii of two chlorine atoms. Therefore the atomic radius of chlorine is ~1Å (or 100 pm, see Fig. 2-27).

The atomic radius changes across the periodic table of elements and is dependent on the atomic number and the electron distribution. Since electrons repel each other due to like charges, the overall size of the atom increases with an increase in the number of electrons in each of the groups (see Fig. 2-27). For example, the radius of a hydrogen atom is smaller than the radius of the lithium atom. The outer electron of lithium is in the $n = 2$ level, so its radius must be larger than the radius of hydrogen which has its outermost electron in the $n = 1$ level. However, in spite of the increase in the number of electrons, the atomic radius decreases when going from left to right across the periodic table. This is a result of an increase in the number of protons for these elements, which all have their valence electrons in the same quantum energy level. Since the electrons are attracted to the protons, the increased charge of the nucleus (more protons) binds the electrons more tightly and brings them closer to the nucleus, causing the overall atomic radius to decrease. For example, the first two elements in the second period of the periodic table are lithium and beryllium.

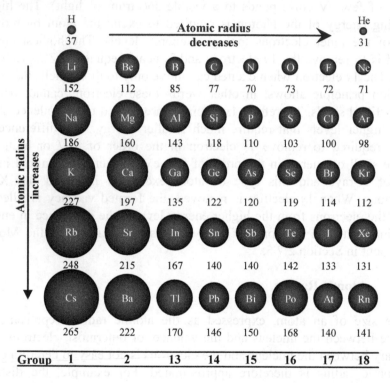

Figure 2-27. Trends of atomic radii (listed in picometers) in the periodic table

The radius of a beryllium atom is 112 pm, which is smaller than that of lithium (152 pm). In beryllium, $Z = 4$, the fourth electron joins the third in the $2s$ level, assuming their spins are anti-parallel. The charge is thus larger and this causes the electrons to be bound more tightly to the nucleus; as a result the beryllium radius is less than the lithium radius. The effect of the increased charge should, however, be seen in the context of the quantum energy levels. For example, cesium has a large number of protons but it is one of the largest atoms. The valence electrons are furthest from the nucleus and the inner electrons shield them from the positive charge of the nucleus;

thus the valence electrons experience a reduced effective nuclear charge and not the total charge of the nucleus. The effect of the increase in the nuclear charge thus only plays a role in the periods from left to right, e.g., from sodium to argon in the third period, since the additional valence electrons (in the same quantum energy level) are exposed to a greater effective nuclear charge along the period.

2.9.4   Ionization Energy

Another important property that shows a trend in the periodic table is the ionization energy (the energy required to remove an electron from an atom). An atom has as many ionization energies as there are electrons. By definition, the *first ionization energy* is the energy required to remove the most outer electron from a neutral atom (Table 2-1):

$$M \rightarrow M^+ + 1e^-$$

The second ionization energy is the energy required to remove the next outer electron from the singly charged ion:

$$M^+ \rightarrow M^{2+} + 1e^-$$

Each successive removal of an electron requires more energy because, as more electrons are removed, the remaining electrons experience a greater effective attraction.

The first ionization potential increases across a period (Fig. 2-28), which is a direct result of the decrease in atomic radius (Fig. 2-27). As the atomic radius becomes smaller the electrons feel a greater attraction from the nucleus. As the force of attraction increases, more energy is required to remove the electrons. The larger nuclear charge in helium ($Z = 2$) that is responsible for the smaller radius (31 pm) results in a higher ionization potential (24.6 eV) compared to that of hydrogen (radius 37 pm and ionization potential 13.6 eV). Lithium, however, has one more electron than helium and this electron is at a higher quantum energy level. The lithium radius is thus greater and the ionization potential is less. The outermost electron in lithium is located in the 2s level, which is outside the 1s level occupied by the first two electrons. The 2s electron is screened by the other two and experiences a charge on the order of one. Thus, the ionization energy of this electron can be estimated to be nearly that of the 2s hydrogen state (which is 3.4 eV). The observed lithium ionization potential, however, is 5.4 eV (see Table 2-1). The value is larger because the outer electron is not perfectly shielded by inner electrons and the effective charge is greater than the assumed value of one. Because lithium has such small ionization energy it is a chemically active element. Next to lithium is beryllium. Due to the larger charge, the radius is smaller (Fig. 2-27) and the ionization potential is thus larger. Next is boron; the first four electrons occupy the $1s$ and $2s$ levels and the fifth electron is in $2p$ level. The increased charge causes the electrons in the new energy level to be more tightly bound, but the new energy level is further away from the nucleus and the valence electron is thus bound with a slightly weaker force. Consequently, the radius is reduced; the ionization potential is also reduced. Although there is an anomaly in the overall trend of ionization potential values across the period, the differences are small (the ionization potentials of beryllium and boron are 9.3 eV and 8.3 eV, respectively). In the next elements leading up to neon, the electrons occupy the $2p$ level (maximum of six electrons). The increasing charge decreases the atomic radius (Fig. 2-27) and ionization potential increases as indicated in Fig. 2-28. The next period starts with sodium. Since the valence electrons of neon fill the $2p$ level, the sodium valence electron can only occupy the higher $3s$ level. This accounts for the larger atomic radius and smaller ionization potential.

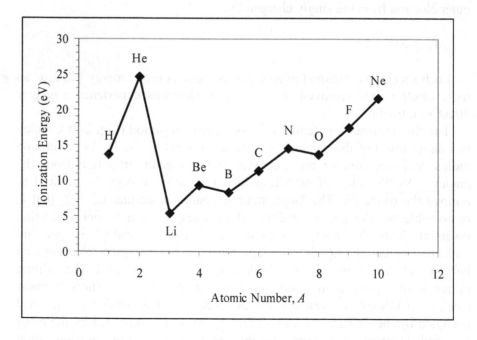

Figure 2-28. Ionization potential of the first ten elements

The small drop in ionization potential of oxygen compared to nitrogen is due to the arrangement of electrons. In nitrogen, two electrons occupy the $1s$ level and two others occupy the $2s$ level. The remaining three electrons occupy the $2p$ level. These three electrons occupy three available and distinct orbitals ($2p$ level accommodates three orientations of the electron's orbital, see Section 2.8.1). This keeps them well separated and reduces the repulsion between them. This in turn makes nitrogen relatively stable with relatively large ionization energy. In oxygen, the fourth electron occupies the $2p$ level and must share one of the orbitals (with opposite spin). These two electrons thus overlap in the orbital they share which increases the repulsion between them and decreases the ionization potential relative to nitrogen.

Examples of the periodic behavior of the elements are evident from the similarities between helium and neon (both very stable, with large ionization potential and small radii), or lithium and sodium (both with very low ionization potential and very large radii).

Example 2.15 First ionization potential

Explain which element from the list has the larger first ionization energy and why: Mg, Na or Al.

Magnesium (Mg), when compared to sodium (Na), has a larger first ionization potential because the first ionization energy tends to increase across a row of the periodic table from left to right (period 3, see Fig. 2-27). Mg also has a larger ionization potential than aluminum (Al) even though Al is to the right of Mg in the periodic table. The electron configurations for Mg and Al are (see Table 2-2)

Mg (12 electrons): $1s^2\,2s^2\,2p^6\,3s^2$

Al (13 electrons): $1s^2\,2s^2\,2p^6\,3s^2\,3p^1$

The outermost electron of Al (in the $3p$ level) is further away from the nucleus than the outermost Mg electrons which are in the $3s$ level. Less energy is thus required to remove the outermost Al electron

2.9.5   Independent Particle Approximation for Electrons

The multi-electron atom is a complicated and complex structure that cannot be exactly modeled. The basic and most common approximation is called the independent particle approximation, known as the IPA. The approximation is derived from the classical theory of the many-body problem such as the solar system in which all forces in between the planets are neglected in comparison to the force from the Sun. This same approach is applied to the atom, assuming the most important force is the force of the nucleus. This is however very poor approximation as the effect of the electrons on each other cannot be neglected. The IPA therefore analyzes each electron independently but the force is the sum of the forces from the nucleus and the force from the average distribution of $Z-1$ electrons. One electron in a multi-electron atom will therefore possess the so-called IPA potential energy, $U(\vec{r})$. It is assumed that the charge distribution affecting each electron is spherically symmetric, thus defining the IPA potential energy to be dependent only on distance from the nucleus, $U(r)$. According to the Gauss's law, any electron outside the charge that is spherically distributed experiences exactly the same force if it is assumed that that same charge, Q, is concentrated at the center of the sphere ($r = 0$)

$$F = k\frac{Qe}{r^2} \tag{2-40}$$

For the electron which is away from the rest of the electrons in an atom, $Z-1$, the total charge Q will be the charge of the nucleus (Ze) reduced by the charge of the rest of the electrons (located inside the radius r). In this extreme case, the charge Q becomes $Ze - (Z-1)e = e$. Thus this one electron experiences the same force as the electron in hydrogen atom, and Eq. (2-40) becomes

$$F = k\frac{e^2}{r^2} \tag{2-41}$$

For the electron located inside a spherical charge the electron will experience no force coming from that charge. This implies the following: the electron close to the nucleus is inside of the charge created by all other electrons; thus it will experience the attractive force from the nucleus (of charge Ze) and will not experience any force from other electrons

$$F = k\frac{Ze^2}{r^2} \tag{2-42}$$

If a force acting on a particle is a function of position only, the potential energy of that particle can be defined as an integral of the force. Thus it follows that when an electron is outside the charge created by all other electrons (at r outside all other electrons) its potential energy is defined by

$$U(r) = \int k\frac{e^2}{r^2}dr = -k\frac{e^2}{r} \tag{2-43}$$

For the electron inside the charge created by all other electrons

$$U(r) = \int k\frac{Ze^2}{r^2}dr = -Zk\frac{e^2}{r} \tag{2-44}$$

The last two equations could be written as follows

$$U(r) = -Z_{eff} k \frac{e^2}{r} \begin{cases} Z_{eff} = Z \\ \text{if } r \text{ (electron) is inside all other electrons} \\ Z_{eff} = 1 \\ \text{if } r \text{ (electron) is outside all other electrons} \end{cases} \qquad (2\text{-}45)$$

Because this potential energy depends only on distance it implies the spherical symmetry of the potential field, and therefore the energy is the same for both spin orientation. For the lowest energy level (like in hydrogen), $n = 1$ and $l = 0$ (Section 2.8.1), the $1s$ shell is twofold degenerate (due to two possible spin orientations) and is the closest to the nucleus. In other words, in many-electron atoms in $1s$ shell the potential energy is large and its numeric value is close to hydrogen-like potential energy; thus $Z_{eff} \sim Z$, giving

$$E_{1s} = -Z^2 E_R \qquad (2\text{-}46)$$

and the most probable radius is a_0/Z.

The next energy level is described with $n = 2$. As explained in Section 2.8.4 the shells $2s$ and $2p$ are located in the region where electrons experience the screening from the $1s$ shell seeing only Z_{eff} that is always smaller than Z. The $2s$ electrons penetrate closer to the nucleus, have lower energy than 2p electrons and experience almost full effect from Z. The energy levels with higher n than 1 are located at increasing distance from the nucleus and these electrons experience the screening effect thus seeing only $Z_{eff}e$. This effective charge is becoming smaller as the orbit quantum number is becoming higher. The most probable radius for such orbits is $\sim n^2 a_0/Z_{eff}$.

2.10 Atomic Parameters

Atomic mass is given in either the absolute unit of *grams* or in a relative unit called the *atomic mass unit* (*u* or *amu*):
- 1 mole of any substance contains 6.02×10^{23} molecules (Avogadro's number), N_a
- The weight in grams of 1 mole of a substance is numerically equal to its molecular weight
- The unified atomic mass unit is exactly one-twelfth of the mass of a C atom (C–12), i.e., the atomic mass of carbon-12 is equal to 12 amu

$$1 \text{ amu} = \frac{m_{C-12}}{12} = 1.661 \times 10^{-24} \text{ kg} = 931.5 \text{ MeV}/c^2 \qquad (2\text{-}47)$$

Example 2.16 Number of atoms

How many ^{12}C atoms are there in 12 g of carbon? What is the mass of one atom of carbon in kg?

Number of atoms in 12 g of carbon is

$$\frac{12 \text{ g}}{(1.661 \times 10^{-24} \text{ g/amu})(12 \text{ amu/atom})} = 6.02 \times 10^{23} \text{ atoms}$$

Because the molar mass of carbon-12 is 12 g, the mass of one atom of carbon 12 can be found by dividing the molar mass by Avogadro's number:

$$\frac{12 \text{ g/mol}}{6.02 \times 10^{23} \text{ atoms/mol}} = 1.993 \times 10^{-23} \text{ g/atom} = 1.993 \times 10^{-26} \text{ kg/atom}$$

The chemical properties of atoms are determined by the distribution of electrons (Section 2.9), and the number of electrons is called the *atomic number* and is usually denoted by Z. The number of protons in an atomic nucleus is also equal to Z, which is a requirement for electrical neutrality. When a neutral atom loses some of its electrons the atom becomes positively charged and is called a *positive ion*. For example, Ca^{2+} is a calcium atom that has lost two of its electrons. An atom can, however, gain electrons and thus become a *negative ion*. For example, Cl^- is a chlorine atom that gained one electron. The *atomic mass number, A*, is an integer that is almost equal to the atomic mass in amu. It is equal to the number of nucleons in the nucleus; that is, it is equal to the sum of the number of protons (Z) and the number of neutrons (N). Atoms (the elements of the periodic table) are denoted as follows:

$$^A_Z X$$

Atoms with the same atomic number Z (for example, ^{35}Ar, ^{38}Ar, ^{40}Ar) are called the *isotopes* of that element (argon). A naturally occurring sample of any element consists of one or more isotopes of that element and each isotope has a different weight. The relative amount of each isotope represents the isotope distribution for that element, and the *atomic weight* is obtained as the average of the isotope weights, weighted according to the isotope distribution.

Example 2.17 Atomic weight

Chromium (atomic weight 51.996) has four naturally occurring isotopes. Three of these are ^{50}Cr with isotopic weight 49.9461 and abundance 4.31%, ^{52}Cr with isotopic weight 51.9405 and abundance 83.76% and ^{54}Cr with isotopic weight 53.9389 and abundance 2.38%. Determine the isotopic weight of the fourth isotope.

$$M^{Cr} = \frac{4.31}{100} M^{50} + \frac{83.76}{100} M^{52} + \frac{2.38}{100} M^{54} +$$

$$\frac{[100 - (4.31 + 83.76 + 2.38)]}{100} M^x = 51.996$$

$M^{53} = 52.9237$ with an abundance of 9.55 %.

Example 2.18 Mass of an atom

Calculate the mass in grams of a ^{52}Cr atom. The atomic mass is 51.94051 amu. A mole contains N_a number of same particles (atoms or molecules); thus

$$M(^{52}Cr) = \frac{52 \ (g/mol)}{6.02 \times 10^{23} \ (atoms/mol)} = 8.638 \times 10^{-23} \ g/atom$$

However, knowing the atomic mass as given in the problem, the more precise mass of the atom is obtained as follows

$$M(^{52}Cr) = \frac{51.9405 \ (g/mol)}{6.02 \times 10^{23} \ (atoms/mol)} = 8.628 \times 10^{-23} \ g/atom$$

Example 2.19 Atom number density

Calculate the molecular weight of water and then determine the atom density of hydrogen in water.

The molecular weight of water is

$$2A_H + A_O = 2 \times 1 + 16 = 18 \quad \text{giving the molecular density of water}$$

$$N(H_2O) = \frac{\rho N_a}{A} = \frac{\left(1 \ g/cm^3\right)\left(6.02 \times 10^{23} \ molecules/mol\right)}{18 \ g/mol}$$

$$= 3.35 \times 10^{22} \ molecules/cm^3$$

The molecular weight of hydrogen is

$$N(H) = 2 \times N(H_2O) = 2 \times \left(3.35 \times 10^{22} \ atoms/cm^3\right)$$
$$= 6.69 \times 10^{22} \ atoms/cm^3$$

APPLICATIONS

Fluorescent Lamp

An atom gains or loses energy by adding or removing energy from the electrons influencing them to move from one orbital to another. Atom is in its neutral state if all electrons are in their lowest possible energy levels. Atoms gain energy by absorbing from photons or by colliding with another particle. As described earlier, when an atom gains energy one or more electrons will move to a higher (more energetic) orbital. Once excited the atoms hold that energy for a very short time, typically in the order of 10^{-12} s. After that time, the electron or electrons will go back to their original energy states, during which process the excess of energy is released in the form of a photon (Section 2.7.3). This emission of photon can be in any part of the light spectrum. The fluorescence lamps operate on that principle. The fluorescence occurs when atom absorbs usually a photon in the ultraviolet range and emits the photon (with the longer wavelength) in the visible part of a spectrum. The transition probability is explained at the end of Chapter 3, Section on APPLICATIONS.

Laser

Laser (light amplification by stimulated emission of radiation) is a device used to produce (amplify) the coherent light (emitted light is in a form of almost perfect sinusoidal waves) as a narrow, low-divergence photon beam, with a narrow wavelength spectrum (called the monochromatic light). The basic principle of the laser device is the transition of the electrons between the orbits in an excited atom and the emission of the photon of a desired frequency. A photon of correct wavelength will stimulate the transition of an electron in an excited atom to return to a lower energy level, thus emitting the photon of the same wavelength as the stimulating (incoming) photon. These two photons will then stimulate two more emissions producing four photons of the same wavelength and the process when continues will produce an amplified stimulated emission of radiation.

From 1960 when laser was invented, it has found numerous applications in science, medicine, engineering, industry, military, information technology and entertainment.

X-rays

The principle of the X-ray excitation is described in Section 2.9.2; the discovery of X-rays is described in Chapter 5; the interaction and effects are summarized in Chapter 6. The X-rays have wide applications in many disciplines: in medicine (medical imaging and cancer treatment), industry (industrial radiography for inspecting particularly the welding sections), homeland security (for example, at the airports for detecting the metal objects), astronomy (X-ray astronomy studying the X-ray emission for celestial bodies in the universe) and biology (X-ray microscopy for imaging the small objects).

PROBLEMS

2.1 Write the electron configuration for potassium, lanthanum, copper and bromine.

2.2 Name the elements whose electron configuration is
(a) $1s^2\,2s^2\,2p^6\,3s^2\,3p^6\,4s^2\,3d^3$
(b) $1s^2\,2s^2\,2p^6\,3s^2\,3p^6\,4s^2\,3d^{10}\,4p^6\,5s^2\,4d^9$
(c) $1s^2\,2s^2\,2p^6\,3s^2\,3p^6$

2.3 How many electrons are in an atom specified by $1s^2\,2s^2\,2p^6\,3s^2\,3p^4$?

2.4 (a) The attractive electrostatic force of the positively charged atomic nucleus forces the negatively charged electron of the hydrogen atom to a circular motion. Write the equation that describes this statement.
(b) Knowing that only orbital radii are allowed for which angular momentum is an integer multiple of $h/(2\pi)$ and using the equation from (a) develop the relation for the allowed radii.

2.5. (a) Express the relation for the frequency of revolution of the electron in hydrogen atom for $n = 1$.
(b) For this case show that $v/c = (ke^2)(2\pi)/hc = 1/137$, which is called the fine structure constant, α.

2.6 Starting from the Bohr's equation for the energy of the nth state of an electron in hydrogen atom, write the equation describing the frequency of light given off when an electron makes a transition from an initial to a final state. From there derive the value for Rydberg constant.

2.7 Calculate the largest velocity, the lowest energy level and the smallest orbit radius for the electron in hydrogen-like atoms. When the orbit becomes infinite what is the value of energy?

2.8 Calculate how many times in a second an electron in hydrogen atom orbiting at the level $n = 30$ goes around the nucleus?

2.9 What is the excited state of sodium atom? What is the excited state of hydrogen atom?

2.10 Knowing that the first excited state of sodium atom is at 2.1 eV above $3s$ level, determine the wavelength and frequency of the photon emitted in the $3p$ -> $3s$ transition.

2.11 The ground state of hydrogen atom has one electron in the $1s$ level with its spin pointing either way. Calculate the energy of the electron in this orbit using Bohr's theory. What is the value of the ionizing energy?

2.12 The ground state of a helium atom has two electrons and both in $1s$ level. How are their spins oriented? The first ionization potential is found experimentally to be 24.6 eV. Calculate the effective charge, Z_{eff}.

2.13 Calculate the value of Rydberg constant for the hydrogen atom taking into account the effect of reduced mass.

2.14 For heavy hydrogenic ions how does the reduced mass change and consequently what is the value of Rydberg constant?

2.15 Calculate the wavelengths of Balmer lines in hydrogen atom.

2.16 Explain departure from Rutherford formula. Give an example.

2.17 Calculate the first ionization potential of hydrogen helium atom.

2.18 Draw possible trajectories of an α particle in Rutherford experiment for different impact parameters and scattering angles.

2.19 For a gold (assuming to have a nuclear radius about 7 fm and an atomic radius of about 0.13 nm) thin foil used in Rutherford experiment estimate its maximum thickness that would not produce the multiple scattering of α particles.

2.20 For an ion of ^{23}Mg write the number of protons, neutrons and electrons.

2.21 Study the Millikan's experiment. Knowing that the oil droplets were produced such to have a radius of 1 μm, and that the voltage between plates positioned at the distance of 0.042 m was recorded whenever the droplets become stationary, using data provided, show that the charge difference is always the integral multiples of 1.602×10^{-19} C. Assume that the density of oil is 900 kg/m^3. The voltage as measured in Millikan's experiment is 391.49 V, 407.80 V, 376.43 V, 337.49 V, 362.49 V, 376.43 V.

2.22 What is the difference between the atomic weight and atomic mass? Give an example.

2.23 Show that the mass of the hydrogen atom is 1.6735×10^{-24} g and that of the oxygen atom is 2.6561×10^{-23} g.

2.24 If naturally occurring carbon consists of 98.892% ^{12}C and 1.108% ^{13}C what is the average mass (in amu) of carbon?

2.25 Calculate the molecular mass of methane (CH_4). What is the percentage by mass of the elements in this compound?

2.26 Using Eq. (2-3) write the computer code to plot the 7.7 MeV α particle's trajectories as a function of impact parameters and angles of deflection. Indicate the points of closest approach.

2.27 Use the Bohr's atomic model and write the computer code to calculate the orbiting velocity of the electron in hydrogen atom (see Example 2.5), helium ion, lithium ion and boron ion. Comment on the results.

2.28 To the computer code developed for Problem 2.27 add the calculation of the time it takes for an electron to complete one revolution in hydrogen atom and ions of helium, lithium and boron. How does the time change, with the orbits moving further away from the nucleus? Comment on the results.

2.29 Write the computer code to reproduce the spectral lines shown in Fig. 2-19.

2.30 Plot the Rydberg energy for hydrogen atom, and first 11 ions from the table of elements. Comment on the results.

2.31 Calculate the ratio of the allowed energies in the helium and lithium ions to that in the hydrogen atom, taking into account the effect of nuclear motion (see Example 2.9).

2.32 Calculate how much energy (in J and eV) does one electron with a principle quantum number of $n = 2$ have.

2.33 Write the electron configuration by explaining the filling of the shells and draw the complete orbital diagram (using the style of Fig. 2-23) for iron Fe.

2.34 In the Rutherford experiment replace the gold foil with the copper foil. For the α particle with the velocity of 1.6×10^9 cm/sec, what is the closest approach to the foil?

2.35 Calculate the ground state hydrogen atom diameter based on Bohr atomic model.

2.36. Demonstrate the Bohr correspondence principle by showing that the frequency of the radiation emitted f_{cl} (classical physics) is equal to the orbital frequency f_{orb} of the electron around the nucleus (Bohr theory).

2.37. Assume the transition from the state n_1 to state n_2 in the Problem 2.37 such that $n_1 - n_2 = 1$ or $= 2$ or $=3$. What is the frequency of the emitted radiation? Comment.

Chapter 3

NUCLEAR THEORY

Basic Principles, Evidence and Examples

The dazzling complexity of the material world can, for almost all purposes, be reduced to a simple trinity: the proton, the electron, and the neutron. The neutron, a component of the nucleus of every atom except that of hydrogen, was the last of the trinity to be discovered, in 1932. Had they all been a little younger, the scientist who uncovered the neutron might have met on the battlefields of World War II. *Brian L. Silver*, ("The Ascent of Science", 1998)

1.   INTRODUCTION

Atomic physics is the science of atoms, their structure and their behavior. To discuss the properties of atoms we need to know about the number of electrons and their configuration (see Chapter 2). In this context the information related to the atomic nucleus is not of great interest except to know that a neutral atom caries Z protons and $A - Z$ neutrons (where Z represents the atomic number and A the atomic mass number). The electrons arrangement is described in the previous chapter.

Nuclear physics is the science of nuclei. Atomic and nuclear physics use similar laws to describe the motion of electrons and the constituents of a nucleus (protons and neutrons). However, innovative approaches had to be developed to describe the forces that hold protons and neutrons in a nucleus. A theoretical understanding of the forces acting inside the nucleus is not yet complete. Nucleons are arranged in the nucleus similar to the electrons in an atom. Both follow the principles of quantum mechanics. Yet, there are important differences: in an atom the nucleus represents a center of Coulomb

T. Jevremovic, *Nuclear Principles in Engineering*,
DOI 10.1007/978-0-387-85608-7_3, © Springer Science+Business Media, LLC 2009

attraction for electrons and the forces between the electrons usually do not play an important role in the atomic model; in the nucleus there is no center of attraction for the nuclei; they are held together by their mutual interaction by the forces less understood than the Coulomb forces.

Since all the electrons are the same and since there are two types of nucleons, protons and neutrons, there are around 100 different atoms but more than 10,000 different nuclei. In this chapter the basic principles and the laws of nuclear theory are presented.

2.   THE NUCLEUS

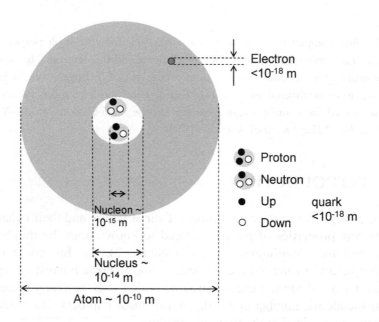

Figure 3-1. Schematic representation of an atom indicating the size of the atom, electron, nucleus, nucleons and quarks

The nucleus of an atom is generally composed of protons and neutrons, collectively called *nucleons*. The nucleus sketched in Fig. 3-1 is currently understood to be a quantum system composed of nucleons of nearly equal mass and the same intrinsic angular momentum (spin) of ½. The proton, a positively charged particle, was discovered by Ernest Rutherford in 1919 (Section 2.5.1). The neutron, an electrically neutral particle, was discovered by the British physicist Chadwick in 1932 (Section 2.5.2). Its presence in the nucleus accounts for the difference between the atomic number and the atomic mass number and also supplies forces that hold the nucleus together.

In addition to its atomic number and atomic mass number, a nucleus is characterized by its size, shape, binding energy, angular momentum and stability.

2.1   Size, Shape and Density of Nucleus

As described in Chapter 2, Section 2.4, Rutherford's experiment showed that it was possible to determine the size of a nucleus by bombarding a gold foil with a beam of α particles. According to the evidence from his experiments, it was understood that an atom has a large radius in comparison to the size of its nucleus. At that time, the α particles were used because they were easily attainable as a product of radioactive decay of unstable nuclei (Chapter 5). Currently, experiments designed to probe the shapes and sizes of nuclei utilize accelerated beams of electrons, protons and α particles. One of the best ways to determine the size of a nucleus is to scatter high-energy electrons from it. From the angular distribution of the scattered electrons, which is dependent on the proton distribution in the nucleus, the shape and an average radius of a nucleus can be determined. Data from these types of experiments indicate that most nuclei have a spherical shape though some (for example those with $Z = 56 - 71$) have ellipsoidal shapes with eccentricities of less than 0.2 (departure from the spherical shape). Since departure from spherical shape is usually minimal, most theoretical models assume that the nucleus is spherical.

The nuclear radius of known elements ranges from 2 fm (He) to 7 fm (U). The radius of any nucleus can be approximated using the *Fermi model*:

$$R = R_0 A^{1/3} \tag{3-1}$$

where A represents the atomic mass number and R_0 is the constant determined from the experiments that depends on the charge distribution of a nucleus. For a medium and heavy nuclei, $R_0 = 0.94$ fm [Henley and Garcia, 2007]. In the older literature the nucleus would be assumed to have a uniform density distribution of charge with $R_0 = 1.2$ fm. The more realistic Saxon–Woods approximation of the nuclear density gives the relation for a nucleus radius that fits better the experimental measurements for most of the nuclei

$$R[\text{fm}] = 1.18A^{1/3} - 0.48 \tag{3-2}$$

Since the volume of a sphere of radius R is proportional to R^3, it follows from Eq. (3-1) that the nuclear volume is proportional to A, i.e., to the total number of nucleons:

$$V = \frac{4}{3}\pi R^3 = \frac{4}{3}\pi R_0^3 A = V_0 A \tag{3-3}$$

Thus, if the volume of a nucleus is proportional to A, it is clear that the volume per nucleon, V_0 is approximately the same for all nuclei. In other words, the density of nucleons (nuclear density) is the same for all nuclei as is the degree of packing of nucleons for all nuclei.

Example 3.1 Nuclear density

Compare the nuclear densities of ^{12}C and ^{235}U.

Knowing that a mass of ^{12}C atom is 12 amu and of ^{235}U atom is 235 amu (see Chapter 2, Section 2.10), it can be shown that the nuclear density is a constant value:

$$V^{^{12}C} = V_0 A = 12V_0 \quad V^{^{235}U} = V_0 A = 235V_0 \quad \text{and} \quad V_0 = \frac{4}{3}\pi R_0^3 \quad \text{giving}$$

$$\rho^{^{12}C} = \frac{M^{^{12}C}}{V^{^{12}C}} = \frac{12\,\text{amu}}{12V_0} = \frac{1.661 \times 10^{-27}\,\text{kg}}{5.13 \times 10^{-45}\,\text{m}^3} = 3.2 \times 10^{17}\,\text{kg/m}^3$$

$$\rho^{^{235}U} = \frac{M^{^{235}U}}{V^{^{235}U}} = \frac{235\,\text{amu}}{235V_0} = \frac{1.661 \times 10^{-27}\,\text{kg}}{5.13 \times 10^{-45}\,\text{m}^3} = 3.2 \times 10^{17}\,\text{kg/m}^3$$

The obtained density inside the nucleus is some 14 orders of magnitude greater than the density of ordinary matter like solids or liquids. For example the density of water at standard temperature and pressure is 1,000 kg/m^3.

Investigation of nuclear size and structure took place in 1950s producing several Nobel Prizes. For example, for his pioneering studies of electron scattering in atomic nuclei and for his discoveries concerning the structure of the nucleons Robert Hofstadter (1915–1990) was awarded the Nobel Prize for physics. He shared the prize with Rudolf Mossbauer. Robert Hofstadter is the father of the cognitive scientist and philosopher Douglas R. Hofstadter best known for his 1980 Pulitzer Prize winning book, *Gödel, Escher, Bach: an Eternal Golden Braid.*

2.2   Equivalence of Mass and Energy

Albert Einstein in his special theory of relativity postulated that the velocity of light in vacuum is the upper limit of speed in the universe. According to his theory, the mass of a moving body is not constant (as classical mechanics would predict) but is a function of velocity. The relation between mass and velocity of a moving body indicates that as the velocity increases, the mass of a body increases:

$$m = m_0 \Big/ \sqrt{1 - \frac{v^2}{c^2}} \qquad (3\text{-}4)$$

where

m – mass of a moving body (also called the variable mass)
m_0 – rest mass of a body (velocity is zero)
v – velocity of a moving body
c – speed of light
The ratio, v^2/c^2 is usually denoted as β^2. Thus

$$m = m_0 \Big/ \sqrt{1 - \beta^2} \qquad (3\text{-}5)$$

Similarly, the relativistic energy of a body moving with velocity v is no more expressed as $mv^2/2$, but

$$E = m_0 c^2 \Big/ \sqrt{1 - \beta^2} \qquad (3\text{-}6)$$

In the relativistic case, velocity increase due to additional energy is smaller than in the non-relativistic case, because the additional energy serves to increase the mass of the moving body rather than its velocity. Equation (3-6) suggests that

- The more energy an object has, the heavier it is.
- The closer the velocity of a moving body is to the speed of light, the larger the force needed to accelerate the body.
- An infinite force is needed to accelerate a material object to the speed of light, which is not physically possible. The only particle that travels at the speed of light is a photon (that has a zero mass). It is also assumed that the graviton moves at the speed of light but there is still no evidence of its existence.
- Mass and energy are equivalent. In other words, all matter contains potential energy by virtue of mass.
- A body at rest ($v = 0 \rightarrow \beta = 0$) (non-relativistic approximation) possesses energy given by the Einstein equation

$$E = m_0 c^2 \quad (v = 0) \qquad (3\text{-}7)$$

In terms of a momentum and mass

$$E^2 = (pc)^2 + (m_0 c^2)^2 \qquad (3\text{-}8)$$

Graphically, the three terms in Eq. (3-8) can be represented as the sides of a right triangle (see Fig. 3-2). The energy–momentum equation is therefore often referred to as the "Pythagorean relation".

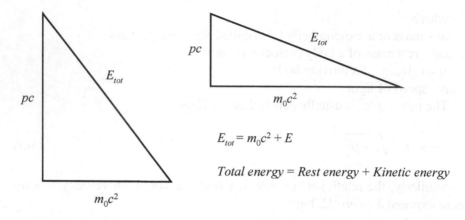

$$E_{tot} = m_0c^2 + E$$

Total energy = Rest energy + Kinetic energy

Figure 3-2. Graphic representation of the relativistic energy–momentum relation: (**a**) when $pc < m_0c^2$ the total energy is mostly rest energy, (**b**) when $pc > m_0c^2$ the total energy is mostly kinetic energy of the moving particle

Example 3.2 Rest energy of nuclear fuel

How much energy can be obtained from 1 g of nuclear fuel?

$$E = m_0c^2 = (1g)(3\times10^{10}\,cm/s)^2 = 9\times10^{20}\,erg = 9\times10^{13}\,J$$
$$= (9\times10^{20}\,erg)(2.78\times10^{14}\,kWh/erg) = 2.5\times10^7\,kWh$$

The result suggests

- A small amount of mass corresponds to a large amount of energy (because the speed of light is large).
- In nuclear reactions an atomic nucleus of initial mass M is transformed into a nucleus of mass M' and the difference in mass is released as energy

$$E = (M - M')c^2 \tag{3-9}$$

Example 3.3 Electron volt (eV)

Show that the energy given to an electron (of charge $e = -1.602 \times 10^{-19}$ coulomb) by accelerating it through 1 volt of electric potential difference called 1 eV is equal to 1.602×10^{-19} J.

The work needed to move one electron through a voltage drop of 1 volt is
$$e\,\Delta V = (-1.602 \times 10^{-19}\,coulomb)(-1\,volt) = 1.602 \times 10^{-19}\,J = 1\,eV$$

Example 3.4 Rest mass of an electron

Calculate the rest energy of an electron in eV and its mass in eV/c^2 if its mass is 9.109×10^{-31} kg.

The rest energy is

$$E = m_0c^2 = (9.109 \times 10^{-31} \text{kg})(3 \times 10^{10} \text{cm/s})^2 = 81.98 \times 10^{-15} \text{J}$$

$$= (81.98 \times 10^{-15} \text{J})\left(\frac{1 \text{eV}}{1.602 \times 10^{-19} \text{J}}\right) = 5.11 \times 10^5 \text{eV} = 0.511 \text{MeV}$$

Thus, $m_0 = 0.511 \text{ MeV}/c^2$.

Example 3.5 Relativistic momentum of an electron

Prove the energy–momentum relation given in Eq. (3-8)

$$E^2 = (pc)^2 + (m_0c^2)^2$$

By definition, momentum can be described as a function of the mass and velocity of a moving body:

$$p = m\upsilon = m_0\upsilon / \sqrt{1-\beta^2} \tag{3-10}$$

Squaring Eq. (3-6)

$$E^2 = (mc^2)^2 = (m_0c^2)^2 / 1 - \frac{\upsilon^2}{c^2} \rightarrow m^2c^4\left(1-\frac{\upsilon^2}{c^2}\right) = m_0^2c^4$$

$$m^2c^4 = E^2 = m_0^2c^4 + m^2c^2\upsilon^2$$

where $p = m\upsilon$, thus giving

$$E^2 = (pc)^2 + (m_0c^2)^2$$

For a massless particle (like a photon) it follows that the total energy depends on its momentum and the speed of light: $E = pc$. This aspect will be discussed in greater detail in later sections.

Example 3.6 Transition of masses

Assume a piece of solid matter initially weighs 6 g. Following a reaction, the mass of the products is one half of the initial mass. Calculate the energy (in J) released in this mass "transition".

From Eq. (3-9)

$$E = \left(M - \frac{M}{2} \right) c^2 = (6 \times 10^{-3} \text{kg} - 3 \times 10^{-3} \text{kg})(3 \times 10^8 \text{m/s})^2 = 27 \times 10^{13} \text{J}$$

2.3   Binding Energy of a Nucleus

Since an atom contains Z positively charged particles (protons) and $N (= A - Z)$ neutral particles (neutrons), the total charge of a nucleus is $+Ze$ where e represents the charge of one electron. Thus, the mass of a neutral atom, M_{atom}, can be expressed in terms of the mass of its nucleus, M_{nuc}, and its electrons, m_e

$$M_{atom} = M_{nuc} + Zm_e \qquad M_{nuc} = Zm_p + (A - Z)m_n \qquad (3\text{-}11)$$

where m_p is the proton mass, m_e the mass of an electron and m_n the mass of a neutron (Appendix 2).

For example the mass of the rubidium nucleus, ^{87}Rb, which contains 37 protons and 50 neutrons, can be calculated as

$$M_{nuc}(^{87}\text{Rb}) = 37 \times 1.007276 + 50 \times 1.008665 = 87.70246 \text{ amu} \qquad (3\text{-}12)$$

The atomic mass, indicated on most tables of the elements, is the sum of the nuclear mass and the total mass of the electrons present in a neutral atom. In the case of ^{87}Rb, 37 electrons are present to balance the charge of the 37 protons. The atomic mass of ^{87}Rb is then

$$M_{atom}(^{87}\text{Rb}) = 87.70246 + 37 \times 0.00055 = 87.72281 \text{ amu} \qquad (3\text{-}13)$$

From the periodic table, the measured mass of an ^{87}Rb atom is found to be $M_{atom}^{measured}(^{87}\text{Rb}) = 86.909187 \text{ amu}$. These two masses are not equal and the difference is given by

$$\Delta m = M_{atom}(^{87}\text{Rb}) - M_{atom}^{measured}(^{87}\text{Rb}) = 0.813623 \text{ amu} \qquad (3\text{-}14)$$

Expanding the terms in Eq. (3-14) shows that the difference in mass corresponds to a difference in the mass of the nucleus

$$\Delta m = M_{atom} - M_{atom}^{measured} = Zm_p + Zm_e + (A - Z)m_n$$
$$- M_{nuc}^{measured} - Zm_e \qquad (3\text{-}15)$$

which reduces to

$$\Delta m = M_{atom} - M_{atom}^{measured}$$
$$= Zm_p + (A-Z)m_n - M_{nuc}^{measured} = M_{nuc} - M_{nuc}^{measured} \qquad (3\text{-}16)$$

Thus, when using the atomic mass values given in the periodic table, the mass difference between the measured and calculated is given with

$$\Delta m = M_{nuc} - M_{nuc}^{measured} = Zm_p + Zm_e + (A-Z)m_n - M_{atom}^{measured} \qquad (3\text{-}17)$$

Notice also that

$$Zm_p + Zm_e = Zm_H \qquad (3\text{-}18)$$

where m_H is a mass of the hydrogen atom.

From this and other examples it can be concluded that the actual mass of an atomic nucleus is *always* smaller than the sum of the rest masses of all its nucleons (protons and neutrons). This is because some of the mass of the nucleons is converted into the energy that is needed to form that nucleus and hold it together. This converted mass, Δm, is called the *mass defect* and the corresponding energy is called the *binding energy* and is related to the stability of the nucleus; the greater the binding energy, the more stable the nucleus. This energy also represents the minimum energy required to separate a nucleus into protons and neutrons. The mass defect and binding energy can be directly related, as shown in Eqs. (3-19) and (3-20)

$$BE = \Delta m \times 931.5 \text{ MeV/amu} \quad \text{or} \qquad (3\text{-}19)$$

$$BE = \left(Zm_p + Zm_e + (A-Z)m_n - M_{atom}^{measured} \right) c^2 \qquad (3\text{-}20)$$

Since the total binding energy of the nucleus depends on the number of nucleons, a more useful measure of the cohesiveness is the *binding energy per nucleon*, E_b

$$E_b = \frac{BE}{A} = \frac{\Delta m(\text{amu}) \times 931.5(\text{MeV/amu})}{A(\text{nucleons})} [\text{MeV/nucleon}] \qquad (3\text{-}21)$$

The binding energy per nucleon varies with the atomic mass number, A, as shown in Fig. 3-3. For example, the binding energy per nucleon in a rubidium nucleus is 8.9 MeV, while in helium it is 7.3 MeV. The curve indicates three characteristic regions:

- *Region of stability* – A flat region between *A* equal to approximately 35 and 70 where the binding energy per nucleon is nearly constant. This region exhibits a peak near *A* = 60. These nuclei are near iron and are called the iron peak nuclei representing the most stable elements.

- *Region of fission reactions* – The heaviest nuclei are less stable than the nuclei near *A* = 60, which suggests that energy can be released if heavy nuclei split apart into smaller nuclei having masses nearer the iron peak. This process is called *fission* (the basic nuclear reaction used in atomic bombs as uncontrolled reactions and in nuclear power and research reactors as controlled chain reactions). Each fission event generates nuclei commonly referred to as fission fragments with mass numbers ranging from 80 to 160. Fission is described in Chapter 7.

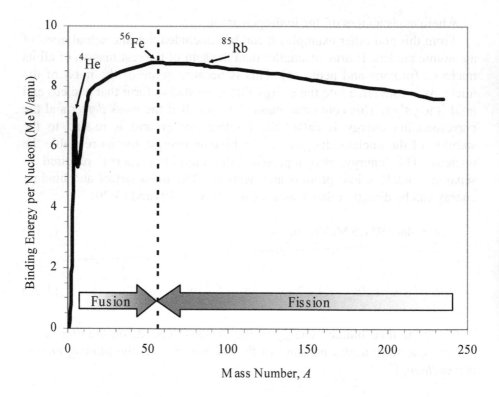

Figure 3-3. Variation of the binding energy per nucleon with the atomic mass number (read also Section 3.1)

- *Region of fusion reactions* – The curve of binding energy suggests a second way in which energy could be released in nuclear reactions. The lightest elements (like hydrogen and helium) have nuclei that are less stable than heavier elements up to the iron peak. If two light nuclei can form a heavier nucleus a significant energy could be released. This process is called *fusion* and represents the basic nuclear reaction in hydrogen (thermonuclear) weapons.

2.4 Stability of the Nucleus

Nuclei that have the same number of protons and different number of neutrons are called *isotopes*. For example, two isotopes of oxygen, $^{16}_{8}O_8$ and $^{17}_{8}O_9$, both have eight protons, but one has eight neutrons while the other has nine. Nuclei with the same mass number such as $^{14}_{6}C_8$ and $^{14}_{7}N_7$ are called *isobars*, while *isotones* are nuclei with the same number of neutrons, for example, $^{13}_{6}C_7$ and $^{14}_{7}N_7$.

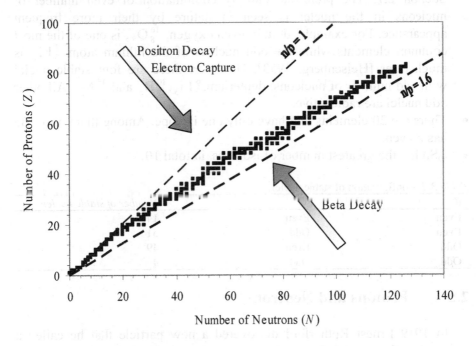

Figure 3-4. Nuclear stability curve (see also Chapter 5)

The naturally occurring elements have the atomic numbers (Z) between 1 and 92 with neutron numbers (N) between 0 and 146. If the number of protons is plotted against the number of neutrons for all nuclides existing in nature as shown in Fig. 3-4, the following tendencies are observed:

- For light nuclei ($A \leq 40$), Z and N are nearly equal. This tendency of $Z \sim N$ is called the *symmetry effect* and is a characteristic of stable nuclei.

As a result of the fundamental similarity between protons and neutrons, an unstable nucleus will transform a proton into a neutron or vice versa in order to reach the stable $Z \sim N$ arrangement.

- For heavier nuclei more neutrons are needed to form a stable configuration and the ratio of N to Z approaches 1.5 for the heaviest nuclei. The tendency for N to be bigger than Z is due to the electrostatic repulsion force acting between the protons. If a nucleus has too many or too few neutrons it is unstable and may spontaneously rearrange to attain stable configuration. Isotopes of atoms with unstable nuclei are called *radioisotopes* and are *radioactive* (see Chapter 5).

- A preference for Z and N to be even is observed in the majority of nuclei. When the numbers of neutrons and protons are both even numbers, the isotopes tend to be far more stable than when they are both odd (Table 3-1). This tendency is the result of a *pairing effect* that is described in Section 2.7. The preferred stability combination of even number of nucleons in the nuclei is seen in nature by their more frequent appearance. For example, doubly even oxygen, $^{16}_{8}O_8$, is one of the most common elements while the odd nucleus of the lithium atom $^{7}_{3}Li_4$ is much rarer [Heisenberg, 1953]. However, there are four stable nuclei with odd number of nucleons: deuterium, $^{6}_{3}Li_3$, $^{10}_{5}B_5$ and $^{14}_{7}N_7$. All other odd nuclei are radioactive.

- There are 20 elements that have only one isotope. Among them only ^{9}Be has Z even.

- $_{50}Sn$ has the greatest number of isotopes, in total 10.

Table 3-1. Configuration of stable nuclei

Z	N	*Number of stable nuclei*
Even	Even	148
Even	Odd	51
Odd	Even	49
Odd	Odd	4

2.5   Protons and Neutrons

In 1919 Ernest Rutherford discovered a new particle that he called a *proton* (the first particle known to be a constituent of every nucleus). He was investigating the effect of α particles interacting with nitrogen gas and noticed signs of hydrogen in the detectors. Rutherford postulated that this hydrogen could have come only from the nitrogen, and therefore that nitrogen contains hydrogen nuclei. At that time it was known that hydrogen had an atomic number equal to 1; thus he suggested that the hydrogen nucleus itself was an elementary particle and he named it proton using the Greek word for "first", *protos*. From this experiment it was understood that the proton carries a positive electrical charge equal in magnitude to the negative charge of an electron because the number of protons in a nucleus was found to be the same as the number of electrons surrounding it for an atom in its neutral state. In 1932 Rutherford's colleague James Chadwick discovered another constituent of the nucleus which he named the *neutron*. Neutron carries no electrical charge and thus can pass through material without being deflected by electrical forces.

Protons and neutrons are approximately equal in mass (each roughly 2,000 times heavier than an electron) and are both composed of up and down quarks whose fractional charges (2/3 and −1/3) combine to produce the 0 or +1 charge of the neutron and proton, respectively (Fig. 3-1).

2.5.1   Protons and Proton Decay

The positive charge of the nucleus of any atom is due to the presence of protons. Every atomic nucleus contains at least one proton and the total number of protons (atomic number) is different for every element. The basic characteristics and constituents of the proton are summarized in Fig.3-5.

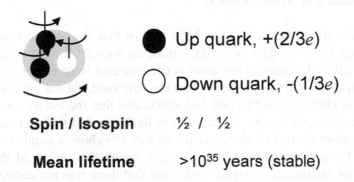

Up quark, +(2/3e)

Down quark, -(1/3e)

Spin / Isospin   ½ / ½

Mean lifetime   >10^{35} years (stable)

Figure 3-5. Nuclear properties of proton

The possibility that the proton may have a finite lifetime has been under experimental investigation for the last decade. The latest experimental evidence suggests that the lower boundary for proton lifetime is over 10^{35} years (many times the present age of the universe, which is estimated to be on the order of 15×10^9 years). How is it possible to detect a time that is longer than the existence of the universe? Obviously it is not possible to watch one proton for 10^{35} years to see if it decays; however, 10^{35} protons can be observed for 1 year with a 50–50 probability that one proton out of 10^{35} will decay. There are two laboratories equipped for this experiment: one in Minnesota and the other in Japan (the Super-Kamiokande). The dominant mode of proton decay is into a positron and a neutral pion: p → e$^+$ + π^0. A positron is an anti-electron, a particle with the same mass and same charge as an electron, but with the opposite charge sign (i.e., a positively charged electron). The pion (or "π meson") is the collective name for three subatomic particles, π^0, π^+ and π^-. The Super-Kamiokande detector has the capability to observe this mode of proton decay. Using a huge pool of water as a source of protons, a proton from either hydrogen or oxygen may decay into a positron and a neutral pion. Upon contact with an electron, the positron is destroyed in a process known as electron–positron annihilation (see Chapter 5). Upon contact, the positron and electron destroy each other, producing two 511 keV photons. The π^0 has a mass of 135 MeV/c^2 and decays into two photons with a very short half-life of 84×10^{-18} s. The experiments look for these electromagnetic showers as an indication of proton decay. To date, proton decay has not been observed at either facility.

The proton is positively charged particle with the charge distributed uniformly around the proton center. The experimental measurements showed that the mean radius of this charge is ~0.8 fm [W. N. Cottingham, D. A. Greenwood, 2001].

2.5.2   Neutrons and Neutron Decay

From the time of Rutherford it has been known that the atomic mass number A of nuclei is more than twice the atomic number Z for most atoms and that almost all of the mass of the atom is concentrated in its center, i.e., at the nucleus. However, it was presumed that the only fundamental particles were protons and electrons. Rutherford had speculated that the nucleus was composed of protons and proton–electron pairs tightly bound together and the fact that an atom neutral in charge required that somehow a number of electrons were bound in the nucleus to partially cancel the charge of the protons. Quantum mechanics, however, indicated that there was not enough energy available to contain electrons in the nucleus (see Section 5.4). An experimental breakthrough came in 1930 when Bothe and Becker bombarded a beryllium target with α particles emitted from a radioactive source. The experiment produced neutral radiation that was observed to be highly penetrating but non-ionizing. In the following years Curie and Joliot showed that when paraffin (a material rich in protons) is bombarded with this neutral radiation it ejects protons with energy of about 5.3 MeV (Fig 3-6). Bothe and Joliot–Curie each explained that the radiation was high-energy gamma rays. This, however, proved to be inconsistent with what was known about gamma ray interactions with matter (Chapter 6).

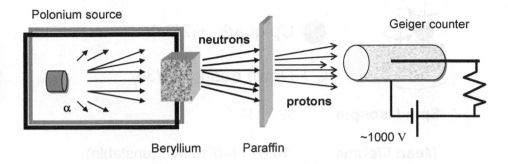

Figure 3-6. Experimental setup that led to the discovery of a neutron in 1932

In 1932 Chadwick performed a number of experiments using different target materials to discover that the emitted radiation was actually a stream of new particles that he named *neutrons*. The discovery proved that there is a neutral particle in the nucleus, but also that there are no free electrons in the nucleus as Rutherford had speculated. Amazingly, once free from the nucleus, neutrons are unstable and decay with a half-life of about 10 min into a proton, an electron and an anti-neutrino.

The α–Be reaction in the experiment shown in Fig. 3-6 was explained by Chadwick

$$_2^4\text{He} + _4^9\text{Be} \rightarrow _0^1\text{n} + _6^{12}\text{C} \tag{3-22}$$

where $_2^4\text{He}$ represents the α particle. He argued that if a photon interacts with a proton in the paraffin target and transfers 100 MeV/c of recoil momentum, the photon itself must have had a momentum of nearly 50 MeV/c, which corresponds to energy of 50 MeV (see Section 2.2). As the energy of the α particles striking the beryllium target was only about 5 MeV, it was impossible that 50 MeV gammas were being emitted. Instead, Chadwick suggested a new particle with approximately the same mass as a proton, which solved the contradiction related to the energy of the assumed photons. In the collision of two particles of equal masses, the incident particle (neutron) can transfer all of its kinetic energy to the target particle (proton). Thus for the observed momentum of 100 MeV/c, the kinetic energy of the neutron was $p^2 / 2m = (100 \text{ MeV}/c)^2 / (2 \times 938 \text{ MeV}/c^2) = 5.3$ MeV. By examining the interactions of neutrons with various materials, Chadwick determined the actual mass of the neutron to be between 1.005 amu and 1.008 amu, or 938 ± 1.8 MeV. The presently accepted value is 939.565 MeV (Fig. 3-7).

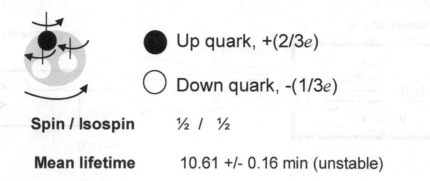

Spin / Isospin   ½ / ½

Mean lifetime   10.61 +/- 0.16 min (unstable)

Figure 3-7. Properties and composition of the neutron

The following is the original note Chadwick sent to *Nature* in February of 1932 describing the arguments for the existence of a neutron. Chadwick was awarded a Noble Prize for his discovery in 1935.

Possible Existence of a Neutron
James Chadwick, *Nature*, p. 312 (Feb. 27, 1932)

It has been shown by Bothe and others that beryllium when bombarded by α-particles of polonium emits a radiation of great penetrating power, which has an absorption coefficient in lead of about 0.3 $(cm)^{-1}$. Recently Mme. Curie-Joliot and M. Joliot found, when measuring the ionization produced by this beryllium radiation in a vessel with a thin window, that the ionization increased when matter containing hydrogen was placed in front of the window. The effect appeared to be due to the ejection of protons with velocities up to a maximum of nearly 3×10^9 cm per sec. They suggested that the transference of energy to the proton was by a process similar to the Compton effect, and estimated that the beryllium radiation had a quantum energy of 50×10^6 electron volts.

I have made some experiments using the valve counter to examine the properties of this radiation excited in beryllium. The valve counter consists of a small ionization chamber connected to an amplifier, and the sudden production of ions by the entry of a particle, such as a proton or α particle, is recorded by the deflexion of an oscillograph. These experiments have shown that the radiation ejects particles from hydrogen, helium, lithium, beryllium, carbon, air, and argon. The particles ejected from hydrogen behave, as regards range and ionizing power, like protons with speeds up to about 3.2×10^9 cm per sec. The particles from the other elements have a large ionizing power, and appear to be in each case recoil atoms of the elements.

If we ascribe the ejection of the proton to a Compton recoil from a quantum of 52×10^6 electron volts, then the nitrogen recoil atom arising by a similar process should have an energy not greater than about 400,000 volts, should produce not more than about 10,000 ions, and have a range in air at N.T.P. of about 1.3 mm. Actually, some of the recoil atoms in nitrogen produce at least 30,000 ions. In collaboration with Dr. Feather, I have observed the recoil atoms in an expansion chamber, and their range, estimated visually, was sometimes as much as 3 mm at N.T.P.

These results, and others I have obtained in the course of the work, are very difficult to explain on the assumption that the radiation from beryllium is a quantum radiation, if energy and momentum are to be conserved in the collisions. The difficulties disappear, however, if it be assumed that the radiation consists of particles of mass 1 and charge 0, or neutrons. The capture of the α-particle by the Be^9 nucleus may be supposed to result in the formation of a C^{12} nucleus and the emission of the neutron. From the energy relations of this process the velocity of the neutron emitted in the forward direction may well be about 3×10^9 cm per sec. The collisions of the neutron with the atoms through which it passes give rise to the recoil atoms, and the observed energies of the recoil atoms are in fair agreement with this view. Moreover, I have observed that the protons ejected from hydrogen by the radiation emitted in the opposite direction to that of the exciting α-particle appear to have a much smaller range than those ejected by the forward radiation. This again receives a simple explanation of the neutron hypothesis.

If it be supposed that the radiation consists of quanta, then the capture of the α-particle by the Be^9 nucleus will form a C^{13} nucleus. The mass defect of C^{13} is known with sufficient accuracy to show that the energy of the quantum emitted in this process cannot be greater than about 14×10^6 volts. It is difficult to make such a quantum responsible for the effects observed.

It is to be expected that many of the effects of a neutron in passing through matter should resemble those of a quantum of high energy, and it is not easy to reach the final decision between the two hypotheses. Up to the present, all the evidence is in favor of the neutron, while the quantum hypothesis can only be upheld if the conservation of energy and momentum be relinquished at some point.

J. Chadwick.
Cavendish Laboratory,
Cambridge, Feb. 17.

2.6 Nuclear Forces

Electrons are held in their orbits around the positively charged nucleus by the electrostatic (Coulomb) force of attraction. Within the nucleus, however, there are only neutral particles (neutrons) and positively charged particles (protons). Therefore the only electrostatic force that acts within the nucleus is a disruptive repulsive force between protons. The gravitational force that acts on all matter regardless of charge is too weak to hold the nucleus, as illustrated with Example 3.7.

Example 3.7 Gravitational force within the nucleus

Use Bohr's atomic model for hydrogen to show that it is impossible to find an atom on the quantum level bound by gravity. Also, show that the gravitational force cannot hold nucleons in the nucleus.

According to Coulomb's law, which has the same form as Newton's universal law of gravity, the electrostatic and gravitational forces acting between an electron of mass m and charge q and a proton of mass M and charge Q are

$$F_{el} = kqQ/r^2 = ke^2/r^2 \quad \text{and} \tag{3-23}$$

$$F_{grav} = GmM/r^2 \tag{3-24}$$

where $G = 6.67 \times 10^{-11}\,\text{Nm}^2/\text{kg}$ and $k = 9.0 \times 10^{-9}\,\text{Nm}^2/\text{C}^2$. According to Bohr's atomic model the energy of an electron in its ground state in H-atom is

$$E_n = -\frac{ke^2}{2a_0}\frac{1}{n^2} \quad (n = 1,2,3,...)\,; \quad a_0 = \frac{\hbar^2}{ke^2 m} = 0.0529\,\text{nm} \tag{3-25}$$

Equation (3-25) can be rewritten as

$$E_n = -\frac{(ke^2)^2 m}{2\hbar^2}\frac{1}{n^2} \tag{3-26}$$

If the force that bounds electrons to the atom was gravitational, Eqs. (3-23), (3-24) and (3-26) yield

$$F_{el} = F_{grav} \implies \frac{ke^2}{r^2} = \frac{GmM}{r^2} \implies ke^2 = GmM$$

$$\implies E_n = -\frac{(GmM)^2 m}{2\hbar^2} \frac{1}{n^2}$$

(3-27)

$$E_1 = -\frac{\left(6.67 \times 10^{-11}\, \text{Nm}^2/\text{kg}\right)^2 \left(9.1 \times 10^{-31}\, \text{kg}\right)^3 \left(1.67 \times 10^{-27}\, \text{kg}\right)^2}{2\left(1.0545 \times 10^{-34}\, \text{Jsec}\right)^2} \frac{1}{1^2} =$$

$$-4.2 \times 10^{-97}\, \text{eV}$$

Recall that according to Bohr's atomic model, the lowest possible energy level corresponds to the ground state for which $n = 1$ and $E_1 = -13.6$ eV.

The ratio of the gravitational and electrostatic forces shown in Eq. (3-28) (which is independent of the distance between the particles) shows that the gravitational force is too weak to overcome the repulsion between the protons and thus hold the nucleons together in the nucleus. In the following equation, the masses are those of the two protons:

$$\frac{F_{grav}}{F_{el}} = \frac{GMM}{ke^2} =$$

$$\frac{\left(6.67 \times 10^{-11}\, \text{Nm}^2/\text{kg}\right)\left(1.67 \times 10^{-27}\, \text{kg}\right)^2}{\left(9.0 \times 10^9\, \text{Nm}^2/\text{C}^2\right)\left(1.6 \times 10^{-19}\, \text{C}\right)^2} = 8 \times 10^{-37}$$

(3-28)

The force that holds the nucleus together is called the nuclear force (or the strong force since it must overcome the electrostatic force of repulsion between the protons) and is the strongest of the four known natural forces (gravitational, electrostatic, nuclear and weak). The force is transferred between nucleons through force carrier particles called *mesons*, π (Fig. 3-8). The exchange of mesons can be understood in analogy to having a ball constantly being thrown back and forth between two people. The strong nuclear force has a very short range, and thus the particles must be extremely close (about 1–2 fm; approximately the diameter of a proton or a neutron) in order for a meson exchange to take place. When a nucleon gets closer than this distance to another nucleon, the exchange of mesons can occur and the particles will bond to each other. When nucleons cannot come within this range, the strong force is considered to be too weak to keep them together, and the competing force (the electrostatic force of repulsion) causes the particles to move apart. Additionally, at distances less than 1 fm, the electrostatic force will overcome the strong nuclear force and the nucleons will repel one another (Fig. 3-8). In other words, when two nucleons are separated by a distance of approximately 1 fm they are bound to each other by the strong nuclear force. Inside of that distance, the electrostatic force becomes dominant and outside of that distance the nuclear force is too weak to bind the nucleons.

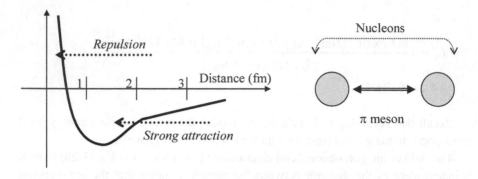

Figure 3-8. Strong nuclear force and force carrier (meson)

As explained in Section 2.3, the mass of any nucleus is always smaller than the sum of the rest masses of its individual nucleons. This difference in mass is a result of the conversion of some mass into the binding energy needed to hold the nucleus together, which is a measure of the strength of the strong nuclear force. This same energy must be applied in order to separate a nucleus into its constituents.

The nuclear density is constant for all nuclei (see Section 2.1) because it is limited by the short-range repulsion. The maximum size of a nucleus (see Chapter 2) is limited by the fact that the attractive force dies away exponentially with distance between the nucleons.

2.7   The Pauli Exclusion Principle and the Symmetry Effect

Nuclei tend to be more stable for nearly equal number of neutrons and protons especially for elements with small atomic mass numbers (see Section 2.4), a phenomenon known as the *symmetry effect*. Protons and neutrons, like electrons, obey the Pauli exclusion principle (described in Chapter 2), which states that no two identical particles, i.e., no two protons or two neutrons, can occupy the same quantum level. This fact results in a preferred nucleus configuration with even number of protons and even number of neutrons. According to this principle a single neutron and a single proton *may* occupy the same quantum level.

The preferred configuration is therefore a multiple of number two. The obvious example is helium nucleus with two neutrons and two protons. This is especially stable structure as well as its extranuclear (electronic) structure containing two electrons. The total binding energy of the helium nucleus is nearly 30 MeV when compared to the binding energy of the deuteron (that has one proton and one neutron therefore valences are not filled to the full capacity) of only 2.2 MeV.

A nucleus, like an atom, can be found in the ground state (the lowest energy level) and in the excited states. The ground state corresponds to the arrangement of all nucleons in their lowest energy levels and according to the Pauli exclusion principle:

- Such arrangements forbid the nucleons to be involved in interactions that would lower their energy, because there are no lower energy states they can move to. Thus the scattering from an incident particle which raises the energy of a nucleon is an allowed interaction, but scattering that lowers an energy level is blocked by the Pauli exclusion principle.

- A dense collection of strongly interacting nucleons would assume the high probability of constant collisions resulting in redirection and perhaps loss of energy for the nucleons. The Pauli principle, however, blocks the loss of energy because only one nuclear particle can occupy a given energy state (with spin 1/2). In this dense collection of matter, all of the low-energy states will fill up first.

It is important to mention that the Pauli exclusion principle is applied only to define the behavior of so-called *fermions*. The fermions are particles which form anti-symmetric quantum states and have half-integer spin: protons, neutrons and electrons. Particles like photons and gravitons (called *bosons*) do not obey the Pauli exclusion principle (they form symmetric quantum states and have integer spin).

The ground state structures of the nuclei of three isobars ($^{12}_{6}C_6$, $^{12}_{7}N_5$ and $^{12}_{5}B_7$) are illustrated in Fig. 3-9 (a). From the arrangement of nucleons, following the Pauli exclusion principle, it can be seen that the seventh neutron in $^{12}_{5}B_7$ must occupy a higher quantum energy level compared to the arrangement in $^{12}_{6}C_6$ and thus the total energy of the $^{12}_{5}B_7$ nucleus is higher. Similarly, the seventh proton in $^{12}_{7}N_5$ must occupy a higher quantum energy level than any of the nucleons in $^{12}_{6}C_6$ and thus has a greater total energy. This example leads to the conclusion that among any set of isobars (nuclei having the same number of nucleons) the nucleus with equal numbers of protons and neutrons will have the lowest total energy. It can be seen from Fig. 3-4 that if a nucleus possesses more energy than the neighboring isobar it will have a tendency to move toward the lowest and most stable energy configuration. This process, as indicated in Fig. 3-4, is called β decay and is explained in Chapter 4.

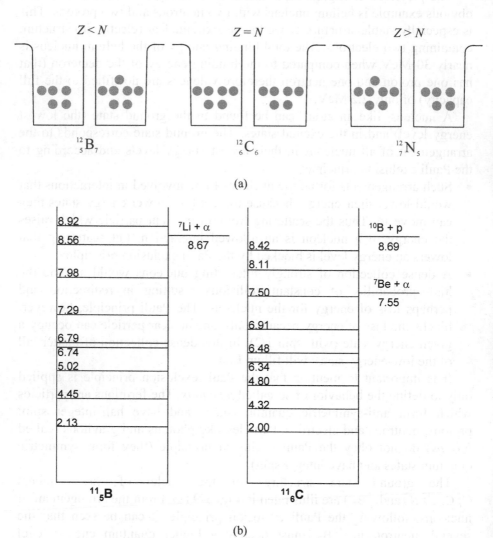

Figure 3-9. (a) Arrangements of protons and neutrons in their ground states for three isobars with the atomic mass number of 12 [Adapted from Taylor, John R., Zafiratos, Chris D., & Dubson, Michael A., 2004]; (b) excited nuclear energy levels (in MeV) in mirror nuclei of boron and carbon; the proton and α separation energies are shown [Adopted from W. N. Cottingham, D. A. Greenwood, 2001]

2.8   Excited States of Nuclei

The ground state of each nucleus as described in previous sections corresponds to the lowest energy level noted as zero where the nucleons are most bound. Almost all nuclei possess the excited states of higher energy than the ground level corresponding to the lower binding energy. The distribution of energy states (distance between and the magnitude of energy levels) vary among the nuclei. In general, the heavier nucleus has more energy levels than the light nucleus. The hydrogen and deuteron have no excited states. The magnitude of the energy level of the first excited state is higher in light nuclei and is also affected by the presence of magic number of nucleons. For example the first excited state of helium nucleus is around 19 MeV. The nuclei such as $^{182}_{73}\text{Ta}, ^{198}_{79}\text{Au}, ^{233}_{88}\text{Ra}$ have over 50 excited states below energy of 1 MeV. The mirror nuclei (in which number of protons in one nucleus equals number of neutrons in other) have similar excited energy levels. Figure 3-9 (b) illustrates the first few energy levels in mirror nuclei of boron and carbon. Nearly equal values of the excited energy levels points to a small difference in Coulomb energies addressing almost same nuclear properties of both nuclei. The excited states are explained using the nuclear shell model (described in Section 3.2). For example, the six neutrons and five protons in $^{11}_{5}\text{B}$ fill first shells, $1s_{1/2}$ and $1p_{3/2}$. Thus, the first excited state would be the so-called single-nucleon excited state in which proton is moved from the shell $1p_{3/2}$ to higher energy level shell $1p_{1/2}$. Many of higher energy states usually correspond to several nucleon excitation levels.

The excited states of a nucleus are obviously not stable. The energies are in the order of MeV for light nuclei and keV for heavy nuclei. Excited states with energies below the lowest value for the break-up of a nucleus will almost always decay to a ground state through the emission of gamma rays or internal conversion. Both processes are described in Chapter 6.

2.9 Independent Particle Approximation for Nucleons

The independent particle approximation (IPA) was introduced in Chapter 2, Section 2.9.5 in studying the electronic structure of atoms. According to the IPA for the electrons, one electron is considered to be independent on all other electrons experiencing the field produced by the average distribution of the remaining number of electrons (i.e., $Z - 1$). Knowing the IPA potential energy values we may determine the energy levels of the nucleons in the nucleus and construct its shell model (Section 3.2).

Since the main force acting among the nucleons in the nucleus is the strong and short-range nuclear force (Section 2.6), the question is would the IPA be applicable to nucleons as well. Figure 3-10 illustrates the IPA

energy potential for nucleons in comparison to the IPA potential energy
for electrons (Chapter 2, Section 2.9.5). In Section 2.1 we learned that
the nucleons are distributed uniformly in a spherical nucleus of radius
defined approximately with Eq. (3-1) and Eq. (3-2). Therefore, the IPA
potential energy of a single nucleon is spherically symmetric as shown in
Fig. 3-10 (b). The radius of the potential well, R, reflects the range of the
nuclear force and is thus always slightly larger than the estimated nuclear
radius. The depth of the IPA potential well is approximately independent
of A and is around −50 MeV. It is important to notice that because the
nuclear force is independent of charge (acting in between the quarks), the
IPA potential is the same for neutrons and protons. However, in the case
of protons there will be an effect of the repulsion Coulomb force acting
between them. The usual approximation to estimate this effect is based
on the assumption that the potential energy of a proton is obtained in the
field of a spherical charge created by the remaining number of protons,
$(Z-1)e$, giving

$$U_{Coulomb}(r) = (Z-1)\frac{ke^2}{r}; \quad U_{Coulomb}(r=0) = \frac{3}{2}(Z-1)\frac{ke^2}{r} \tag{3-29}$$

The Coulomb potential energy peaks at the center of a nucleus; its
value is around 1–2 MeV for light nuclei (and thus could be neglected)
and around 30 MeV for heavy nuclei (see α decay in Chapter 6) and thus
must be taken into account. The total IPA potential energy for heavy
nucleus is therefore a sum of the contributions from the Coulomb and
nuclear forces.

The IPA potential energy distributions for a proton and a neutron as
a function of a distance from the center of a nucleus are shown in Fig.
3-11. As expected at very large distances the proton potential energy is
zero. As proton approaches the nucleus the Coulomb repulsive force
increases the proton potential energy. When proton reaches the surface
of a nucleus the strong nuclear force become dominant and binds the
proton to the nucleus and creates the negative energy potential. When a
neutron approaches the nucleus there will be no Coulomb force
repulsing it from the nucleus. That is why the potential energy of a
neutron remains zero as long as neutron reaches the surface of a
nucleus. At this point the strong attractive nuclear force starts to act and
binds the neutron to a nucleus creating as proton, a negative potential
well.

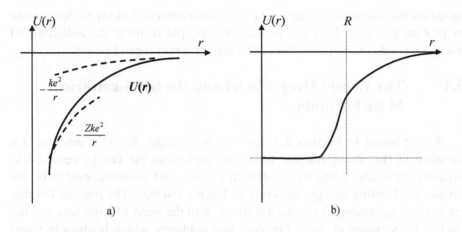

Figure 3-10. (**a**) IPA potential energy for electron; (**b**) IPA potential energy for nucleons
[Adapted from Taylor, John R., Zafiratos, Chris D., & Dubson, Michael A., 2004]

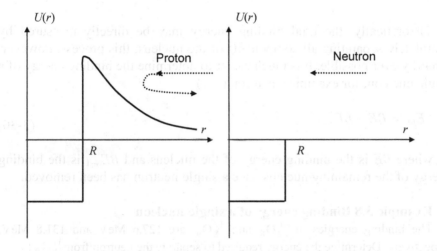

Figure 3-11. The IPA potential energy for a proton and a neutron

3. NUCLEAR MODELS

There are two classes of particle models used to describe the nucleus: the liquid drop model belongs to the class of *collective particle models* that describe the nucleus as a whole entity assuming the nucleons interact strongly in the interior of the nucleus and thus have a small mean free path; the shell nuclear model belongs to the class of *independent particle models* based on the assumption that set of quantum numbers defines nucleons state of motion and that the Pauli exclusion principle restricts the collisions of nucleons inside the nucleus, thus allowing for larger mean free path.

3.1   The Liquid Drop Model and the Semi-empirical Mass Formula

As explained in Section 2.3, the binding energy, *BE*, of a nucleus is a measure of the strong nuclear force and represents the energy required to separate the nucleus into its constituent protons and neutrons; evidently, the greater the binding energy, the more stable the nucleus. The nuclear binding energy is large enough to cause a difference in the mass of a nucleus and the sum of the separate masses of protons and neutrons, which is given in terms of the binding energy as discussed in Section 2.3, Eq. (3-20)

$$BE = \left(Zm_p + Zm_e + (A - Z)m_n - M_{atom}^{measured} \right)c^2$$

Theoretically, the total binding energy may be directly measured by completely separating all components of the nucleus; this process, however, is hardly ever possible. It is much easier to determine the binding energy of a single nucleon, for example a neutron (E_{bN})

$$E_{bN} = BE - BE_{N-1} \tag{3-30}$$

where *BE* is the binding energy of the nucleus and BE_{N-1} is the binding energy of the remaining nucleus once a single neutron has been removed.

Example 3.8 Binding energy of a single nucleon

The binding energies of $^{16}_{8}O_8$ and $^{17}_{8}O_9$ are 127.6 MeV and 131.8 MeV, respectively. Determine the energy required to separate the neutron from $^{17}_{8}O_9$:

$$E_{bN} = BE - BE_{N-1} = 131.8 - 127.6 = 4.2 MeV$$

The liquid drop model was proposed by Bohr and Wheeler to explain the structure and shape of the nucleus. In this model, a nucleus is described in analogy with a drop of incompressible liquid. This is a crude model that does not explain all of the properties of nuclei, but can easily account for the spherical shape of most nuclei and explain the process of fission (see Chapter 6). The liquid drop model is based on the assumption that the nucleus could be approximated with a homogeneous mixture of nucleons that interact strongly with each other and maintain a spherical geometry due to surface tension. Mathematical analysis of the model produces a semi-empirical equation that can be used to predict the binding energy of a nucleus as a function of Z, N and A. The formula is also called the Weizsaecker semi-empirical formula and contains the following five terms:

- *Volume energy – the BE of a nucleus is proportional to the number of nucleons, A*: There are two reasons for this tendency. First, the strong nuclear force as discussed in Section 2.6 acts only within a very small distance between nucleons (up to ~ 2 fm; Fig. 3-12), which means that each nucleon is bound to only a fraction of other nucleons. The second reason is that the nuclear density as described in Section 2.1 is approximately constant for all nuclei, which means that the degree of packing of nucleons is also nearly equal for all nuclei, and consequently each nucleon is bound to roughly the same number of neighboring nucleons. Thus the average *BE* for each individual nucleon is the same in all nuclei and the total binding energy is proportional to the total number of nucleons, A. Additionally, the total number of nucleons, and therefore the total *BE*, is proportional to the volume of the nucleus. This is expressed mathematically using a *volume term*

$$BE \propto BE_{vol} = a_{vol} A \qquad (3\text{-}31)$$

where a_{vol} is a constant such that $a_{vol} > 0$.

- *Surface energy – the surface nucleons tend to reduce the total BE of a nucleus because they do not have neighbors on all sides:* Nucleons near the surface of a nucleus are less tightly bound because they only feel the nuclear force from inside the nucleus having fewer neighboring nucleons (Fig. 3-12 (a)). Thus the binding energy curve (Fig. 3-3) increases with mass number at small mass numbers. The number of surface nucleons is proportional to the surface area of a nucleus, which is related to the total number of nucleons by the Fermi model

$$4\pi R^2 = 4\pi R_0^2 A^{2/3} \qquad (3\text{-}32)$$

Thus the total *BE* of a nucleus is reduced by a factor proportional to $A^{2/3}$

$$BE \propto BE_{vol} + B_{surf} = a_{vol} A - a_{surf} A^{2/3} \qquad (3\text{-}33)$$

where a_{surf} is a constant such that $a_{surf} > 0$.

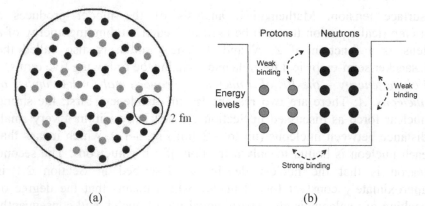

(a)   (b)

Figure 3-12. (**a**) Distribution of nucleons in the interior of a nucleus (each nucleon is bound only to the nucleons within the short range of ~ 2 fm). Keep in mind that light nuclei are not so similar to a liquid drop; (**b**) binding of protons and neutrons in an asymmetric nucleus
[Adapted from Rydin, 1977]

- *Coulomb energy – the electrostatic (Coulomb) repulsion forces between protons tend to disrupt the nucleus and thus reduce the total BE of a nucleus*: This reduction is equal to the potential energy of the total nuclear charge and tends to be largest at the high mass number; assuming that the total charge of Ze is uniformly distributed within the sphere it follows

$$U_{Coulomb} = \frac{3}{5} \frac{k(Ze)^2}{R_0 A^{1/3}} = a_{Coulomb} \frac{Z^2}{A^{1/3}} \qquad (3\text{-}34)$$

The total BE of a nucleus is thus further reduced by a factor proportional to $Z^2/A^{1/3}$

$$BE \propto BE_{vol} + B_{surf} + B_{Coulomb} = a_{vol}A - a_{surf}A^{2/3} - a_{Coulomb}\frac{Z^2}{A^{1/3}} \qquad (3\text{-}35)$$

The exact value of $a_{Coulomb}$ is dependent on the real shape of the nucleus and the approximation of Eq. (3-35) is thus valid for spherical nuclei.

- *Asymmetry term – due to the symmetry effect the total BE of a nucleus tends to decrease as the quantity $Z - N$ increases*: As discussed in Section 2.7 a nucleus with $Z = N$ has the highest binding energy (and greatest stability). As the nuclei depart from this equality (as $Z - N$ increases), the total BE decreases. The extra neutrons or extra protons lie at the higher quantum energy levels and are therefore less tightly bound (Fig.3-12 (b)). This phenomenon is accounted for mathematically as follows

$$BE \propto BE_{vol} + B_{surf} + B_{Coulomb} + B_{symm} =$$

$$a_{vol} A - a_{surf} A^{2/3} - a_{Coulomb} \frac{Z^2}{A^{1/3}} - a_{symm} \frac{(Z-N)^2}{A} \qquad (3\text{-}36)$$

- *Pairing term – the pairing effect (described in next section) affects the total BE of a nucleus*: As described in Chapter 2, a nucleus having an even number of both protons and neutrons tends to have the greatest stability, and therefore greater *BE*. This preference requires a correction to the total *BE*

$$BE \propto BE_{vol} + B_{surf} + B_{Coulomb} + B_{symm} + B_{pair} =$$

$$a_{vol} A - a_{surf} A^{2/3} - a_{Coulomb} \frac{Z^2}{A^{1/3}} - a_{symm} \frac{(Z-N)^2}{A} + \delta a_{pair} \frac{1}{A^{1/2}} \qquad (3\text{-}37)$$

where a_{pair} is a constant and δ is dependent upon nuclear configuration as shown in Table 3-2.

From the Weizsaecker semi-empirical formula it can be concluded that the total binding energy of a nucleus depends primarily on A, which is why *BE* is usually discussed in terms of binding energy per nucleon (Section 2.3). It can be also understood why the binding energy per nucleon decreases above mass number of 60. As the mass number (Z) increases the Coulomb force increases so the repulsion term in Eq. (3-37) becomes larger in magnitude requiring more neutrons to hold the nucleus. The extra neutrons then bring asymmetry to the binding of nucleons that as a result prevents adding more neutrons to a nucleus.

Table 3-2. Correction term for the semi-empirical binding energy formula

δ	Z	N
1	Even	Even
0	Even or odd	Odd or even
-1	Odd	Odd

The five constants of the formula are determined empirically so as to fit as many experimentally measured binding energies as possible. For those nuclides that are difficult to measure experimentally, the following set of values is generally used to approximate the binding energy (all values are given in MeV and are based on Taylor, John R., Zafiratos, Chris D., & Dubson, Michael A. 2004):

$$a_{vol} = 15.75 \quad a_{surf} = 17.8 \quad a_{Coulomb} = 0.711 \quad a_{symm} = 23.7 \quad a_{pair} = 11.2$$

Example 3.9 Prediction of nuclear binding energy from the Weizsaecker semi-empirical formula

Calculate the total binding energy for $^{87}_{37}Rb_{50}$ and compare with the measured value of 757853.053 ± 2.487 keV.

$$BE \propto (15.75)(87) - (17.8)(87)^{2/3} - 0.711\frac{(37)^2}{(87)^{1/3}} - 23.7\frac{(37-50)^2}{87} + 0$$

$$\Rightarrow BE \sim 755.05 \text{ MeV}$$

The liquid drop model also permits the development of a semi-empirical formula for the prediction of nuclear masses:

$$M(Z,A) = Zm_p + (A-Z)m_n - a_{vol}A + a_{surf}A^{2/3} + a_{Coulomb}Z^2A^{-1/3} +$$

$$a_{symm}\frac{(Z-A/2)^2}{A} - \frac{(-1)^Z + (-1)^N}{2}a_{pair}A^{-1/2}$$

or

$$M(Z,A) = 0.99389A - 0.00081Z + 0.014A^{2/3} + 0.083\frac{(A/2-Z)^2}{A} +$$

$$0.000627\frac{Z^2}{A^{1/3}} + \Delta$$

where the values of Δ are given in Table 3-3.

Table 3-3. Correction term for the semi-empirical nuclear mass formula

Δ	Z	A
$-0.036/A^{3/4}$	Even	Even
$+0.036/A^{3/4}$	Odd	Even
0		Odd

Example 3.10 Prediction of a nucleus mass from the Weizsaecker semi-empirical formula

Calculate the nuclear mass for $^{87}_{37}Rb_{50}$ and compare with the measured value of 86.9091835 ± 0.0000027 amu.

$$M(Z,A) = 0.99389(87) - 0.00081(37) + 0.014(87)^{2/3} + 0.083 \cdot \frac{(87/2-37)^2}{87} +$$

$$0.000627 \cdot \frac{(37)^2}{(87)^{1/3}} - 0.036\frac{1}{(87)^{3/4}} = 86.94609 \text{amu}$$

The Weizsaecker semi-empirical formula for the prediction of masses is quadratic in Z. Thus, for an A odd nuclei the plot of $M(Z,A)$ as a function of Z gives a parabola indicating only one stable isobar, while for A even there are two parabolas due to the pairing energy as defined in Table 3-2 with numerous isobars. The trend is shown in Chapter 5, Fig. 5-17.

For a fixed A value it is possible to find the number of protons for which $M(A,Z)$ has a minimum:

$$\left[\frac{\partial M(A,Z)}{\partial Z}\right]_{A=const} = 0 \quad \Rightarrow \quad Z_{M(A,Z)|_{min}} = \frac{A}{2}\frac{m_n - m_p + a_{symm}}{a_{Coulomb}A^{2/3} + a_{symm}}$$

The stability is therefore obtained if the number of protons is smaller than the number of neutrons. This equation when plotted against the number of neutrons will represent the valley of stable nuclei as given in Fig. 3-4.

3.2   The Shell Model

An alternative to Bohr and Wheeler's liquid drop model of the nucleus is the *shell model*, according to which the various nucleons exist in certain energy levels within the nucleus (Fig. 3-9) and move independently of each other. According to the shell model the nucleons are not free particles but subject to the potential well whose shape and spatial extension are similar to the nuclear density distribution, Fig. 3-13(a). Since the nuclear force has a short range, nucleon experiences the potential energy proportional to the number of nucleons in its vicinity which corresponds to nuclear density. Such a potential well consists of a series of single particle energy levels whose relative spacing depends on the depth, size and shape of the potential well. In 1949, M. G. Mayer (USA, 1906–1972) and H. D. Jensen (Germany, 1907–1973) et al. independently proposed an average potential to reproduce the nuclear magic numbers.

Each nucleon is identified by its own set of quantum numbers similar to the electrons in their orbits. In this manner, nuclear energy levels containing successively 2, 8, 20, 50, 82 and 126 nucleons exhibit a very high level of stability due to completely filled energy levels. The next set of magic numbers that would represent the super-heavy nuclei (not yet discovered) are expected to be for neutrons 184, 196, 271 and 318, and for protons 114, 126 and 164. Likewise the atoms that have a full valence electron shell, nuclei containing the *magic numbers* of nucleons are inert in the nuclear sense. For example, these nuclei do not readily react when bombarded with neutrons. Table 3-4 compares the atomic and nuclear closed-shell numbers. Nuclei which have both a magic number of neutrons and protons are particularly stable and are labeled "doubly magic":

$$\ce{^4_2He_2} \quad \ce{^{16}_8O_8} \quad \ce{^{40}_{20}Ca_{20}} \quad \ce{^{48}_{20}Ca_{28}} \quad \ce{^{208}_{82}Pb_{126}}$$

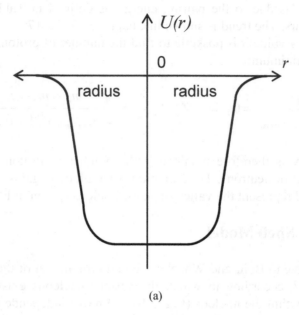

(a)

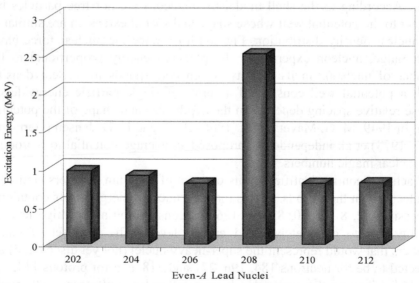

(b)

Figure 3-13. (a) Nucleon potential energy [Adapted from Born, 1969] ; (b) first excited energy level in even-*A* nuclei of lead

Table 3-4. Atomic and nuclear closed-shell numbers

Atomic closed-shell numbers	2, 10, 18, 36, 54, 86
Nuclear closed-shell numbers (magic numbers)	2, 8, 20, 28, 50, 82, 126

The existence of the magic numbers and the shell structure of a nucleus are confirmed through many observations:

1. Nuclei with a magic number of neutrons or protons tend to have more stable isotopes. For example, (a) tin ($_{50}$Sn) has 10 stable naturally occurring isotopes: ^{112}Sn (0.97%), ^{114}Sn (0.65%), ^{115}Sn (0.34%), ^{116}Sn (14.54%), ^{117}Sn (7.68%), ^{118}Sn (24.22%), ^{119}Sn (8.58%), ^{120}Sn (32.59%), ^{122}Sn (4.63%), ^{124}Sn (5.79%); (b) isotones with $N = 82$ have seven stable isotopes.

2. Isotopes with the magic number of neutrons and protons tend to be more abundant in nature. The high isotope abundance is directly related to the high binding energy therefore magic number of nucleons.

3. The stable elements at the end of the naturally occurring radioactive series (see Chapter 5) all have a magic number of neutrons or protons. The thorium, uranium and actinium series decay to lead, which has a magic number of 82 protons, while the neptunium series ends with bismuth, which has a magic number of 126 neutrons.

4. The neutron absorption cross section (see Chapter 7) for nuclei with a magic number of neutrons is much lower than that for the neighboring isotopes. The closed neutron shells increase the stability of the nuclei making them less likely for neutron interactions.

5. The binding (or the separation) energy of the last neutron in nuclei with a magic number plus one drops rapidly when compared to that of a nucleus with a magic number of neutrons. This means that the nucleus having a magic number of neutrons is more tightly bound.

Example 3.11 Separation energy of the last neutron

Compare the separation energy of the last neutron in ^{40}Ca$_{20}$ with its binding energy per nucleon. What is the separation energy of the last neutron when one neutron is added to the nucleus of ^{40}Ca$_{20}$. The masses are $m(^{40}$Ca$_{20}) = 39.9625912$ amu; $m(^{39}$Ca$_{19}) = 38.9707177$ amu; $m_n = 1.008665$ amu; $m_p = 1.007825$ amu; $m(^{41}$Ca$_{21}) = 40.9622783$ amu.

The separation energy of the last neutron in ^{40}Ca$_{20}$ is

$$E_{BEn} = (38.9707177 + 1.008665 - 39.9625912) \times 931.5 \text{ MeV} = 15.6 \text{ MeV}$$

The *BE* per nucleon is (see Sections 2.3 and 3.1)

$$BE = \left(Zm_p + Zm_e + (A-Z)m_n - M_{atom}^{measured}\right)c^2$$
$$= (20 \times 1.007276 + 20 \times 0.00055 + 20 \times 1.008665 - 39.9625921)c^2$$
$$= 352.3 \text{ MeV}$$
$$\Rightarrow BE/40 = 8.8 \text{ MeV}$$

The separation energy of the last neutron is almost twice the average nucleon BE in the doubly magic isotope of calcium-20. When one neutron is added, according to the Pauli principle and the shell structure of the nucleus, the nucleus must begin to fill a new shell. The separation (or the binding) energy of that neutron is thus much less

$$E_{BEn} = (39.9625912 + 1.008665 - 40.9622783) \times 931.5 \text{ MeV} = 8.3 \text{ MeV}$$

6. The excitation energy from the ground energy level to the first excited energy level is greater for the closed-shell nuclei. An interesting observation is that the discrete energy levels of the electrons in an atom are measured in eV or keV, while the energy levels of a nucleus are on the order of MeV. Nuclei, like atoms, tend to release energy and return to the ground state following the excitation. The excitation energies for the even-A nuclei of lead are shown in Fig. 3-13(b). The histogram indicates that the required energy is dramatically larger for the nucleus with a magic number of neutrons. This means that the nuclei having a magic number of neutrons are more tightly bound.

APPLICATIONS

Nuclear Resonance Fluorescence

The resonance fluorescence (sometimes referred to as the optical) is a process in an atom or molecule in which a photon emitted from that atom or molecule has the same frequency as the absorbed photon (see the description on fluorescence lamp at the end of Chapter 2). As described below the frequency of the emitted photon is actually slightly different from the frequency of the absorbed photon due to the recoil of the atom. The photon absorbed by an atom or a molecule will cause its electron to move to a higher orbit (energy level) from which it will fall back to its original orbit, emitting a photon of the same energy as the absorbed one (because the transition in both directions involves the same two orbits). The direction of the photon emission is random.

The nuclear resonance fluorescence (NRF) is analogue to the optical resonance fluorescence except it involves the transition of nucleons in the nucleus. The basic principle of the NRF is described as follows: each nucleus has a unique number of nucleons located at the specific energy levels according to the shell model described in Section 3.2; when the nucleus is excited (means the energy is added) after a short period of time it will decay into its ground state by emitting one or few gamma rays of energy specific to that nucleus thus giving the nucleus-specific NRF signature. The NRF technique requires a monoenergetic source of photons of energy usually between 2 MeV and 8 MeV (the region in which the energy lines in a majority of nuclei of interest are measured and known; in addition these photons are powerful enough to penetrate few centimeters of steel) that is directed to the target under the interrogation. The detected scattered photons are of discrete energies corresponding to the spectra that are characteristic to a particular nucleus. Therefore, the NRF is a unique technique that is used to identify potential threats by accurately detecting the hidden nuclear materials, explosives, toxic materials and weapons of mass destruction. However, the detection of the emitted NRF photons is not so simple.

The main difference between the NRF and the optical resonance fluorescence is described in the following example. Figure 3-14 shows the two energy levels of an electron in an atom, the ground level denoted as 1 and the first excited state denoted as 2. The energy difference between these two energy levels (electron orbits; review Chapter 2) is therefore

$$E_0 = E_2 - E_1 \tag{3-38}$$

This energy difference between the two energy levels of an electron in an atom is distributed between the emitted photon of energy E (during the electron transition from the state 2 to state 1) and recoiling of that atom of energy E_r. Thus we may write

$$E_0 = E_\gamma + E_r \tag{3-39}$$

From the last two equations we may see that the energy of the emitted photon is slightly smaller than the energy of the excitation (or the energy of the incoming photon)

$$E_\gamma = E_0 - E_r \tag{3-40}$$

where for the optical resonance fluorescence $E_r \ll E_0$ giving that the emitted photon has the same energy as the absorbed photon. The emitted photon and the recoiling atom carry equal momentum and are emitted (moved) in opposite directions as sketched in Fig. 3-14. From the conservation of momentum and energy it therefore follows

$$p_r = p_\gamma \quad \rightarrow \quad E_r = \frac{p_r^2}{2M} = \frac{p_\gamma^2}{2M} = \frac{E_\gamma^2}{2Mc^2} \quad \leftarrow \quad E_\gamma = p_\gamma c \qquad (3\text{-}41)$$

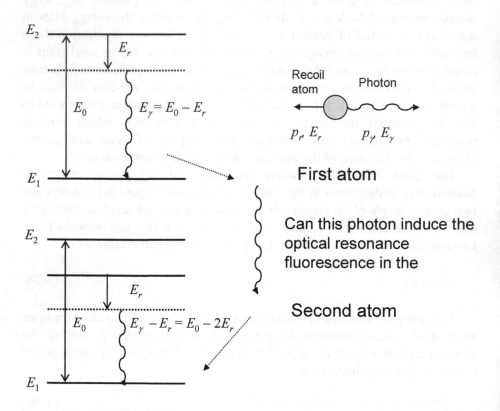

Figure 3-14. Schematics of the optical resonance fluorescence

The question is would the emitted photon of the energy reduced by the recoiling atom energy be able to induce the same transition in a second atom? If we say that the energy difference between the exact energy of the absorbed photon (equal to the energy difference between the two states in atom) is ΔE which is equal to E_r, then the atom that absorbs this emitted photon will produce the transition of energy

$$\Delta E = 2E_r = \frac{E_\gamma^2}{Mc^2} \qquad (3\text{-}42)$$

As it is described in Chapter 7 based on the explanations of the uncertainty principle derived in Chapter 4, it is known that every energy level of an electron in an atom or nucleon in a nucleus has intrinsic width, Γ, that is estimated with the following relation

$$\Gamma = \hbar / \tau \tag{3-43}$$

where $\hbar$ represents the reduced Planck constant (see Chapter 4, Section 1) and τ denotes the mean life time of that energy level [obviously the ground state would have an infinite mean life time]. If the width of the energy level is larger than the deficiency in photon energy (as we defined ΔE), the optical resonance fluorescence is possible. This can be understood from the following example: for an atom of atomic mass number 92 and expected energy of an electron transition in the order of 1 eV

$$\Delta E \left(A = 92 \right) = 2 E_r = \frac{E_\gamma^2}{Mc^2} = \frac{\left(1\text{eV} \right)^2 \left(1.6 \times 10^{-19} \text{J/eV} \right)}{92 \left(1.67 \times 10^{-27} \text{kg} \right) \left(3 \times 10^8 \text{m/s} \right)^2}$$

$$\cong 1.2 \times 10^{-11} \text{eV}$$

Assuming the average mean time of electron energy level is in the order of 10^{-8} s

$$\Gamma \left(A = 92 \right) = \frac{\hbar}{\tau} = \frac{1.05 \times 10^{-34} \text{Js}}{10^{-8} \text{s}} \cong 10^{-7} \text{eV}$$

By comparing the numeric values of the last two equations it is easily understood that since $\Gamma \gg \Delta E$, the optical resonance fluorescence is possible with the photons of certain deficiency in their energy compared to the energy levels difference (caused by the recoiling of the atom), Fig. 3-15. However, is that true for the NRF as well? The same example can be used to illustrate the energy deficiency and the state width for the nucleus. In the case of nucleus the energy levels differ on average in the order of 100 keV. The mean life time of the nucleon transition is shorter than that for electrons and could be assumed to be in the order of 10^{-10} s. Thus for the NRF transition the same photon emitted from the nucleus will not be able to induce another same transition in a second nucleus because $\Gamma \ll \Delta E$

$$\Delta E(A=92)=2E_r=\frac{E_\gamma^2}{Mc^2}=\frac{(100\,\text{keV})^2(1.6\times10^{-19}\,\text{J/eV})}{92(1.67\times10^{-27}\,\text{kg})(3\times10^8\,\text{m/s})^2}$$

$$\cong 0.1\,\text{eV}$$

$$\Gamma(A=92)=\frac{\hbar}{\tau}=\frac{1.05\times10^{-34}\,\text{Js}}{10^{-10}\,\text{s}}\cong10^{-5}\,\text{eV}$$

For the same reason the detection of such photons is difficult. However, there are methods that can induce the strength of the detection. Here are the two most common methods:

1. Doppler broadening (that is introduced in Chapter 7, Section 4.8) increases the intrinsic width of the energy levels and thus compensates for the recoiling energy loss. By heating the target the energy level widths are broadened due to thermal motion of the atoms. If T denotes the absolute temperature of the atom, the optical resonance fluorescence width becomes

$$\Gamma(T)=\frac{E_0}{c}\sqrt{\frac{2kT}{M}}=\frac{E_\gamma}{c}\sqrt{\frac{2kT}{M}} \qquad (3\text{-}44)$$

For the example as discussed earlier ($A=92$), the optical resonance fluorescence width increases according to Eq. (3-44) to $\sim7\times10^{-5}$ eV and the NRF width becomes nearly 0.1 eV and thus comparable to ΔE.

2. Rotation of the source at high velocity that affects the Doppler shift. For example, for the motion of the source in the direction toward the detector, the frequency of the emitted photon is increased by the factor $f_0 \upsilon/c$ increasing the energy of the emitted photon (f_0 is the frequency of the absorbed photon and the υ is the velocity of the rotation of the source). The required velocity to compensate for the recoiling nucleus energy loss is obtained from the following relation

$$\frac{hf\upsilon}{c}=\frac{E_\gamma f}{c}=\frac{E_\gamma^2}{Mc^2} \quad \rightarrow \quad \upsilon=\frac{E_\gamma}{Mc} \qquad (3\text{-}45)$$

For the NRF example of $A=92$ and $E_\gamma=100$ keV the required velocity would be around 350 m/s.

The Mössbauer Spectroscopy

Mössbauer discovered in 1957 the phenomenon related to the resonant and recoil-free emission and absorption of photons by atoms bound in solids. For this discovery he has received the Noble Prize in 1961.

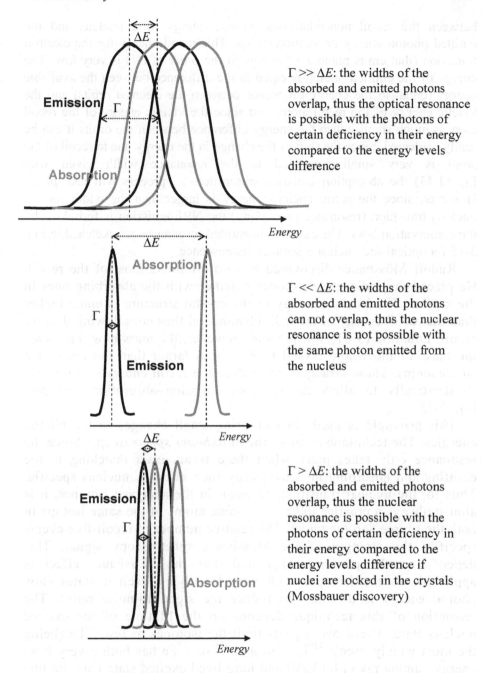

$\Gamma \gg \Delta E$: the widths of the absorbed and emitted photons overlap, thus the optical resonance is possible with the photons of certain deficiency in their energy compared to the energy levels difference

$\Gamma \ll \Delta E$: the widths of the absorbed and emitted photons can not overlap, thus the nuclear resonance is not possible with the same photon emitted from the nucleus

$\Gamma > \Delta E$: the widths of the absorbed and emitted photons overlap, thus the nuclear resonance is possible with the photons of certain deficiency in their energy compared to the energy levels difference if nuclei are locked in the crystals (Mossbauer discovery)

Figure 3-15. Schematics of the Mössbauer effect

We showed with deriving Eq. (3-41) that when atom emits a photon it must recoil to conserve the momentum; the same equation gives the relation between the recoil non-relativistic kinetic energy of a nucleus and the emitted photon energy or its momentum. This recoil energy for the electron transition (that emits photons of energy in the order of eV) is very low. The energy of the emitted photon is equal to the difference between the available energy (equal to the energy difference between the electron orbits) and the kinetic energy of the recoil atom, and since the kinetic energy of the recoil atom is much smaller than the energy difference between the orbits it can be easily neglected. Since the shift or the change in the energy due to recoil of the atom is very small compared to the resonance width given with Eq. (3-43) the absorption–emission–absorption–... process will take place. However, since the recoil nucleus energy is larger than the width for the nucleon transition (resonance probability) the NRF emission is forbidden by the conservation laws. The cascades in emitted resonances are sketched in Fig. 3-15 for optical and nuclear resonance fluorescence.

Rudolf Mössbauer discovered how to eliminate most of the recoil. He placed the radioactive isotopes together with the absorbing ones in the crystal. The binding energy of the crystal structure is much higher than the recoil energy of the individual nuclei thus not allowing them to recoil. Equation (3-41) still is valid, however, the mass now represents the mass of the whole crystal ($\sim 10^{20}$ times larger than the mass of a single atom). These arrangements reduce the recoil energy of a nucleus so drastically to allow the resonant emission–absorption cascades, Fig. 3-15.

This principle is used in measuring small changes in the photon energies. The technique is called the *Mössbauer spectroscopy*. Since the resonance only takes place when there is an exact matching in the emitting and absorbing nucleus energy such effect is nucleus specific. Thus for the nuclear transitions to occur in the analyzed isotopes, it is almost always a practice to have the same atoms of the same isotope in both the source and the target. The relative number of recoil-free events specifies the strength of the Mössbauer spectroscopy signal. This depends on the photon energy and thus the Mössbauer effect is applicable to the isotopes with very low lying excited states (low photon energy is desirable to reduce the signal-to-noise ratio). The resolution of this technique depends on the lifetime of the excited nucleus state. These two aspects limit the isotopes to few, ^{57}Fe (being the most widely used), ^{129}I, ^{119}Sn and ^{121}Sb. ^{57}Fe has both a very low-energy gamma ray (14.4 keV) and long-lived excited state with the line width of 5×10^{-9}ev.

PROBLEMS

3.1. If the radius of a nucleus is given by Eq. (3-1), calculate the density of nuclear matter in g/cm^3 and in $nucleons/fm^3$. Assume the mass of a nucleon is 1.67×10^{-24} g.

3.2. Use Eq. (3-1) to calculate the radius of ^3H, ^{60}Co and ^{239}Pu.

3.3. Show by expanding $[1 - (\upsilon/c)^2]^{1/2}$ in powers of $(\upsilon/c)^2$ that the kinetic energy can be written as

$$E = \frac{1}{2} m_0 \upsilon^2 + \frac{3}{8} m_0 \frac{\upsilon^4}{c^2} + ...$$

Does non-relativistic formula $m_0 \upsilon^2 / 2$ overestimate or underestimate the kinetic energy of particle with the rest mass m_0 and speed υ?

3.4. An electron and proton are each accelerated from rest by a total potential of 500 million volts (500 MeV). Calculate the increase in mass and fractional increase in mass of each of these particles as well as their final speeds.

3.5. Calculate the total binding energy and the binding energy per nucleon for $^{32}P_{17}$ (atomic mass = 31.975697 amu).

3.6. Calculate how much energy would be absorbed or released if two atoms of ^{12}C were fused together to create one atom of $^{24}Mg_{12}$ (atomic mass = 23.985042 amu).

3.7. Calculate the amount of energy needed to dissociate one atom of ^{12}C into three atoms of ^4He (atomic mass = 4.002603 amu).

3.8. Calculate the magnitude of the Coulomb and gravitational potential energy between adjacent protons using the radius of a nucleus as the separation distance. Compare these two energies with the binding energy per nucleon of $^{32}P_{17}$. Use the Eq. (3-1) to calculate the radius of a nucleus.

3.9. How much energy is required to remove a proton from $^{40}Ca_{20}$ (atomic mass = 39.962589 amu)?

3.10. Find the energy released in the reaction: $^{238}_{92}$ U $\rightarrow$ $^{234}_{90}$ Th + $^{4}_{2}$ He. M($^{238}_{92}$U) = 238.050786 amu, M($^{234}_{90}$Th) = 234.043583 amu, M($^{4}_{2}$He) = 4.002603 amu.

3.11. The radius of a heavy nucleus is $\sim 10^{-12}$ cm. When the velocity of neutron becomes large enough that $\lambda/2\pi$ is of the same order of magnitude as the nuclear radius, the neutron can be diffracted about the nucleus, what is known as *shadow scattering*. Show that at neutron energy of $\sim$0.21 MeV this effect becomes important [$h = 6.62 \times 10^{-27}$ erg-s; neutron mass $= 1.675 \times 10^{-24}$ g].

3.12. Show that on average 200 MeV is released when one atom of ^{235}U fissions by capture of a thermal neutron? In fission usually two fission fragments are released. Use Fig. 3-3 to estimate the average binding energy of fission fragments. What is the binding energy of a captured neutron?

3.13. Calculate the wavelength of a proton needed to excite an electron in Li^{2+} from the state $n = 2$ to the state $n = 5$.

3.14. Show that the minimum mass using the semi-empirical formula (3-37) is obtained for

$$Z(min\ mass) = \frac{A}{2 + 0.015A^{2/3}}$$

3.15. The energies at which non-relativistic expressions for energy and momentum of electrons and protons are in error by 1% are considered to be threshold values for relativistic approach. Show what is the condition for which the relativistic momentum departs this 1% from the non-relativistic expressions.

3.16. For the 1% threshold for momentum as defined in Problem 3.15 determine what is the kinetic energy of electron, proton and α particle. [*Hint*: The relativistic kinetic energy of a particle is $E = mc^2 - m_0c^2$; the non-relativistic particle kinetic energy can be found in Problem 3.3]

3.17. Use Eq.(3-5) to write a computer code to calculate and plot the dependence of a mass of a moving body and its energy.

3.18. Calculate the nuclear density of the smallest and the largest nucleus in the periodic system of elements. Comment on the results.

3.19. Calculate the separation energy of last neutron in first 10 nuclei in the periodic system of elements. Comment on the results.

3.20. Select an even-A nuclei and plot the excitation energy of the first excited state. Compare with Fig. 3-13(b) and discuss the differences.

3.21. Compare the size of the nuclei (calculated using the Fermi model) and the size of the corresponding atoms across the periodic system of elements. What is your observation? How would you explain the change in nucleus size analyzing the number of nucleons, nuclear forces, Coulomb forces and nuclear density?

3.22. Show that when two deuterons react they form tritium with the net gain in the binding energy of the system of 4.02 MeV.

3.23. Calculate the binding energy per nucleon of ^{238}U. What is an approximate gain in the binding energy of the system if ^{238}U splits into two equal nuclei? What would be the corresponding amount of energy released in this reaction? Compare with the values discussed in Chapters 7 and 8.

3.24. Using the solution of Eq. (3-37) for minimum mass show that for light stable nuclei $Z \sim A/2$ and give few examples. Also, show that for stable heavy nuclei $Z \sim A^{1/3}/0.015$.

3.25. Calculate the mass defect and binding energy per nucleon of a ^{63}Cu nucleus (the actual mass of a copper nucleus is 62.91367 amu).

3.26. Derive the relativistic equation for the kinetic energy of a moving particle, i.e., $E = mc^2 - m_0c^2$.

3.27. Based on the derivation in 3.26 show that the total energy of a moving body is given by Einstein formula, $E_{tot} = mc^2$.

NOTE: Some of the problems listed are adopted from the website developed by Dr. C. N. Booth, http://www.shef.ac.uk/physics/teaching/phy303/

Chapter 4

DUALITY OF NATURE

Basic Principles, Evidence and Examples

Light and matter are both single entities, and the apparent duality arises in the limitations of our language. It is not surprising that our language should be incapable of describing the processes occurring within the atoms, for, as has been remarked, it was invented to describe the experiences of daily life, and these consist only of processes involving exceedingly large numbers of atoms. Furthermore, it is very difficult to modify our language so that it will be able to describe these atomic processes, for words can only describe things of which we can form mental pictures, and this ability, too, is a result of daily experience. Fortunately, mathematics is not subject to this limitation, and it has been possible to invent a mathematical scheme – the quantum theory – which seems entirely adequate for the treatment of atomic processes; for visualisation, however, we must content ourselves with two incomplete analogies – the wave picture and the corpuscular picture. (Quantum Theory, 1930), *Werner Heisenberg (1901–1976)*

Newton's and Schrödinger's equations both describe the motion. Newton's laws describe the motion of macroscopic objects and are the foundation of classical physics. Schrödinger's equations describe the motion of quantum particles giving the solutions in a form of wave functions and probabilities. This chapter is all about the quantum particle–wave duality and the explanations on the quantum laws with examples and descriptions of experimental evidence.

T. Jevremovic, *Nuclear Principles in Engineering*,
DOI 10.1007/978-0-387-85608-7_4, © Springer Science+Business Media, LLC 2009

1.   PLANCK'S THEORY OF QUANTA

In 1900 Max Planck developed the theory that energy is absorbed and emitted in small energy packets that he called *quanta*. The size of quanta of low-frequency (red) light is smaller than the size of quanta of high-frequency (violet) light. In 1905, Albert Einstein published his famous paper on the photoelectric effect postulating the quantum nature of light (for which he received the Nobel Prize in 1921). According to Einstein's theory (described in Chapter 6), light is composed of particles (which he called *photons*) such that a beam of light is analogous to a stream of bullets. Thus, ultraviolet light (UV, 10^{16} Hz) consists of a stream of photons each having 100 times the energy of infrared light photons (IR, 10^{14} Hz). That is why UV light can cause skin cancer, while IR has no significant effect on the skin.

The energy of an atom, as discussed in Chapter 2, can be increased only in discrete values, just as American money cannot be counted in units less than cents. The energy of the quanta (E) is proportional to the frequency (f) of oscillation of the light wave. Therefore the total energy can be equal only to an integer number of quanta (similar to the fact that one can have 3 dollars and 20 cents, but one cannot have 3 dollars and 20.5 cents). The size of the energy quantum defined by Planck is given by

$$E = nhf \quad n = 0,1,2,3,... \tag{4-1}$$

where h is the Planck's constant equal to 6.626×10^{-34} Js determined (by Planck) to fit the best experimental data related to the black body radiation (Section 1.1). Equation (4-1) indicates that the quantum of photon energy varies with its frequency (recall that the quantum of charge is the same for all charges, e).

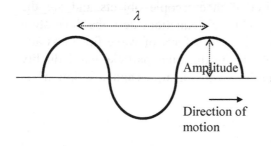

λ : wavelength indicating the distance between two wave crests (the longest are the radio waves)

f : frequency describing how many crests pass a given point each second

υ : velocity [$= \lambda / f$]

E : energy [$= hf = hc / \lambda$]

Figure 4-1. Definition of a wave

The importance of this idea is explained further in Sections 1.1, 1.4, and 1.5. It is important to note that increasing the intensity of a light source increases the rate at which photons are emitted. If the frequency of emitted light has not changed, the energy of the emitted photons has also not changed. The relationships between frequency, wavelength, velocity, and energy of a light wave are sketched in Fig. 4-1.

1.1 Black Body Radiation

At the beginning of the 20th century it was known that heat causes the molecules and atoms of matter to oscillate, and that any body with a temperature greater than absolute zero radiates some energy. It was also observed that the intensity and frequency distribution of the emitted radiation depended on the detailed structure of the heated body. The model analyzed for more than 40 years in order to explain the dependence of emitted radiation energy on wavelength was the "black body" model (Fig. 4-2).

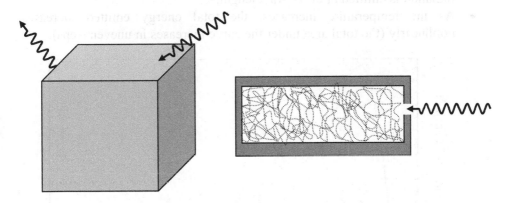

Figure 4-2. Black body

A black body is a hypothetical object that absorbs 100% of all radiation that it is exposed to. At normal temperatures, such a body reflects no radiation and thus appears to be perfectly black. When heated, the radiation emitted by a black body is called black body, or cavity, radiation. It is not dependent upon the type of incident radiation. In practice, no material has been found to exhibit the exact properties of this model. A black body may be thought of as a furnace with a small hole in the door through which heat energy can enter from the outside. Once inside the furnace, the heat is entirely absorbed by the inner walls, which may emit radiation to be absorbed by another part of the furnace wall or or to escape through the hole. The radiation that escapes from a black body may have any wavelength.

The black body radiation spectrum (Fig. 4-3), which represents the intensities of each of the wavelengths of radiation emitted from a black body as a function of the energy of radiation, indicates that

- A black body radiates energy at every wavelength, while energy decreases exponentially as wavelength increases.
- A black body emits most of its radiant energy at a peak wavelength. For example, at 5,000 K the peak wavelength is about 5×10^{-7}m (500 nm) which is in the visible light (yellow–green) region.
- At each temperature a black body emits a standard amount of energy represented by the area under the curve. A hotter body thus emits radiation with shorter wavelengths. For example, black bodies at higher temperatures are blue, and those at lower temperatures are red.
- As the temperature increases, the peak wavelength emitted by a black body decreases and begins to move from the infra-red to the visible end of the spectrum. Since none of the curves cross the *x*-axis, it follows that radiation is emitted at every wavelength.
- As the temperature increases, the total energy emitted increases nonlinearly (the total area under the curve increases in uneven steps).

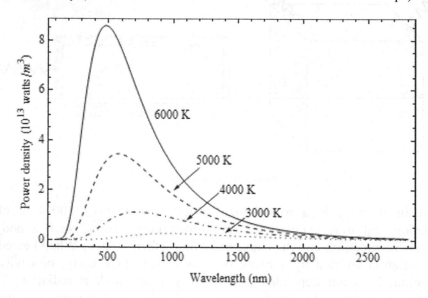

Figure 4-3. Black body radiation spectrum

1.2   Wein's Displacement Law

When the temperature of a black body increases, the overall emitted energy increases, and the peak of the radiation curve moves to shorter wavelengths (Fig. 4-3) as defined by the Wein's displacement law

$$\lambda_{max} \, T = 2.8898 \text{x} 10^{-3} \, mK \tag{4-2}$$

where λ_{max} is the wavelength at which the energy of the emitted radiation is maximum, and T is the temperature in *Kelvin*.

Wein's displacement law is used to evaluate the temperatures of any radiant object whose temperature is far above that of its surroundings (such as stars, for example). Wilhelm Wein was awarded the Nobel Prize in Physics in 1911 for his work in optics and radiation.

Example 4.1 Wein's displacement law

Use the Wein's displacement law to calculate the temperature (in K) of a star whose maximum wavelength is 3.6×10^{-9}m (an X-ray star).

From Eq. (4-2) it follows $\lambda_{max} \, T = 2.8898 \text{x} 10^{-3} \, mK \quad T = 8.03 \times 10^{5} \, K$

1.3   The Stefan–Boltzmann Law

According to the Stefan–Boltzmann law the energy, E, radiated by a black body per unit time and unit area (energy flux or emissive power) is proportional to the fourth power of the *absolute* temperature, T

$$E = \sigma T^4 \tag{4-3}$$

where σ is the Stefan–Boltzmann constant equal to 5.67×10^{-8} Wm^{-2}K^{-4}.

The Stefan–Boltzmann law gives the total energy that is emitted at all wavelengths from a black body (which corresponds to the area under the black body radiation spectrum, Fig. 4-3), and it explains the increase in the height of the curves with temperature. The increase in energy is very abrupt, since it is proportional to the fourth power of the temperature.

Example 4.2 Stefan–Boltzmann law

Calculate the ratio of radiated energy from the Sun to that of the Earth assuming their temperatures are 6,000 and 300 K, respectively. Determine the maximum wavelengths of the emitted radiation.

According to the Stefan–Boltzmann law, Eq. (4-3) it follows that

$E_{Sun} = 5.67 \times 10^{-8}$ Watts m^{-2} K^{-4} $(6,000 \text{ K})^4 = 7.3 \times 10^7$ Watts m^{-2}

$E_{Earth} = 5.67 \times 10^{-8}$ Watts m^{-2} K^{-4} $(300 \text{ K})^4 = 459$ Watts m^{-2}

$E_{Sun} / E_{Earth} = (6,000 \text{ K})^4/(300 \text{ K})^4 = 1.6 \times 10^5$

According to the Wein's displacement law it follows that

Sun: $\lambda_{max} = 2.8898 \ 10^{-3} \text{ mK}/6,000 \ K = 0.48 \ \mu\text{m}$

Earth: $\lambda_{max} = 2.8898 \ 10^{-3} \text{ mK}/300 \ K = 9.6 \ \mu\text{m}$

1.4   The Rayleigh–Jeans Law

At the beginning of the 20th century a major problem in physics was to predict the intensity of radiation emitted by a black body at a specific wavelength. Wein's displacement law (described in Section 1.3) could predict the overall shape of the black body spectrum, but at long wave–lengths the predictions disagreed with experimental data. Rayleigh and Jeans developed a theory that the radiation within a black body is made up of a series of standing waves. They argued that electromagnetic radiation was emitted by atoms oscillating in the walls of the black body. The oscillating atoms emit radiation that creates a standing wave moving back and forth between the walls. Their formula is shown in Fig. 4-4.

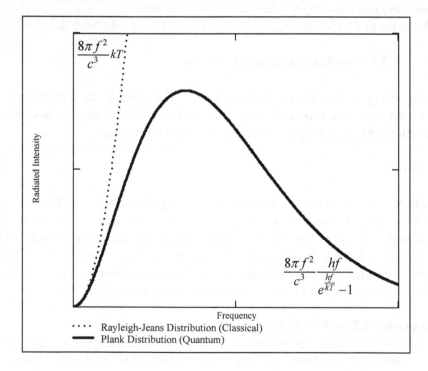

Figure 4-4. Rayleigh–Jeans law of black body radiation

The Rayleigh–Jeans formula agreed with the experimental data for the long wavelengths, but in the region of short wavelengths the disagreement with measured values was extreme. According to the Rayleigh–Jeans formula, the radiation intensity becomes infinite as wavelength approaches zero. When compared to the radiation energy distribution as a function of temperature and wavelength of the emitted light (Fig. 4-3), it can be observed that the peak wavelength was not predicted by the Rayleigh–Jeans law. This failure to account for the decrease in energy emitted at long wavelengths (the UV wavelengths) is known as the *ultraviolet catastrophe*.

At a meeting of the German Physical Society in October of 1900, Max Planck presented his theory that radiation is emitted in discrete portions, or quanta, as given with Eq. (4-1), and showed that his formula fit all experimental data.

1.5 Planck's Law

Planck explained that the oscillating electrons, of the surface atoms of a black body, emit radiation according to Maxwell's laws of electromagnetism. At that time, classical mechanics predicted that such radiation could have any value of energy.

Planck postulated that the energy is emitted or absorbed only in discrete amounts because the frequencies of the oscillating electrons could have only specific discrete values. Since the energy of electromagnetic radiation is proportional to frequency ($E = hf$), it too can be available only in discrete amounts as given with Eq. (4-1) that defines Planck's law, the basic law of quantum theory. According to this law, the energy of electromagnetic waves is restricted to quanta radiated or absorbed as a whole with magnitude proportional to frequency. Any individual wave would be restricted to posses no more than kT of energy thus restricting that any standing wave in a black body could not have energy quanta larger than kT. The value kT is defined by the classical thermodynamics and represents the mean total energy of a wave. Further details on black body radiation are not of great importance for the topics covered in this book.

2. THE WAVE–PARTICLE DUALITY

The ultimate belief and tendency in modern physics is toward a large overview that will incorporate all laws of nature into one unified theory. This theory would bring together the laws of the subatomic world and laws of galaxies and everything in between; a concept that Einstein called the *ideal limit of knowledge*. Kepler, Galileo, Copernicus, and Newton were the first to develop the theory of the universe, according to which the universe was infinite in all directions and light travelled at infinite or near infinite speed. With the 20th century came Einstein's theories of quantum mechanics, and the understanding of physics from the macro–world to the subatomic realm changed.

A *quantum* is a discrete quantity and *mechanics* is the study of motion; thus quantum mechanics describes a nature to consist of small, discreet parts (quanta), and is applied to describe events on the subatomic scale. Newtonian (classical) physics is applicable to the macro-world, but is not applicable to the subatomic realm. Newton's laws are based on every day observations and predict *events* such as a ball's trajectory or the velocity of celestial bodies. Quantum mechanics is based on subatomic experiments and predicts *probabilities*. Subatomic phenomena cannot be observed or detected directly, as an atom or subatomic particle cannot be seen by the same means as macro objects. Using Newton's laws of motion, the future or the past of a moving object can be easily predicted given initial conditions. For example, if the present positions and velocities of the Earth, the Moon, and the Sun are known, it is possible to predict where the Earth has been, or will be, in relation to the Moon and the Sun. For example, the space program would not be possible without Newtonian calculations of the movements of spacecraft relative to the movements of the Earth and Moon.

The ability to predict the future and the past, based on knowledge of the present and Newton's laws of mechanics, suggests that from the moment the universe was created and set into motion, everything that was to happen within the universe was already determined. However, according to quantum mechanics, it is not possible to know enough about the present to make a complete prediction of the future. A prediction or observation of the subatomic world requires a decision as to which aspects must be known, because the laws of quantum mechanics forbid precise knowledge of more than one of them at the same time.

This section introduces the basic aspects of the quantum mechanics concept, describes the evidence for the wave–particle duality nature of subatomic constituents, explains the uncertainty principle, and gives a brief introduction to the Schrödinger wave equation.

The scope of the presented theories is directly related to applications in nuclear engineering disciplines.

2.1   De Broglie's Hypothesis

The development of quantum mechanics began in 1900 with Planck's study of black body radiation (see Section 1.1). Planck found that the energy of oscillation of electrons that produce the radiation is absorbed and emitted in discrete amounts, quanta, given by $E = nhf$, where n is an integer value and h is Planck's constant which value was determined from the black body radiation spectra. However, Planck was not able to explain *why* the energy would be quantized, because at that time radiated energy was considered to be *wave like*. This theory was derived from the Thomas Young's double-slit experiment (see Section 2.2) that in 1803 demonstrated the interference pattern of light.

In 1905, Einstein explained the photoelectric effect that proved Planck's discovery of quanta and showed not only that energy absorption and emission are quantized, but also that the energy of light itself is quantized. With this explanation he introduced a new concept of light; theorizing that light quanta are *particle like* (photons), and that light, therefore, behaves as a series of particles. This was a confrontation to classical physics, and the two sets of (repeatable *by demand*) experiments (the photoelectric effect and the double-slit experiment) were proving different natures of light.

The idea that light could behave as a wave *and* as a particle created a new question: does an electron have particle–wave properties, or, in a larger frame, is the dual particle–wave nature of light a property of all material objects as well? The answer to this question was given by Prince Louis de Broglie in his Doctoral thesis of 1923 in which he argued that all material objects can behave, like light, both as a particle and as a wave at the same time. Equation (4-1)

$$E = hf$$

was difficult to apply to particles with finite mass. It describes a total, kinetic or total relativistic energy (as all are identical) of light. However, the relationship of momentum, p, to wavelength

$$p = h/\lambda \tag{4-4}$$

is valid for any particle or material object.

De Broglie suggested (without any experimental evidence) that for any particle with non-zero mass (such as electrons, protons, or bowling balls) moving with momentum p, there is an associated wave of wavelength λ related to its momentum as

$$\lambda = h/p \tag{4-5}$$

The wavelength of a moving particle calculated from this equation is called the *de Broglie wavelength*.

Example 4.3 De Broglie wavelength

Calculate the de Broglie wavelength for (1) an electron moving at 3.0×10^6 m/s ($= 0.01c$) and (2) a 1,000-kg car traveling at 100 km/h.

(1) $\lambda = h/p = 6.6 \times 10^{-34}$ Js/$[(9.11 \times 10^{-31}$ kg$) (3.0 \times 10^6$ m/s$)] = 2.4 \times 10^{-10}$ m

Since the wavelength of the electron is comparable to atomic dimensions, the effect of its wave nature is important.

(2) $\lambda = h/p = 6.6 \times 10^{-34}$ Js/$[(1,000$ kg$) (100 \times 10^3$ m/$3,600$ s$)] = 2.4 \times 10^{-38}$ m

The wave character of the car is much smaller than the car itself; hence the wave-like motion of the car (or of any macro object, for that matter) is not evident.

Using the analogy of sound waves (known to vibrate at discrete frequencies when confined to a finite region such as an organ pipe), de Broglie argued that the quantization of electron energy in an atom can be explained as the quantization of electron–wave frequency (for an electron confined inside the atom), which would explain the quantization of the angular momentum of an electron in a hydrogen atom.

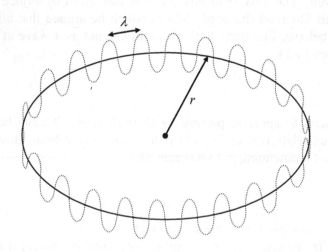

Figure 4-5. Electron wave in Bohr's atom according to the de Broglie representation

De Broglie pictured the electron wave oscillating along the circular orbit of Bohr's atom (Fig. 4-5) such that its circumference is equal to the finite integral number of wavelengths

$$2\pi r = n\lambda, \qquad n = 1, 2, 3, \dots \tag{4-6}$$

Replacing the wavelength of the electron wave with the de Broglie relation of Eq. (4-5), it follows

$$rp = \frac{nh}{2\pi} \qquad\qquad (4\text{-}7)$$

For a circular orbit, rp represents the angular momentum, L. Combining last equations leads to

$$L = \frac{nh}{2\pi} = n\hbar, \qquad n = 1, 2, 3, \dots \qquad\qquad (4\text{-}8)$$

the Bohr quantization condition (read Chapter 2).

2.2   Double-Slit Experiment

In 1803, the British physicist, physician, and Egyptologist, Thomas Young (known for deciphering the Rosetta stone), carried out a very simple but unique experiment known as the *double–slit experiment*. The goal of this experiment was to understand the nature of light. He analyzed the pattern created by light while passing through two slits (either through one or both). In order to understand the nature of such an experiment, we start with a brief review of the properties of waves.

The best analogy for understanding the wave property of light is to consider the water waves created at the entrance of a harbor. For example, if the entrance of the harbor is wide enough the waves move straight through it. This is because the distance between the crests, the wavelength, is smaller than the size of harbor entrance as illustrated in Fig. 4-6 (a). However, if the entrance of the harbor is small (smaller than or equal to the wavelength), the waves spread out into semicircles, a phenomenon called the *diffraction* of waves as illustrated in Fig. 4-6 (b).

It was assumed that because light is a wave it should behave the same way when passing through slits of sizes smaller or larger than its wavelength. If light passes through the cut-out screen as shown in Fig.4-7(a), it will behave as the ocean waves passing through the large harbor entrance, because the opening in the slit is large compared to the wavelength. When the slit opening is small (Fig. 4-7 (b)) the light diffracts, and there is no sharp boundary between the bright and the dark area at the screen where the image is projected.

Knowing this, Thomas Young developed an experiment as shown in Fig. 4-8. He analyzed the patterns, that are created by the light on the wall screen depending on the size of the slits and whether one or both are opened. Young observed that when one slit was closed the image obtained at the wall indicated light diffraction (like that shown in Fig. 4-7 (b)). However, when both slits were opened, the expected image of a simple sum of the light waves did not appear. Instead, the pattern showed bands of light and dark areas (Fig. 4-8), a phenomenon called the light *interference*.

Because the spacing of the maxima and minima in the interference pattern depends on light wavelength, changing the wavelength (color) of light will change the location and number of bright and dark bands on the screen. If the distance between the slits is increased, more bands of light will be created on the screen.

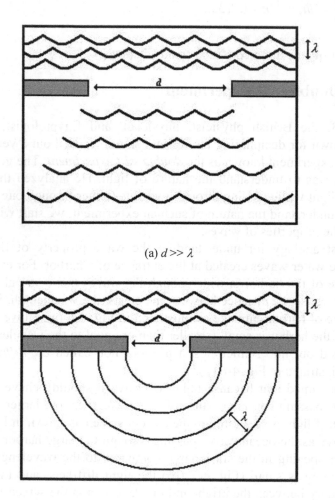

(a) $d \gg \lambda$

(b) $d \sim \lambda$

Figure 4-6. Patterns that water waves create while passing through a harbor entrance of different sizes: (**a**) the entrance is larger than the wavelength of water waves and (**b**) the entrance is smaller than the wavelength of water waves

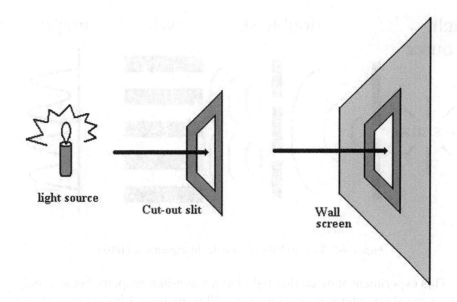

(a) Size of the slit larger than light wavelength

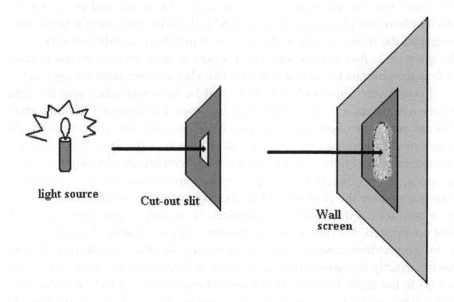

(b) Size of the slit comparable to light wavelength

Figure 4-7. Patterns created by light passing through slits of various sizes

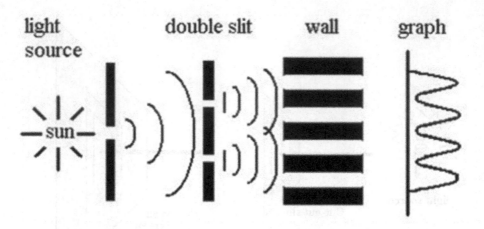

Figure 4-8. Thomas Young's double slit experiment (1803)

This experiment showed that light has a wave-like property because only waves can show interference. However, 100 years later, Einstein proved that light also exhibits the properties of a particle. Assuming a light is a stream of particles, the double-slit experiment can be analyzed in a very interesting way. When both slits are opened, photons *"fired from the light gun one at the time"* will hit the screen (wall) at particular areas, and there will be places where the photons will never land (otherwise there would be no dark areas and the image would be the same as when there is only one slit). Now, the question is *how* the photons know where to land, and where not to land, or *how* the photons know that there are two slits and that both are opened?

If one of the slits is closed, there will be no interference and the dark bands will disappear (the whole wall becomes illuminated including areas that are dark when both slits are opened). When only one slit is opened, the future of the event can be easily predicted because the laws governing the phenomenon are known, as well as the initial conditions (the origin of light, its speed, and its direction). Using Newton's laws of motion it is possible to determine where the photons will land on the wall. The initial conditions, for each case (one slit or two), are identical. However, in the case when both slits are opened, Newton's laws of motion will give exactly the same results as in the previous case, which will be wrong. In other words, two photons having exactly the same initial conditions, in two different experiments, will not go to the same location. In the second experiment, it can be understood that a wave pattern is created on the screen by a large group of photons. In that pattern it is not possible to know where a single photon will land. All that can be known is the probability of finding a single photon in a given location. *What determines WHERE a single photon will land?*

According to quantum mechanics there is only a probability which guides a photon to a particular area. The experiment can be viewed once again considering the light as a particle, as a wave or as a wave–particle:

Light as a particle – if light is considered as a stream of particles we may ask why the photons avoid making spots on certain areas of the screen when both slits are open. Every particle has two opportunities, two slits to pass through. These two opportunities interfere with each other since the image obtained at the wall shows the bands of light and dark areas. The interference can be explained by saying that the particles are controlled in such a way that each particle passes through a slit alone. The particles do not bump into each other and two particles never pass through one slit at the same time. The next question is how to explain this interference using quantum mechanics.

Light as a wave – if light is considered to behave as a wave it can reach both slits at the same time, which a particle cannot do. The wave can then break up into two waves and each would pass through each slit individually. This phenomenon is seen when a real wave (at a harbor, for example) comes to two openings (like the space between piers). Two waves can travel separate paths, go through separate slits, and reach the wall where they can interfere with each other. Waves are made of moving hills and valleys (Fig. 4-1); if at some point on the wall the valley of one wave meets the hill of another wave, these two waves cancel out at that point. This easily explains the light and dark bands at the wall when both slits are opened. If one slit is closed, then there would be no reason for the wave to split into two parts and the wave will reach the screen unimpeded. This consideration seems to solve the problem by stating that the possibilities always interfere with each other if an object behaves as a wave. It could be concluded that there were no particles in the stream of light and that the stream was simply a wave. However, when the waves arrive at the screen they do not land everywhere like waves reaching the beach shore, indicating that light does not always behave like a wave.

Light as a particle and a wave – Waves reach the screen in a series of points. Since the real waves cannot do that, it can be concluded that a particle always leaves a track while traveling as a wave through the space. This statement can explain that waves are particles and that particles are the waves. *Such behavior of particles when confronted with two or more possibilities is called wave–particle duality* (Fig. 4-9). Although it is still not known why subatomic matter behaves in this way, the laws of quantum mechanics can explain the phenomena in many applications (lasers, microchips, photocells, nuclear reactors, long-range deep-space communication devices, transistors and materials at very low temperatures).

High probability that
particle is here

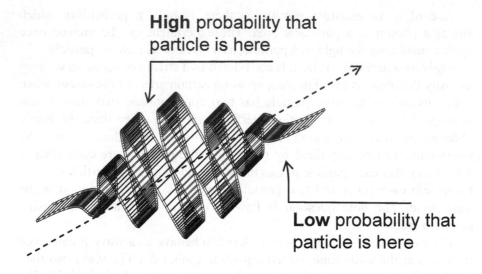

Low probability that
particle is here

Figure 4-9. Wave-particle duality

2.3   Experimental Evidence for the Wave–Particle Duality

The de Broglie's hypothesis as discussed in Section 2.1, explained that matter waves were not evident in the macroscopic world because the wavelength was much smaller than the size of the objects. His equation indicated that because the size of the subatomic particles is smaller than the wavelengths of their associated waves the wave properties are noticeable. For example, for a non-relativistic particle of mass m and kinetic energy $E = p^2/2m$, the de Broglie wavelength can be expressed as

$$\lambda = \frac{h}{\sqrt{2mE}} \qquad\qquad (4\text{-}9)$$

This relation clearly indicates that the particles of lower mass have longer wavelengths. It follows that the particle–wave behavior of the lightest known particle, the electron, should be easy to detect. By expressing the kinetic energy of an electron in eV and placing its mass of 9.109×10^{-31} kg into Eq. (4-9), the de Broglie wavelength for an electron is

$$\lambda = \sqrt{\frac{1.5}{E}} (nm) \qquad\qquad (4\text{-}10)$$

For example, an electron with energy 1.5 eV has a wavelength of 1 nm, while an electron with energy 15 keV has a wavelength of 0.01 nm. Since the distances between the atoms in crystalline structures of solid matter are in the order of electron wavelengths for electron energies in the range of eV to keV, electrons are expected to be diffracted by crystal lattices. In 1926, just 2 years after de Broglie presented his hypothesis, C. J. Davisson and L. H. Germer, at the Bell Telephone Laboratories, were able to verify the wave property of electrons in crystal diffraction experiments. Davison used electrons with energy of 54 eV and wavelength of 0.167 nm which were diffracted from a nickel-coated surface. In 1927 G. P. Thomson used the electrons with energy of approximately 40 eV and wavelength 0.006 nm to demonstrate a diffraction by micro-crystals. In the Davisson-Germer experiment the electrons were of low energy and thus they did not penetrate very far into the crystal. To analyze the experimental data and show the evidence of the wave nature of electrons, it is sufficient to assume that the diffraction took place in the plane of atoms at the surface of the nickel. From an independent X-ray diffraction data available at that time, it was known that the spacing between the rows of atoms in a nickel crystal was 0.091 nm. Therefore, according to Bragg's law (knowing that the maximum angle of diffraction was measured to be $\theta = 65°$), the wavelength of the diffracted electron was 0.165 nm (Fig. 4-10). This value, when compared to the de Broglie wavelength of 0.167 nm, provides strong evidence for wave-like behavior of the electrons.

Example 4.4 Electron diffraction from crystal planes

Electrons accelerated through a voltage of 100 V are diffracted from a crystal with a plane distance of $d = 2 \times 10^{-10}$ m. Calculate the electron scattering peaks for the first three orders of diffraction.

The electron wavelength is obtained as follows

$$qV = \frac{p^2}{2m} \quad \rightarrow \quad p = \sqrt{2mqV} \quad \rightarrow$$

$$\lambda = \frac{h}{p} = \frac{h}{\sqrt{2mqV}} = \frac{6.626 \times 10^{-34} \, Js}{\sqrt{2(9.1 \times 10^{-31} \, kg)(1.6 \times 10^{-19} \, C)(100V)}} = 1.228 \times 10^{-10} \, m$$

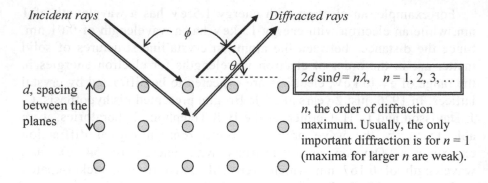

Figure 4-10. The Bragg's law

Using Bragg's law (Fig. 4-10) the scattering peaks for the first three orders of diffractions are

$$2d \sin \theta = n\lambda \quad \rightarrow \quad \theta = \sin^{-1} \frac{n\lambda}{2d} = \sin^{-1}(0.30699n) \quad \rightarrow$$

$$\theta(n=1) = 17.88° \quad \theta(n=2) = 37.88° \quad \theta(n=3) = 67.07°$$

Following these first experiments of particle diffraction many more were carried out to confirm the wave–particle duality of protons, neutrons, atoms and molecules. Additionally, the double-slit interference experiment was performed with electrons (1989), neutrons (1991) and even atoms (1991) and molecules (1999). The double-slit interference experiments with electrons demonstrated that a very weak source of electrons (only one electron passing through the slits at any given time) generated the pattern of waves on the screen. Such experiments showed that particles of matter are not classical solid particles with well-predicted and defined trajectories, but that they behave as waves whenever there is a choice of more than one possibility (such as in the double-slit experiment). In other words, in the double-slit experiment every particle is given two trajectories. In a wave form the particle travels along both trajectories arriving at a random point on the screen causing the interference pattern. In all experiments with all types of particles, the pattern consists of bands with a spacing of $\lambda L/a$, where a is the separation between the slits, L is the distance between the slits and the screen, and λ is the de Broglie wavelength (Fig. 4-8).

Example 4.5 Single slit diffraction

For the red light (660 nm) impinging on slits of width $a = 0.05$ mm and $a = 0.2$ mm placed $L = 1$ m away from the screen, determine the angular separation θ between the center line and the first minimum of the resulting diffraction pattern and its distance from the first maximum, y (Fig. 4-11).

$$a\sin\theta = n\lambda, \quad n=1 \quad \rightarrow \quad \theta = \sin^{-1}\frac{\lambda}{a} \quad \rightarrow$$

$$\theta(0.05\ \text{mm}) = 0.76°, \theta(0.2\ \text{mm}) = 0.19°$$

$$y = L\tan\theta \quad \rightarrow \quad y(0.05\ \text{mm}) = 13.2\ \text{mm}, \quad y(0.2\ \text{mm}) = 3.3\ \text{mm}$$

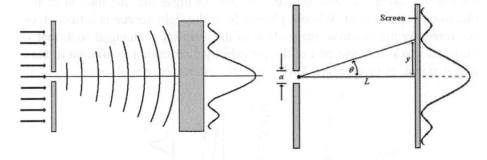

Figure 4-11. Single-slit diffraction pattern

2.4   The Uncertainty Principle

The double-slit experiment considering photons as both particles and waves is discussed in Section 2.2. In order to introduce the uncertainty principle associated with the subatomic realm, we will analyze this experiment using the electrons as shown in Fig. 4-12 (a). In the experiment it is assumed that all electrons coming from the "gun" have nearly the same energy. The electrons behave in the same way as photons (as discussed in Section 2.3) and produce interference pattern on the screen. The question is can we "watch" each of the electrons to see their trajectories and thus understand which slit they go through and at which point on the screen they land. In order to do this, two modifications are made to the experimental setup: (1) a light source is placed directly behind the slit screen, and (2) a detector or an array of detectors is placed on the wall. As each electron passes through a slit, the light emitted from the source is reflected such that the observer may determine which slit the electron passed through. The detectors then indicate where the electron hit the wall. Thus every electron that arrives at the screen is placed into one of the two categories: those that passed through slit one, and those that passed through slit two. From the number of events recorded in each category, we obtain the probabilities, P'_1 and P'_2, respectively, of each event. The distribution of these probabilities as a function of distance from the centerline is shown in Fig. 4-12 (b). When the light source is in place we "watch" each electron passing through the screen; the expected result is each electron passes through only one slit regardless of the number of open slits. When both slits are open, the probabilities of each event simply add to determine the total probability, P'_{12} = $P'_1 + P'_2$, and no interference is observed. However, if the light source is removed, interference is once again observed (P_{12}) as in the original experiment. The conclusion is that the observation of the electron trajectories using the light source somehow changes the distribution of the electrons at the screen. When a photon from the light source is reflected, or scattered, by the electron, the motion of the electron is changed such that it may fall into a different part of the probability distribution (a minima instead of a maxima, or vice versa). This is why the interference is observed.

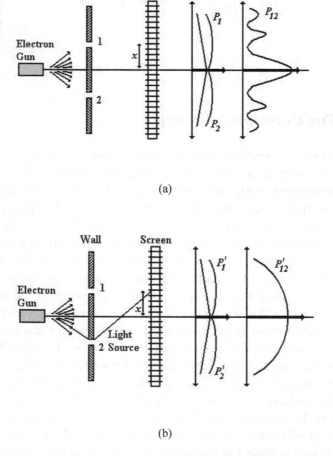

(a)

(b)

Figure 4-12. (**a**) Double-slit experiment with electrons and (**b**) the results of the experiment when electron trajectories are "watched" by shining a light on the electrons

The next question then becomes what will happen to the probability distribution if the intensity of the light source is reduced? Remember, however, that when the intensity is reduced, the energy (quanta) of the photons is not changed; only the rate at which they are emitted from the source is reduced. Thus when the source of light is dim, we may not see each electron trajectory and must now record the events into three categories: those electrons that passed through slit one, those that passed through slit two, and those that were not seen but were registered at the detector. The probabilities of these events are P'_1, P'_2, and P_{12}, respectively. Those electrons that are not seen cause interference, which is understandable because when we do not see the electron, it means there were no photons to disturb its trajectory (to change the direction and speed of motion), and when we do see it, there is a photon disturbing it (changing the direction and the speed of motion). We now come naturally to the last question: Is there any way we can see the electrons without disturbing them?

The momentum carried by a photon is $p = h/\lambda$, implying that in order to disturb the electrons only slightly, not the intensity, but the frequency of the light source should be changed. Using a light of lower frequency (for example the red light), i.e., the longer waves, the interference pattern at the screen will be restored. If the interference pattern is restored, however, we are no longer able to see where the electron hits the screen. The conclusion from this experiment is that it is impossible to design an apparatus such that it can be used to distinguish which slit the electron passed through without disturbing the electrons enough to destroy the interference pattern. This introduces the basic characteristic of subatomic wave–particle duality, known as the *uncertainty* in measuring more than one of the particle's parameters. In this experiment there was a reason why we have chosen a light source (photons) to monitor the trajectories of the electrons. In the macro world we are able to see the objects due to the reflection of light from them. For example, a lamp in the night emits photons that travel through space and interact with the surrounding objects. The reflected photons travel back toward our eyes where we detect the image. Therefore, the interaction of the photons with the objects around us is the core of the process; if an object can be observed, it must have undergone interactions with light (photons). In the macro world, where Newtonian physics applies, the interactions of photons with measured objects are ignored since such interactions will not affect the motion of macro objects (a table in a room will not move when the light is turned on). Thus in the macro world the act of measurement does not affect the object being measured. From the double-slit experiment, however, we have seen that the motion of subatomic particles *is* affected by observation and measurement. The very important conclusion is that every measurement taken on a quantum scale has an effect on the system.

Werner Heisenberg, after earning his Doctoral degree in Munich, Germany, worked with Bohr and Born in the emerging field of quantum mechanics. He developed the *Heisenberg Uncertainty Principle* as an explanation for the uncertainty in measuring parameters of the subatomic particles. The principle states

(uncertainty in position) × (uncertainty in velocity) > h/m or

$$\Delta x \times \Delta \upsilon \geq \frac{\hbar}{2m} \tag{4-11}$$

where m is the mass of the particle and h is the Planck's constant. The uncertainty in a position, Δx, represents the error made in measuring the position of a particle, and the uncertainty in velocity, $\Delta \upsilon$, refers to the error made in measuring the velocity of that particle. Thus, if we choose to measure the position of a particle we will introduce an uncertainty in the velocity and vice versa. That is why the product of these two measured values is not equal to zero. In other words, as the uncertainty in one variable becomes smaller and smaller, the uncertainty in the other becomes larger and larger in order to maintain a constant product. At the conceptual limit, if we could know the exact location of a particle in the subatomic realm, we would not know anything about its velocity; or, if we knew the exact speed of the particle, we would not be able to know where the particle is. The uncertainty principle applies not only to position and velocity but also to all parameters of a subatomic particle. For example, if we want to measure the energy of a quantum system we will need a certain amount of time, Δt, to take the measurement. During this time the energy of the system may change, ΔE, without our knowledge. The uncertainty principle describing the relation between the energy of a quantum system and the time needed to measure is given by

$$\Delta E \times \Delta t \geq \frac{\hbar}{2} \tag{4-12}$$

The principle states that if a particle has a definite energy ($\Delta E = 0$), then Δt must be infinite. In other words, a particle with definite energy is localized in the same region for all time. Such states are called the stationary states corresponding to Bohr's stationary orbits as discussed in Chapter 2. If a particle does not remain in the same state forever, Δt is finite and therefore

ΔE is not zero, and the energy of a particle must be uncertain. An example of this condition is an unstable atom or a nucleus. As mentioned previously, an unstable atom or nucleus will eventually rearrange in order to reach a stable condition, thus Δt is finite and the energy of an unstable atom or nucleus has a minimum uncertainty given as

$$\Delta E \approx \frac{\hbar}{2\Delta t} \tag{4-13}$$

Finally, the uncertainty principle can be expressed in terms of particle momentum, p, and particle position, knowing that $p = m\upsilon$

$$\Delta x \times \Delta p \geq \frac{\hbar}{2} \tag{4-14}$$

Example 4.6 Uncertainty in the position and energy of an electron confined in a nucleus and an atom

If the size of an atom is 10^{-10} m, and the size of a nucleus is 10,000 times smaller, calculate the momentum and energy of an electron confined in an atom and in a nucleus. Compare these to the binding energy of an electron in a hydrogen atom.

For an electron confined in the nucleus

$$r_{nucleus} = 1 \times 10^{-14}\,\text{m} \quad \therefore \quad \Delta x = 1 \times 10^{-14}\,\text{m}$$

$$\Delta x \Delta p \geq \frac{\hbar}{2} \quad \Rightarrow \quad \Delta p = \frac{\hbar}{2\Delta x} = \frac{6.626 \times 10^{-34}\,\text{Js}/2\pi}{2 \times 10^{-14}\,\text{m}} = 5.27 \times 10^{-21}\,\text{kgm/s}$$

$$E = \frac{(\Delta p)^2}{2m} = \frac{\left(5.27 \times 10^{-21}\,\text{kgm/s}\right)^2}{2\left(9.1 \times 10^{-31}\,\text{kg}\right)} = 1.53 \times 10^{-11}\,\text{J} \times \frac{1\text{eV}}{1.6 \times 10^{-19}\,\text{J}}$$

$$= 9.55 \times 10^7\,\text{eV}$$

The energy of an electron localized in a volume comparable to that of a nucleus is very large when compared to the binding energy of the electron in a hydrogen atom (see Chapter 2). This implies clearly that an electron cannot be localized to such a small volume in the atom. However, for an electron confined in an atom

$$r_{nucleus} = 1 \times 10^{-10}\,\text{m} \quad \therefore \quad \Delta x = 1 \times 10^{-10}\,\text{m}$$

$$\Delta x \Delta p \geq \frac{\hbar}{2} \quad \Rightarrow \quad \Delta p = \frac{\hbar}{2\Delta x} = \frac{6.626 \times 10^{-34}\,\text{Js}/2\pi}{2 \times 10^{-10}\,\text{m}} = 5.27 \times 10^{-25}\,\text{kgm/s}$$

$$E = \frac{(\Delta p)^2}{2m} = \frac{\left(5.27 \times 10^{-25}\,\text{kgm/s}\right)^2}{2\left(9.1 \times 10^{-31}\,\text{kg}\right)} = 1.53 \times 10^{-19}\,\text{J} \times \frac{1\text{eV}}{1.6 \times 10^{-19}\,\text{J}} = 0.955\text{eV}$$

The energy of an electron localized in a volume comparable to that of an atom is comparable to the binding energy of the electron in a hydrogen atom.

Example 4.7 Exchange of quanta between two electrons

As explained in Chapter 2, Section 2.6, the electromagnetic force between charged particles is assumed to be mediated by photons, or the energy quanta (as explained by the Planck's law). In other words the exchange of the photons between say two electrons explains the electromagnetic interaction between them. Assume that the collision between two electrons is considered in the center of mass system and that the collision is elastic (meaning the energy does not change after the collision). What is the maximum distance the photon would be able to travel between the two colliding electrons?

Assuming the collision to be elastic, the energies of the colliding electrons do not change, thus $E_1' = E_1$ and $E_2' = E_2$. The total available energy before the collision is $E = E_1 + E_2$. The total energy at the moment of photon emission of energy E_γ is $E = E_1 + E_2 + E_\gamma$. This means, that at that moment, the energy of the system is not conserved. This is allowed by the Heisenberg Uncertainty Principle based on Eq. (4-13). Thus the duration of that "moment" in which the energy is not conserved within an amount of ΔE is $\hbar / 2\Delta E$. A photon of energy $E_\gamma \cong \Delta E = \hbar \omega$ cannot be observed if it exists for a time shorter than $\hbar / 2\hbar\omega = 1 / 2\omega$. Such a photon will therefore travel the distance of (velocity x time), $c / 2\omega$. If the frequency of the photon is small, the distance of the electromagnetic interaction becomes very large; the Coulomb force can theoretically act at the infinite distance because it is $1/r^2$ dependent on the distance between the particles.

3.   SCHRÖDINGER EQUATION

3.1   Interpretation of Quantum Mechanics

A quantum system is divided into two parts: (a) the observed system and (b) the observing system. For example, the observed system in the double-slit experiment is a photon. The observing system represents the environment that surrounds the observed system including the experimenter (observer). The observed system travels according to a physical law called the Schrödinger wave function. This wave function refers to probabilities, e.g., the probability of finding a subatomic particle in one location rather than another (Fig. 3-20). In the macroscopic world it is intuitive that every event exists in three dimensions and in time. For example, a wave function associated with two particles will be written in six spatial dimensions (three for each particle). If the wave function represents the probability associated with 20 particles, it will exist in 60 spatial dimensions. Thus when an experiment with subatomic particles is carried out, their multi-dimensional reality is reduced to three dimensions in order to be compatible with our macroscopic world. The wave-particle duality, that employs the concept that an entity simultaneously possesses localized (particle) and distributed (wave) properties, has been introduced in order to account for observations in experiments with subatomic particles. The dominant view of this approach is that quantum probabilities become determinate by the act of measurement. Thus it is said that the wave function is collapsed when an observer looks at the system. In the double-slit experiment, according to classical physics, a photon emitted from the light source travels from the source to the slit, passes through the slit, and travels to the screen where it is detected. Thus its location at the screen can be determined. However, according to quantum mechanics, there is no real particle that travels between the source and the screen. There is only a wave function and the probability that the photon will pass through one slit or the other. The photon is detected *only when the observer looks at the screen*. In other words, the quantum reality is not described until an act of measurement takes place, at which point the wave function collapses to a single possibility.

In the autumn of 1927, the *5th Solvay Conference* was held in Brussels, Belgium. The conclusion of this meeting became known as the Copenhagen interpretation of quantum mechanics. During this Conference Bohr and Einstein conducted their famous debate:

> "*I shall never believe that God plays dice with the world*!" questioning the probabilistic nature of quantum theory. And Bohr's answer: "*Einstein, stop telling God what to do!*"

The Copenhagen interpretation of quantum mechanics, for the first time, acknowledged that a complete understanding of reality lies beyond the capability of a rational thought.

In 1957 Hugh Everett, John Wheeler, and Neill Graham proposed another explanation of the quantum wave functions called the many worlds interpretation of quantum mechanics. According to this theory, the wave function is real, and all possibilities that it predicts are real. This theory can also be demonstrated by analyzing the double-slit experiment. Suppose that when a photon goes through slit one, you run up the stairs. When a photon goes through slit two you run down the stairs. According to the Copenhagen interpretation of quantum mechanics, these two possibilities are mutually exclusive because it is not possible for you to run up and down the stairs at the same time. However, according to the many worlds interpretation of quantum mechanics, at the moment the wave function "collapses" the universe splits into these two worlds. In one of them you run upstairs, and in another one you run downstairs. There are two editions of you doing different things at the same time; but each of these two editions is unaware of the existence of the other. These two editions of you will never meet, as these worlds remain forever separated branches of reality.

3.2　　Standing Waves

De Broglie's hypothesis (Section 2.1) that each material object has a wave property opened new developments in quantum mechanics. It immediately pointed to a much more natural way of understanding atomic phenomena than Bohr's model of the atom. Bohr's model of hard, spherical electrons that orbit the nucleus at specific distances and specific energy levels, emitting photons by jumping between the orbits, explained the spectrum of simple atoms. However, it did not explain *why* each shell contains a certain number of electrons or *how* electrons move between the shells. Austrian physicist Erwin Schrödinger postulated that electrons are not spherical objects, but rather *patterns of standing waves*.

The standing wave can be explained in analogy with a rope tied to one pole at one end and then flicked sharply upward and downward from the other end, forming a hump, or a wave, that appears to travel between the two ends. By sending a series of waves down the rope, a pattern of standing waves as shown in Fig. 4-13 is generated. The simplest pattern is that of a single standing wave, shown in Fig. 4-13 with $n = 1$. This pattern is formed by the superposition of two waves traveling in opposite directions. In reality, it is not the rope that is moving, but the *pattern*; these stationary patterns are called standing waves. Regardless of the length of the rope, the rope will always show a pattern of a whole number of standing waves (i.e., one wave, two waves, three waves, etc.) that must divide the rope evenly into whole sections. The number of wavelengths or half-wavelengths that will fit along the rope is determined by its length. The first three possible standing waves are shown in Fig. 4-13. The lowest frequency is called the fundamental frequency or the first harmonic, and the higher frequencies are called overtones. Integer multiples of the fundamental frequency (the first harmonic) are labeled as the second harmonic, third harmonic, etc.

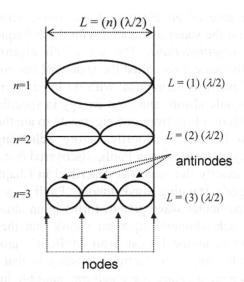

Figure 4-13. Standing waves and electron's orbits around the nucleus (see also Fig. 4-5)

In 1925 Schrödinger proposed that standing waves of subatomic particles are "quantized", similarly to the waves in the rope. For example, an electron orbiting a nucleus must travel a certain distance around the nucleus, which can be thought of as the length of the rope, therefore, only a whole number of standing waves, never a fraction of one, can form the length of orbiting electrons. Schrödinger developed the complex equation (called the Schrödinger equation) to describe the quantum wave function of subatomic particles. The equation can be solved exactly only for the simplest structure, the one-proton and one-electron structure of atomic hydrogen. The reason that the Schrödinger equation cannot be solved exactly for an atom which contains more than one electron is a mathematical problem that also appears in other areas such as astronomy: there is no exact solution to the equations describing the motion of more than two mutually interacting bodies. No exact solution of the Schrödinger equation is possible for any of the atoms heavier than hydrogen, but methods of successive approximations can be used to obtain solutions. The solution of the Schrödinger equation gives the energies an electron should have in an atom. Since light is emitted or absorbed by an atom when an electron moves from one permitted location to another, knowledge of the energies of the various levels available to an electron also gives the emission and absorption spectra of the atom.

The electrons are confined to the space surrounding the nucleus in similar to the way in which the waves of a guitar string are controlled within the string. As the tightness of the string forces it to vibrate with specific frequencies, an electron also can only vibrate with specific frequencies (see

Chapter 2). In the case of an electron, these frequencies are called the *eigenfrequencies*, and the states associated with these frequencies are called the *eigenstates* or *eigenfunctions*. The set of all eigenfunctions for an electron form a mathematical set called the *spherical harmonics*. There is an infinite number of spherical harmonics, with no in-between states. Thus an atomic electron can only absorb and emit energy in specific quanta. It does this by making a quantum leap from one eigenstate to another. This term has been introduced in Chapter 2. Shortly before Schrödinger's discovery, another Austrian physicist, Wolfgang Pauli, discovered that no two electrons in an atom can be exactly the same (as described in Chapters 2 and 3). In terms of Schrödinger's standing wave theory, Pauli's exclusion principle means that once a particular wave pattern forms in an atom, it excludes all others of its kind. Schrödinger's equation shows that there are only two possible wave patterns in the lowest orbit of Bohr's atomic model, and therefore there can be only two electrons existing in that orbit (1*s* level). There are eight different standing wave patterns possible in the next energy level; therefore there can be only eight electrons, and so on. Although Schrödinger was sure that electrons were standing waves, he was not sure *what was waving*. He was however convinced that something was waving and that he called the *"psi"* (Ψ), a *quantum wave function*.

3.3   Quantum Waves

The quantum wave thus introduced by Schrödinger and described by the 23rd letter of Greek alphabet, Ψ, is to be understood more as a mathematical formalism of the physical phenomena in the quantum world of subatomic particles. The quantum wave is not located anywhere in the space like a real wave and it possesses no energy. The square of the amplitude of an ordinary wave gives its energy. However, the square of the amplitude of a quantum wave at certain position in space represents the probability that a particle (a localized quantum of energy) with that wave property will be located at that particular spatial coordinate. Since the quantum wave does not carry energy it is not directly detectable; quantum particles are (Section 2.2). The presence of the quantum wave is identified after many particle events (Section 2.2). According to this interpretation, the quantum wave described with the quantum function is $\Psi(\vec{r}, t)$. The probability of finding the particle at time t in the volume element $d^3 r$ is defined as

$$\left| \Psi\left(\vec{r}, t\right) \right|^2 d^3\vec{r} \rightarrow \text{probability of finding the particle} \qquad (4\text{-}15)$$

As Fig.4-9 indicated, the particle could be found anywhere in space, however it is most likely to be found where the probability of its wave function is large. By integrating the Eq. (4-15) over all the possible positions in the space, the integral will give the probability of finding the particle somewhere in the universe. Since the particle will certainly be found somewhere, the integral must be equal to unity

$$\int_{space} \left| \Psi\left(\vec{r},t\right)\right|^2 d^3\vec{r} = 1 \tag{4-16}$$

For a particle moving along one dimension (most simplified case as will be discussed in proceeding sections) the Eq. (4-16) takes the following form

$$\int_{-\infty}^{\infty} \left| \Psi(x,t)\right|^2 dx = 1 \tag{4-17}$$

3.4 General Characteristics of the Quantum Wave Function

Each particle is represented by a wave function, $\Psi(x,t)$, which is obtained by solving the Schrödinger equation

$$i\hbar \frac{\partial \Psi(x,t)}{\partial t} = -\frac{\hbar^2}{2m} \frac{\partial^2 \Psi(x,t)}{\partial x^2} + U(x,t)\Psi(x,t) \tag{4-18}$$

where i is the square root of negative 1, $\hbar$ is the Plank's constant divided by 2π, and $U(x,t)$ is the potential energy field. This equation plays the role of Newton's law of conservation of energy in classical mechanics. The equation predicts the probability of future behavior of dynamic quantum subatomic systems; given a large number of events, it predicts the distribution of probabilities. As defined with the Eq. (4-15), the square of the wave function represents the probability amplitude for finding a particle at a given point in space at a given time. In order to represent a physically observable system, the wave function must satisfy the following constraints: it must be a solution to the Schrödinger equation, it must be normalized implying that the total probability over all x is unity as described in Section 3.3, it must be a continuous function of x, and it must have a continuous slope. Basic steps in developing the fundamental quantum mechanics equation are as follows:

1. *Conservation of energy*:

$$E = E_k + U \tag{4-19}$$

where E is the total energy, E_k ($mv^2/2 = p^2/2m$) is the kinetic energy, and U is the potential energy of a particle with mass m.

2. *De Broglie hypothesis*: for a free particle with momentum p, the wavelength, $\lambda = h/p$. If $k = 2\pi/\lambda$ is a wave number, then it follows

$$p = \frac{h}{\lambda} = \frac{hk}{2\pi} \tag{4-20}$$

Combining the Eqs. (4-19) and (4-20), the kinetic energy of a free particle to which a de Broglie wave is associated, can then be written as

$$E_k = \frac{mv_2}{2} = \frac{p^2}{2m} = \frac{h^2 k^2}{4\pi^2 2m} \tag{4-21}$$

which bears a clear resemblance to the kinetic energy term (first term) of the Schrödinger equation.

3. *Continuous solution*: a solution concerning the location or state of motion of a particle should not show discontinuity (the particle can not appear and disappear at different locations in a system).

4. *Single-valued solution*: a solution should give only one probability for the particle to be in a specific location at a specific time.

5. *Linear solution*: solution must be linear in order to assure that de Broglie waves will have the superposition property expected for waves.

The following sections will explore the time-independent Schrödinger equation where the time-dependent solution is assumed to be separable in the following form

$$\Psi(x,t) = \psi(x)f(t) \tag{4-22}$$

The partial derivatives in the Schrödinger equation may now be transformed into two absolute derivatives under the assumption that the potential field is not a function of time

$$ih\frac{1}{f(t)}\frac{df(t)}{dt} = E$$

$$-\frac{\hbar^2}{2m}\frac{d^2\psi(x)}{dx^2} + U(x)\psi(x) = E\psi(x) \tag{4-23}$$

A separation constant, E, is introduced representing the particle energy state. The time-dependent portion has the following general solution

$$f(t) = Ce^{-iEt/\hbar} \qquad (4\text{-}24)$$

The remaining, time-independent, portion will be now analyzed for some specific potential energy distributions.

3.5 Wave Function for a Particle in an Infinite Well

The simplest case to analyze is a single particle in an infinite well, for which the potential is defined as shown in Fig. 4-14

$$U(x) = \begin{cases} 0, & \text{for } 0 \le x \le a, \\ \infty, & \text{otherwise} \end{cases} \qquad (4\text{-}25)$$

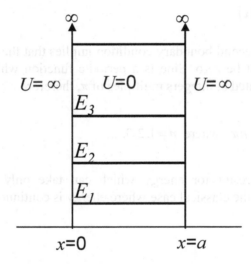

Figure 4-14. Infinite square well potential distribution

For this case, the wave function takes a zero value *outside* of the well because the infinitely high walls confine the particle within. Therefore, inside the well, the time-independent Schrödinger equation reduces to

$$-\frac{\hbar^2}{2m}\frac{d^2\psi(x)}{dx^2} = E\psi(x) \qquad (4\text{-}26)$$

or, equivalently

$$\frac{d^2\psi(x)}{dx^2} = -k^2\psi(x), \text{ where } k \equiv \frac{\sqrt{2mE}}{\hbar} \qquad (4\text{-}27)$$

It is clear that the general solution is the same as that for the classical simple harmonic oscillator

$$\psi(x) = A\sin(kx) + B\cos(kx) \quad \text{for} \quad 0 \le x \le a \qquad (4\text{-}28)$$

where A and B are constants to be determined using boundary conditions for wave function continuity at the boundaries

$$\psi(0) = \psi(a) = 0 \qquad (4\text{-}29)$$

Imposing the first boundary condition implies that B must be zero and the solution reduces to

$$\psi(x) = A\sin(kx) \qquad (4\text{-}30)$$

Imposing the second boundary condition implies that the argument of the sine function must be zero. Sine is a periodic function which takes a zero value when evaluated at integers multiples of π, therefore

$$ka = \frac{\sqrt{2mE}}{\hbar} = n\pi, \text{ where } n = 1,2,3,\dots \qquad (4\text{-}31)$$

This gives a result for energy which can take only discrete values (Fig. 4-14) unlike the classical case where energy is continuous

$$E_n = \frac{n^2\pi^2\hbar^2}{2ma^2} \qquad (4\text{-}32)$$

The full solution for this case is the product of the time-dependent and time-independent components

$$\Psi(x,t) = \psi(x)f(t) = A\sin\left(\frac{n\pi x}{a}\right)e^{-iEt/\hbar} \qquad (4\text{-}33)$$

where the two constants have been combined into a single constant, A.

In order to determine the constant A, we must normalize the integral of the square of the wave function to unity, as explained in Section 3.3.

$$\int_{-\infty}^{\infty}|\Psi(x,t)|^2\,dx = \int_{-\infty}^{\infty}\Psi(x,t)\Psi^*(x,t)dx = 1 \qquad (4\text{-}34)$$

where Ψ^* is the complex conjugate of the wave function. Expanding the conjugate product reveals that the time-dependent terms cancel each other leaving only a simple integral with respect to x. Evaluating this integral gives $A^2 = 2/a$, and the full solution is, therefore

$$\Psi(x,t) = \sqrt{\frac{2}{a}}\sin\left(\frac{n\pi x}{a}\right)e^{-i\left(\frac{n^2\pi^2\hbar}{2ma^2}\right)t} \quad \text{for} \quad 0 \le x \le a \qquad (4\text{-}35)$$

This wave function may now be used to determine the probability that the particle will be located at any position x and time t within the infinite well.

3.6   A Wave Function for a Free Non-Relativistic Particle

In this case, the particle is assumed to be totally free, namely $U(x) = 0$ for all x. Upon insertion of this potential distribution, the time-independent Schrödinger equation takes the same form as inside the infinite square well and has a solution in the following form

$$\psi(x) = A\sin(kx) + B\cos(kx) \quad \text{for all } x \qquad (4\text{-}36)$$

At this point in solving the infinite well problem, we applied the appropriate boundary condition which determined the allowed values for k and therefore, E. However, in this case there are no such boundary conditions to restrict the value of the k because the particle is totally free. However, upon closer inspection of the definition of k, the allowed energies may be found to be

$$E = \frac{\hbar k}{2m} \qquad (4\text{-}37)$$

which is precisely the same as the energy predicted by the de Broglie hypothesis for a free particle. It is important to note that due to the lack of a restriction on k, the energy values are not quantized, meaning that a free particle can possess any energy value. Another interesting note about this solution, is that it is not normalizable, namely the integral of $|\Psi(x,t)|^2$ is infinite. Thus, the wave function can never be normalized and the wave function obtained here may not be used to predict probabilities.

3.7   Tunneling Phenomena

Subatomic particle tunneling (passage) through a classically forbidden potential wall is one of the most important demonstrations of the validity of the quantum theory. Examples of such processes are the α decay, fission and other processes involving particles crossing the Coulomb barrier. To illustrate the phenomena of quantum tunneling we will assume that the potential barrier is approximated with the rectangle as shown in Fig. 4-15 with the barrier "height" of U_0 and thickness a, such that

$$U(x) = \begin{cases} 0 & \text{for} \quad -\infty < x < 0 \\ U_0 & \text{for} \quad 0 < x < a \\ 0 & \text{for} \quad a < x < \infty \end{cases} \qquad (4\text{-}38)$$

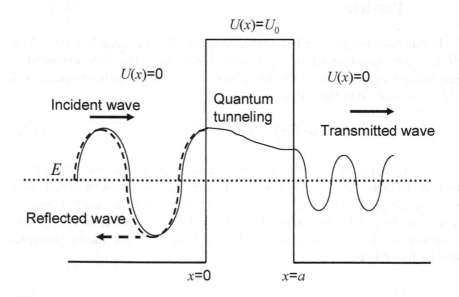

Figure 4-15. A rectangular potential well of height U_0 and thickness a illustrating the quantum tunneling: the real part of the wave function representing particle of energy smaller than the barrier height approaches from the left and tunnels through the barrier

According to classical physics, every particle (represented with the quantum wave) approaching the potential barrier shown in Fig. 4-15 having the energy lower than U_0 will be reflected back. Quantum physics shows that there will be a probability of such a particle to be transmitted through the high potential barrier. The behavior of a particle of a given mass m in the potential well is described with Eq. (4-23). In order to calculate the probabilities for a particle reflection and transmission through the potential well shown in Fig. 4-15, we will consider the stationary state that is described with the wave function of the form given with Eq. (4-27) representing the solution of Eq. (4-26). Thus we have to find the solution of the stationary equation for various regions along the direction x

- *Left from the barrier* the potential energy is zero and the wave function satisfies the following equation

$$\frac{d^2\psi_I(x)}{dx^2} = -k^2\psi_I(x) \quad \text{with} \quad k^2 = \frac{2mE}{\hbar^2} \tag{4-39}$$

The solution for the incoming and reflected wave of the intensities $|A_i|^2$ and $|A_r|^2$ is given with

$$\psi_I(x) = A_i e^{ikx} + A_r e^{-ikx} \tag{4-40}$$

- *The wave function inside the potential barrier* takes different forms depending on the particle energy. For $E > U_0$, a classically allowed situation, the wave function satisfies the following equation

$$\frac{d^2\psi_{II}(x)}{dx^2} = -k_0^2\psi_{II}(x) \quad \text{with} \quad k_0^2 = \frac{2m(E - U_0)}{\hbar^2} \tag{4-41}$$

with the general solution of the form

$$\psi_{II}(x) = A e^{ik_0 x} + A' e^{-ik_0 x} \tag{4-42}$$

For $E < U_0$, a classically forbidden situation, the wave function satisfies the following equation

$$\frac{d^2\psi_{II}(x)}{dx^2} = \chi^2\psi_{II}(x) \quad \text{with} \quad \chi^2 = -\frac{2m(E - U_0)}{\hbar^2} \tag{4-43}$$

with the general solution of the form

$$\psi_{II}(x) = B e^{-\chi x} + B' e^{\chi x} \tag{4-44}$$

where B and B' are the arbitrary constants. The solution includes both the negative and the positive terms, because the wave function cannot become infinite in the region in which x does not become infinite.

– *Right from the barrier* the potential energy is zero and the wave function satisfies Eq. (4-39). The solution is obtained for the transmitted wave of intensity $|A_t|^2$ as follows

$$\psi_{III}(x) = A_t e^{ikx} \tag{4-45}$$

because it is assumed that there is no source of particles in the region right from the barrier. The only source of particles (waves) streaming toward the barrier is from its left.

For all three regions the wave function must satisfy the conditions of continuity and also to be a single valued (as explained in Section 3.4). Thus the continuity conditions are

$$
\begin{aligned}
&\psi_I(x=0) = \psi_{II}(x=0) \\
&\left.\frac{d\psi_I(x)}{dx}\right|_{x=0} = \left.\frac{d\psi_{II}(x)}{dx}\right|_{x=0} \\
&\psi_{II}(x=a) = \psi_{III}(x=a) \\
&\left.\frac{d\psi_{II}(x)}{dx}\right|_{x=a} = \left.\frac{d\psi_{III}(x)}{dx}\right|_{x=a}
\end{aligned}
\tag{4-46}
$$

Combining the solutions given with Eqs. (4-43), (4-44), (4-45), and the continuity conditions specified with Eq. (4-46), we can derive the expressions for the intensities of the reflected and transmitted waves and determine the corresponding probabilities. The probability that the particle is reflected, P_r, and the probability that the particle is transmitted, P_t, are defined as the particle flux ratios

$$
P_r = \frac{|A_r|^2}{|A_i|^2} \qquad\qquad P_t = \frac{|A_t|^2}{|A_i|^2}
$$

$$
P_r = \frac{\sinh^2(\chi a)\left(1 + k^2/\chi^2\right)^2}{\left(1 + k^2/\chi^2\right)^2 \sinh^2(\chi a) + (2k/\chi)^2}
\tag{4-47}
$$

$$
P_t = \frac{(2k/\chi)^2}{\left(1 + k^2/\chi^2\right)^2 \sinh^2(\chi a) + (2k/\chi)^2}
$$

The summation of these two probabilities must be equal to unity to assure the conservation of particle flux. We are however interested to determine the probability of quantum tunneling, P_t, for the particles of energy lower than the barrier height as shown in Fig. 4-15. The following equation is valid under the assumption of the wide barrier, i.e., $\exp(-2\chi a)<<1$ in which case $B'<<B$ thus giving

$$ P_t = \frac{|A_t|^2}{|A_i|^2} \cong \left[\frac{16k^2\chi^2}{\left(k^2+\chi^2\right)^2}\right]e^{-2\chi a} = \left[\frac{16E\left(U_0-E\right)}{U_0^2}\right]e^{-2a\frac{\sqrt{2m}}{\hbar}\sqrt{U_0-E}} \qquad (4\text{-}48) $$

The wave function inside the wide barrier is dominated by the exponentially decaying term in Eq. (4-44), while the probability of tunneling through such a barrier is proportional to $\exp(-2\chi a)$. The application of this formula is found in alpha decay, and two other examples are given at the end of this chapter.

3.8 Hydrogenic Wave Functions

As mentioned before, the Schrödinger equation is analytically solvable only for the simplest case of the hydrogen atom consisting of a single proton and a single electron. In this case, the only relevant force is the attractive force between these two particles. However, when more particles are introduced into the system, like for instance a helium atom, the problem becomes increasingly more difficult. There are now additional force terms representing the repulsion of like particles as well as additional attractive terms for the newly introduced particles. You can imagine how complex the system would be if one were to try and solve this problem for a high-Z atom such as uranium.

There are however some special cases other than hydrogen where a solution may be obtained. These are known as hydrogenic atoms, literally meaning atoms which are *like* hydrogen. The simplest example is that of a singly ionized helium atom. In this case, the nucleus contains two protons as compared to the one in hydrogen and there in only one orbital electron. The actual solution of such cases is outside the realm of this text, but a list of recommended literature is provided if a more in-depth study is desired.

3.9 Quantazation of Angular Momentum

The physical interpretation of the quantum number n for the electron in an atom is explained in Chapter 2; it determines the energy level and the distance of an electron from the nucleus. The Schrödinger equation predicts the radial probability of finding an electron at certain distance from the nucleus and is in agreement with the Bohr model. The average radius varies with n^2, meaning that an electron at orbit $n = 1$ is away from the nucleus at the average distance equal to the Bohr radius a_0, at the orbit $n = 2$ electron will be four times farther from the nucleus, for $n = 5$ it will be 25 times farther from the nucleus, and so forth. This section describes the physical meaning of the l and m quantum numbers.

As described in the Appendix 5 the general definition of the angular momentum is given with a product of mass, size, and speed of rotation

Angular momentum $=$ (mass) $\times$ (radius) $\times$ (rotation speed)

or it can be written as follows

$$\vec{L} = \vec{r} \times \vec{p} \tag{4-49}$$

where $\vec{r}$ is the position vector of the particle (or the object) and $\vec{p}$ is its linear momentum. The direction of $\vec{L}$ is perpendicular to the plane of the orbit. For example, the orbital angular momentum describes the motion of the Earth around the Sun, while the intrinsic angular momentum of the Earth describes its rotation about the axis. The planet in the solar system can have any shape of its orbit from almost circular (like the Earth's orbit) to a very elongated one (like the orbits of the comets). These orbits differ in the value of their angular momentum, L. As shown in Fig. 4-16, the orbit with the largest L is of a circular shape, while the orbit with zero L is a thin and long ellipse.

The quantum theory differs from this example based on classical physics.

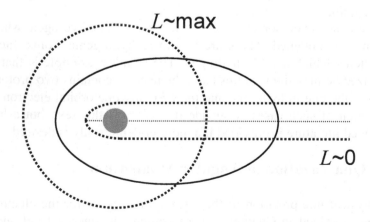

Figure 4-16. Interpretation of the angular momentum of the planets in solar system

According to quantum theory the magnitude of the angular momentum (the length) for an electron is different from the ones obtained in Bohr'model (see Chapter 2, Section 2.8.2) and is defined as a quantum mechanical vector

$$\left|\vec{L}\right| = L = \sqrt{l(l+1)}\hbar \tag{4-50}$$

Thus the orbital motion of an electron in a state with $l = 1$ has the length of the angular momentum equal to $\sqrt{2}\hbar$, with $l = 2$ it will be equal to $\sqrt{6}\hbar$, and so forth. Now, consider for example the quantum state defined with $n = 3$. According to Bohr's model, the angular momentum of such an electron will have only one value and equal to $3\hbar$. However, since the state $n = 3$ can have three values for the quantum number l, there will be three different values for the angular momentum, $0, \sqrt{2}\hbar, \sqrt{6}\hbar$.

Vector $\vec{L}$ has three components in the three-dimensional space. The Schrödinger equation provides the values for the vector $\vec{L}$ components. Since the z axis is the axis of reference in the spherical coordinate system, the z component of the vector $\vec{L}$ is analyzed. In addition, according to quantum mechanics we may have an exact value for only one of the three components of the angular momentum vector. The other components are uncertain according to the Heisenberg Uncertainty Principle. Figure 4.17 shows a schematic representation of the quantized values of angular momentum along the z-direction. This component has a restricted value defined as a multiple of magnetic quantum number and reduced Planck constant as follows

$$L_z = m\hbar \quad m = 0, \pm 1, \pm 2, ..., \pm l \tag{4-51}$$

Until now, only the analysis of the angular momentum or the effect of the electron (or a planet) orbiting around the center was considered. However, Earth also orbits around its own axis, and this is found to be true for subatomic particles as well. This orbiting is called an intrinsic spin. Therefore, the total angular momentum is defined as a sum of two terms and for a general orbiting body can be written as follows

$$\vec{J} = \vec{L} + \vec{S} = \vec{r} \times \vec{p} + I\vec{\omega}$$

where I represents the body's moment of inertia and ω is the angular velocity of spinning around its own axis. The magnitude of the spin is found to have similar form as the magnitude of the angular momentum

$$\left|\vec{S}\right| = L = \sqrt{s(s+1)}\hbar \tag{4-52}$$

As discussed in Chapter 2, Section 2.8, the spin quantum number has a fixed value of ½ . Therefore, the electron's spin S has fixed (constant) magnitude

$$\left|\vec{S}\right| = L = \sqrt{s(s+1)}\hbar = \frac{\sqrt{3}}{2}\hbar \tag{4-53}$$

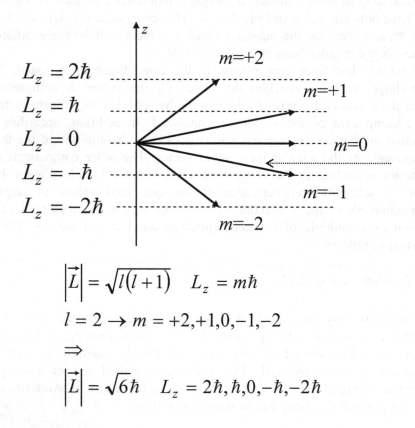

$$\left|\vec{L}\right| = \sqrt{l(l+1)} \quad L_z = m\hbar$$

$$l = 2 \rightarrow m = +2,+1,0,-1,-2$$

$$\Rightarrow$$

$$\left|\vec{L}\right| = \sqrt{6}\hbar \quad L_z = 2\hbar,\hbar,0,-\hbar,-2\hbar$$

Figure 4-17. Interpretation of the quantum angular momentum [example is given for l =2]

That is why the spin is also described as intrinsic angular momentum of the electron. The possible values of the z-component of the electron spin are

$$S_z = \frac{1}{2}\hbar \quad \text{or} \quad S_z = -\frac{1}{2}\hbar \tag{4-54}$$

In other words, the electron spin can be "up" or "down" spin. The experimental evidence for the electron spin is found indirectly through the magnetic moment existing for any rotating charge. Figure 4-18 shows the classical magnetic momentum due to the circular motion of a charged body.

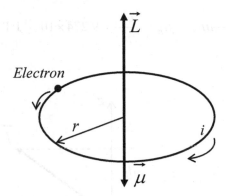

Figure 4-18. Interpretation of the magnetic moment [example is given for *electron*]

The magnetic moment is a vector, $\vec{\mu}$, with the magnitude equal to the product of the circulating curent, i, and the area enclosed by the orbital loop, $A = \pi r^2$. Its direction is defined by the right-hand rule; in case of negative charge, it has the opposite direction from the angular momentum direction. According to the Heisenberg Uncertainty Principle, the exact knowledge of the angular, and therefore magnetic, moment are not allowed. But, as we already shown, we may determine the z-projections of these two vectors as depicted in Fig. 4-19.

Using the Bohr's atomic model (Chapter 2, Section 2.6) which assumes that an electron orbits around the nucleus in a circular orbit, the current it generates is determined as the ratio of charge and time as it completes the loop of a given radius. According to the Example 2.5, the time is given as $2\pi r / v$. The velocity is determined based on the electron energy (i.e., the principal quantum number) or we may write $v = p/m_e$, because we also know that $\left| \vec{L} \right| = r m_e v = rp$. Therefore, it follows

$$\mu = i\left(\pi r^2\right) = \frac{-e}{T}\left(\pi r^2\right) = \frac{-e}{2\pi r / v}\left(\pi r^2\right) = \frac{-e}{2m_e / p}r = \frac{-e}{2m_e}\left| \vec{L} \right| \qquad (4\text{-}55)$$

or

$$\vec{\mu} = \frac{-e}{2m_e}\vec{L} \qquad (4\text{-}56)$$

The negative sign due to the electron charge (that is negative) indicates that the angular and magnetic moment has opposite directions as shown in Fig. 4-19. The z-component of the magnetic moment is defined as a function of the Bohr magneton, μ_B

$$\mu_z = \frac{-e}{2m_e} L_z = -m\mu_B \quad \mu_B = \frac{e\hbar}{2m_e} = 9.274 \times 10^{-24} \, \text{J/T} \quad (4\text{-}57)$$

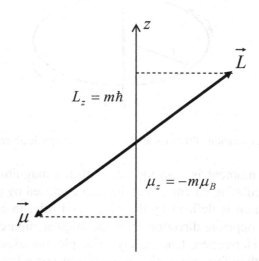

Figure 4-19. The z-components of the angular and magnetic moment vectors for the electron orbiting around a nucleus

Besides the orbital magnetic moment an electron has also the spin magnetic moment that can be shown to be equal to

$$\vec{\mu}_s = \frac{-e}{m_e} \vec{S} \quad (4\text{-}58)$$

APPLICATIONS

From Quantum to Classical Physics or *Where* is the Connection between Microscopic and Macroscopic World

The most intriguing question is to explain the change from the quantum to classical physics laws and the transition from the microscopic to the macroscopic world. Quantum physics was developed to explain the subatomic phenomena; Newton's laws are valid only in the macroscopic world. What makes the transition in between, and how does something subatomic, driven with the laws of uncertainty, become macroscopic and certain? To illustrate that transition and explain the inter-connections, we will use the example of the infinite square well and the probability of finding the particle in that well. The numerical example, given at the end of this chapter, shows the plots of the probability functions corresponding to the first seven states for the particle trapped in an infinite well. For the first state for example the probability function shows that a particle could be found anywhere in the well in contrast to classical physics that would specify only one position. When increasing the quantum number states, the probability distribution requires more nodes with the distance between them becoming smaller such that they are practically smeared together for very high quantum number (Fig. 4-20). The position of the particle will approach an average and the measurement will show an even distribution in the well, thus approaching the classical physics predictions. In other words, when the quantum number of a system is very high, it behaves classically. This is called the correspondence principle or classical limit, and it was formulated by Niels Bohr in 1923.

Scanning Tunneling Microscope

A scanning tunneling microscope produces three-dimensional images of the metallic and semiconductor surfaces at the atomic level. The technique, developed in 1981, is very useful for characterizing surface unevenness and defects, and determining the size and conformation of surface aggregates. The basic principle of how the scanning tunneling microscope operates is sketched in Fig. 4-21; an extremely fine metal probe (the tip size being that of one single atom) is placed very close to the sample (distance of only an atom's diameter) so that the tunneling of electrons between the probe and the sample is induced. Thus, if the tip is in contact with the sample, the electron tunneling will take place, and the "tunneling current" will be recorded. The tunneling current increases with decreased distance between the tip and the sample surface. This change of tunneling current results in atomic resolution of the surface for each position of the tip that moves over the sample surface thus producing the surface image.

In Chapter 6, Section 5.2.1, from the photoelectric effect explanation, it follows that a minimum energy required to eject an electron from the surface of a metal is in order of a few eV. In the Chapter 2 we learned that an electron is kept in the cloud surrounding the nucleus by the attractive force and thus we can imagine that the electron resides in an attractive potential energy field. When two metal surfaces are placed close to each other, there will be two low-level potential energy fields forming the potential well similar to the one shown in Fig. 4-15. Because the electrons are quantum particles, they will tunnel through the potential barrier with the probability based on Eq. (4-48).

$$P_t \cong e^{-2a\frac{\sqrt{2m_e}}{\hbar}\sqrt{U_0-E}} \qquad (4\text{-}59)$$

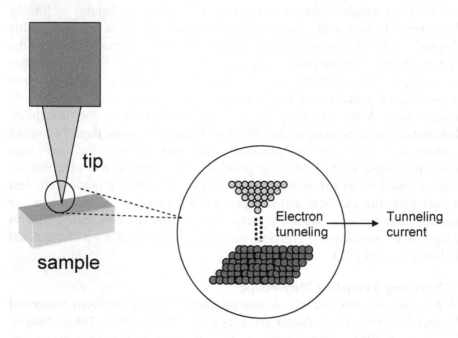

tip

Electron tunneling

Tunneling current

sample

Figure 4-21. Principle of scanning tunneling microscope

where a represents the distance between the metal surfaces, m_e is the electron mass, the most energetic electron coming from the metal surface has energy E, and the U_0 is the potential barrier height at the metal surface. It can be seen that the tunneling probability depends on the distance between the surfaces such that if the distance changes by a small value, Δa, the fractional probability illustrating the sensitivity of the scanning tunneling microscope becomes

$$\frac{\Delta P_t}{P_t} \cong -2\Delta a \frac{\sqrt{2m_e(U_0-E)}}{\hbar} \qquad (4\text{-}60)$$

The easiest way to understand the application of this phenomenon is to calculate this fractional tunneling probability. Assuming that the typical energy required to eject an electron from the metal surface is around 4 eV, which represents the difference between the potential barrier and the electron energy, the penetration parameter can be obtained

$$\chi = \frac{\sqrt{2m_e(U_0 - E)}}{\hbar} = \frac{\sqrt{2m_e c^2(U_0 - E)}}{\hbar c} =$$

$$\frac{\sqrt{2(0.511 \times 10^6 \, \text{eV})(4\text{eV})}}{1.97 \times 10^{-7} \, \text{eVm}} \cong 10^{10} \, \text{m} \tag{4-61}$$

Equation (4-60) illustrates how large the sensitivity of the scanning tunneling microscope is: for example if the distance between the surfaces changes by 0.001 nm, the change in the probability is in the order of 2 % with the penetration parameter obtained with Eq. (4-61) [Phillips, 2003].

Gamow Energy and Fusion

The center of the stars is made of a highly dense, ionized gas of electrons, protons, and light nuclei. The temperature of this gas is around 10^7 K in the center of the Sun. The high temperature and high density of the gas creates the conditions for the nuclear processes to take place in which protons (hydrogen nuclei) are fused to create helium and release energy. The size of the Sun is so large that the energy released at the center of the Sun takes about 50 million years to reach the surface of the Sun, going through a numerous processes of absorptions and re-emissions during that time. In order to understand the source of this energy, and the conditions for the fusion reaction, we will analyze two protons approaching each other at the center of Sun. The fusion event probability is proportional to the probability of protons tunneling through the potential barrier. The kinetic energies of these two protons are of the order of 1 keV or more precisely

$$E = kT \cong 0.86 \, \text{keV} \tag{4-62}$$

The potential energy of two protons changes with the distance between them, Fig. 4-22; at large distances it is described by the repulsive Coulomb potential

$$U(r) = \frac{e \cdot e}{4\pi\varepsilon_0 r} \tag{4-63}$$

where ε_0 is the electrical permittivity of space, e is the proton charge, and r is the distance between the two protons. At small distances, in the range of the strong nuclear force of the order of 2 fm, the potential energy is attractive creating the net Coulomb barrier of around 1 MeV in height. Thus, when two protons of energy close to 1 keV approach each other at the center of the Sun they see the barrier of around 1 MeV to cross in order to be able to fuse.

If protons were not the quantum particles, the Sun would not shine. According to classical physics there is a well-defined distance of closest approach (Rutherford scattering), r_c, for protons of energy E, given with

$$E = \frac{e \cdot e}{4\pi\varepsilon_0 r_c} \qquad (4\text{-}64)$$

According to quantum physics these two protons are described with Eq. (4-23). However, we will write this equation in three-dimensions and solve it for the potential barrier sketched in Fig. 4-17

$$-\frac{\hbar^2}{2\mu}\nabla^2\psi\!\left(\vec{r}\right) + U(r)\psi\!\left(\vec{r}\right) = E\psi\!\left(\vec{r}\right) \qquad (4\text{-}65)$$

where μ is the reduced proton mass of $m_p/2$ [$=m_p m_p/(m_p+m_p)$] [Phillips, 2003]. Assuming the spherical symmetry without angular dependence, the wave function has the following form

$$\psi\!\left(\vec{r}\right) = w(r)/r \qquad (4\text{-}66)$$

with $w(r)$ satisfying

$$-\frac{\hbar^2}{2\mu}\frac{d^2 w(r)}{dr^2} + U(r)w(r) = E \qquad (4\text{-}67)$$

For the three-dimensional barrier of constant height U_0 and width r_c-2 fm, $w(r)$ decays exponentially (in the classically forbidden region) as (see Section 3.7 to understand the solution)

$$w(r) \propto e^{\chi r} \qquad \chi^2 = -\frac{2\mu}{\hbar^2}(E - U_0) \qquad (4\text{-}68)$$

The probability that protons will tunnel from r_c to ~ 2 fm distance between them is

$$P_t = \frac{\left|w(r=2\text{fm})\right|^2}{\left|w(r=r_c)\right|^2} \cong \left|e^{-\chi(r_c-2fm)}\right|^2 \qquad (4\text{-}69)$$

For the Coulomb barrier shown in Fig. 4-22, the wave function in the region 2 fm $< r < r_c$, is approximated with

$$w(r) \propto e^{\chi r} \quad \chi^2 = -\frac{2\mu}{\hbar^2}\left(E - \frac{e \cdot e}{4\pi\varepsilon_0 r}\right) \tag{4-70}$$

in which case the penetration parameter depends on distance r. Using Eq. (4-68) we obtain the approximate probability of proton tunneling

$$P_t = \frac{|w(r = 2\text{fm})|^2}{|w(r = r_c)|^2} \cong \left| e^{-\int_{2\text{fm}}^{r_c} \chi dr} \right|^2 \tag{4-71}$$

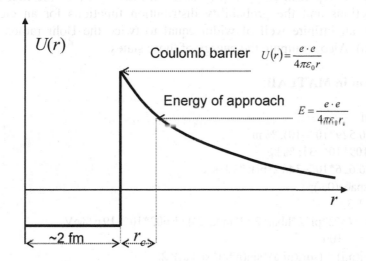

$U(r)$

Coulomb barrier $U(r) = \dfrac{e \cdot e}{4\pi\varepsilon_0 r}$

Energy of approach $\quad E = \dfrac{e \cdot e}{4\pi\varepsilon_0 r_c}$

r

~2 fm $\quad r_c$

Figure 4-22. Potential energy profile for two protons with separation distance equal to r

Assuming that $r_c \gg 2$ fm, Eq. (4-71) reduces to a known relation for fusion probability for two protons (for which $Z = 1$) defined as a function of Gamow energy, E_G

$$P_t \cong e^{-\sqrt{E_G/E}} \quad E_G = \left(\frac{e \cdot e}{4\pi\varepsilon_0 \hbar c}\right)^2 2\pi^2 \mu c^2 = 2\mu c^2 \left(\pi\alpha Z_p Z_p\right)^2 \tag{4-72}$$

where $Z_p = 1$ and α is the fine structure constant ~1/137. Thus for the case of two protons approaching each other, the Gamow energy is 493 keV. And finally the Eq. (4-72) gives the probability of about one in three billion that the two protons with thermal energies of ~ 1 keV will fuse by tunneling through the potential Coulomb barrier of ~ 1 MeV, at the center of Sun where the temperature is ~ 10^7 K

$$P_t \cong e^{-\sqrt{E_G/E}} = e^{-\sqrt{493/1}} = e^{-22} \sim 3 \times 10^{-10} \tag{4-73}$$

Because this probability of quantum tunneling is so low, the stars shine for billions of years.

NUMERICAL EXAMPLE

Wave Function States in an Infinite Square Well

As described in Section 3.5, the wave function of a particle in an infinite square well is represented by Eq. (4-23). At time = 0, plot the first seven wave functions and the probability distribution functions for an electron trapped in an infinite well of width equal to twice the Bohr radius (r_0 = 0.0529 nm). Also, compute the energy of these states.

Solution in MATLAB:

```
clear all
a = 2*(0.529*10^-10); % m
m = 9.109*10^-31; % kg
hbar = 6.626*10^-34/(2*pi); % J*sec
x = linspace(0,a);
for n = 1:3
    E(n) = (n^2*pi^2*hbar^2/(2*m*a^2))/(1.602*10^-19); %eV
    for i = 1:100
        phi(n,i) = (sqrt(pi/a)*sin(n*pi*x(i)/a))^2;
    end
    subplot(3,1,4-n)
    plot(x/a,phi(n,:),'k')
    xlabel('Distance Normalized to a')
    ylabel('Probability Density')
end
disp(E)
```

PROBLEMS

4.1. Calculate the minimum de Broglie frequency of the neutron that is capable of exiting an electron in He^+ from the ground state to the state $n = 3$. What is the wavelength of the X-ray emitted when electron falls back to its ground state?

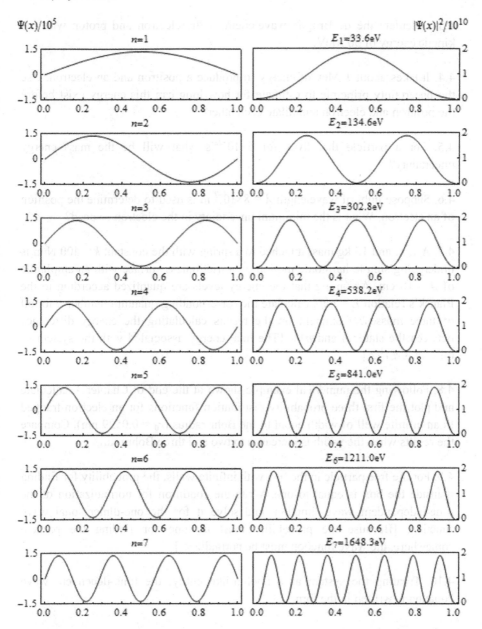

Figure 4-20. Probability distributions for first seven states on an electron trapped in an infinite well of width twice the Bohr radius

4.2. Starting from the Bohr's atomic model show that the energy (kinetic plus potential) corresponding to the circular orbit of the mass m in a three-dimensional harmonic oscillator potential is $n\hbar\sqrt{k/m}$.

4.3. Calculate the de Broglie wavelength of the electron and proton with the kinetic energy of 50 MeV.

4.4. It takes about 1 Mev of energy to produce a positron and an electron. Use the uncertainty principle to estimate for how long can this energy exist before the positron and electron annihilate each other?

4.5. For a particle that lives for 6×10^{-22} s what will be the mass–energy uncertainty?

4.6. Suppose light of wavelength $\lambda = 8 \times 10^{-7}$ m is used to determine the position of an electron. What is the minimum uncertainty in the electron's speed?

4.7. A 1, 5, and 15 kg mass attached to a spring with the constant $k = 400$ N/m is undergoing simple harmonic motion on a frictionless surface with an amplitude of $A = 10$ cm. Assuming that the energy levels are quantized according to the Planck's relation $E = nhf$, calculate the corresponding quantum number n for all of three masses? Comment on the results calculating the energy difference between the states n and $n+1$. [The total energy associated with the system of mass m is $kA^2/2$.]

4.8. Following the numerical example shown at the end of Chapter 4, calculate and plot the first three probability distribution functions for an electron trapped in an infinite well of width equal to the Bohr radius ($r_0 = 0.0529$ nm). Compare the results when the width is increased to two and three Bohr radii.

4.9. For the free particle in the box with infinite walls, the probability for finding it inside the box is equal to one. Write the condition for normalization of the time-independent wave function, and solve it for the one-dimensional wave function. [Because the probability must be one for finding the particle somewhere, the wave function must be normalized.]

4.10. Treating the system as a photon-like entity, the time-dependent wave equation is written in the form

$$\frac{\partial^2 E}{\partial x^2} = \frac{1}{c^2} \frac{\partial^2 E}{\partial t^2}$$

Assuming the solution to be

$E(x,t) = E_0 \cos(kx - \omega t)$ show that the energy of a photon is $E = pc$.

4.11. The Schrödinger time-dependent one-dimensional equation for an electron can be written in the following form

$$-\frac{\hbar^2}{2m}\frac{\partial^2\Psi(x,t)}{\partial x^2}+U(x,t)\Psi(x,t)=i\hbar\frac{\partial\Psi(x,t)}{\partial t}$$

Assuming the potential energy to be constant ($U = U_0$), and that the solution of the above wave equation can be expressed as

$$\Psi(x,t) = Ae^{i(kx-\omega t)} = A[\cos(kx-\omega t)+i\sin(kx-\omega t)]$$

show that the total energy of the electron is a summation of potential and kinetic energy

$$\hbar\omega = U_0 + \frac{\hbar^2 k^2}{2m}$$

4.12. A particle of energy E that is smaller than the height of the barrier (potential energy U_0), according to classical mechanics it is forbidden to penetrate inside the region (see the figure below). The wave function associated with the free particle must be continuous, and thus there is a finite probability that the particle will tunnel through the barrier (Fig. 4-23).

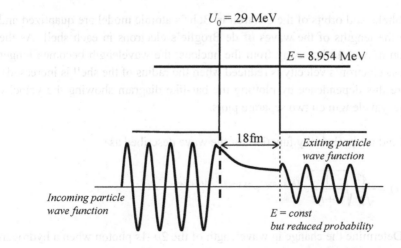

Figure 4-23. Potential well for free particle wave penetration

For a particle approaching the barrier, a wave function for the free particle can be used. However when a particle reaches the barrier, it must satisfy the Schrödinger equation in the form

$$-\frac{\hbar^2}{2m}\frac{\partial^2 \Psi(x)}{\partial x^2} = (E - U_0)\Psi(x) \quad \text{which has a solution}$$

$$\Psi(x) = Ae^{-\alpha x} \quad \alpha = \sqrt{\frac{2m(U_0 - E)}{\hbar^2}}$$

Calculate the tunneling probability for the α particle described in Example 5.6 in Chapter 5.

4.13. The resolving power of an electron microscope is assumed to be equal to the wavelength of the used light. Calculate the required kinetic energy of electrons in order to be able to "see" an atom. The required resolving power is 10^{-11} m.

4.14. Based on the wave-particle duality in one dimension (the de Broglie's relation between wavelength and momentum and the Planck's relation between frequency and energy) show that the Schrödinger equation for free particle (like an electron for example)

$$-\frac{\hbar^2}{2m}\frac{\partial^2 \Psi(x,t)}{\partial x^2} = i\hbar\frac{\partial \Psi(x,t)}{\partial t} \quad \text{has the following solution}$$

$$\Psi(x,t) = Ae^{i(kx-\omega t)} = Ae^{i\left(\frac{2\pi}{\lambda}-\omega t\right)}$$

4.15. Shells and orbits of the electron in Bohr's atomic model are quantized and so are the lengths of the waves of de Broglie's electrons in each shell. As the electron moves further away from the nucleus, the wavelength becomes longer (because electron's velocity is reduced when the radius of the shell is increased). Capture this dependence by plotting the bar-like diagram showing the velocity and the wavelength on two separate plots.

4.16. Find the probability function for the wave described as

$$\Psi(x,t) = \frac{A}{\sqrt{2}}\sqrt{\frac{1}{a+i\eta}}e^{i(k_0 x-\omega_0 t)}e^{-(x-\beta t)^2/4(a+i\eta)}$$

4.17. Determine the change in wavelength of the 2p–1s photon when a hydrogen atom is placed in a magnetic field of various intensities: 3T, 5T, 10T, and 30T. Comment on the results.

4.18. Use Planck's law to write a computer code that will calculate the black body radiation as a function of light frequency.

4.19. Plot the dependence of de Broglie wavelength and the velocity of motion for an electron, proton, and a 1,000 kg body moving at non-relativistic speeds.

4.20. Plot the Eq. (4-1). What can you conclude from the slope of the plot?

4.21. Derive the following relation for the wavelength of an electron:

$$\lambda = \frac{h}{m_0 c} \frac{\sqrt{1 - v^2/c^2}}{v/c}$$ where $h/m_0 c$ is the Compton wavelength for the

electron. Calculate its value.

Chapter 5

RADIOACTIVE DECAY
Radioactivity, Kinetics of Decay, Examples

It can be thought that radium could become very dangerous in criminal hands, and here the question can be raised whether mankind benefits from knowing the secrets of Nature, whether it is ready to profit from it or whether this knowledge will not be harmful for it.

The example of the discoveries of Nobel is characteristic, as powerful explosives have enabled man to do wonderful work. They are also a terrible means of destruction in the hands of great criminals who are leading the people towards war. I am one of those who believe with Nobel that mankind will drive more good than harm from new discoveries. *Pierre Curie* (1859–1906), 1903 Nobel Prize address.

1. INTRODUCTION

Nuclides exist in two main forms, stable and unstable. A nuclide is considered to be stable if there is no proof of its spontaneous transformation into another nuclide. The probability of transformation is characterized by the half-life, which is defined as the time needed for half of the starting amount of an unstable nuclide to transform. Elements above lead are all unstable and have very long half-lives (order of 10^8–10^{10} years) compared to the age of the atom (assumed to be formed some 10 billion years ago).

If, for example, a stable nucleus of ^{59}Co with 27 protons and 32 neutrons receives one neutron (which must possess an energy of 7.5 MeV), the newly formed nucleus ^{60}Co is artificial (does not exist in nature), unstable, and in an

T. Jevremovic, *Nuclear Principles in Engineering*,
DOI 10.1007/978-0-387-85608-7_5, © Springer Science+Business Media, LLC 2009

excited state. The instability is caused by the addition of a new particle (with its associated energy) that requires rearrangements of nucleons inside the nucleus. The process nuclei undergo in order to return to the ground state is called the radioactive decay. In the case of ^{60}Co the radioactive decay scheme is sketched in Fig. 5-1.

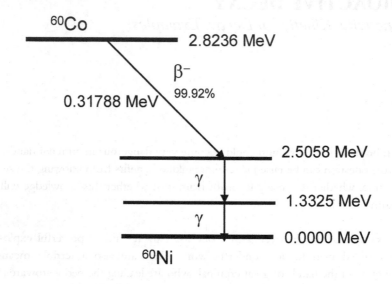

Figure 5-1. Radioactive decay scheme for ^{60}Co

None of the artificially created isotopes are stable; all are radioactive and decay with half-lives in the range of microseconds to years. However, based on the quantum mechanics (see Chapter 4) it is expected that it is possible to create isotopes in a new "island" of stability (Fig. 5-2). The most stable nuclides are those whose protons and neutrons close the shells (energy levels). These are the nuclides with a magic number of protons, neutrons or both. The next proton magic number is 114 (beyond those already known to exist). The number of neutrons needed to overcome the proton–proton repulsion in such a nucleus is estimated to be at least 184 and perhaps as many as 196. It is not easy to bring together two nuclei that would give both the correct number of protons and the necessary number of neutrons with a half-life long enough to be detected. Element 114 was experimentally observed in 1998 at Russia's Joint Institute for Nuclear Research in Dubna. Their very complex experiments showed the possibility to create short-lived heavy nuclei around the new "island" of stability. By accelerating atoms of ^{48}Ca into a target of ^{244}Pu, atoms of element 114 (with a nuclear weight of 289) were detected through their decay into element 112 as follows

$$^{244}_{94}\text{Pu} + {}^{48}_{20}\text{Ca} \rightarrow {}^{289}_{114}\text{Uuq} + 3\ {}^{1}\text{n} \qquad 175 \text{ neutrons}$$

$$^{244}_{94}\text{Pu} + {}^{48}_{20}\text{Ca} \rightarrow {}^{288}_{114}\text{Uuq} + 4\ {}^{1}\text{n} \qquad 174 \text{ neutrons}$$

The lifetimes of the elements 114 and 112 are 30 s and 280 ms, respectively.

This chapter focuses on the laws of physics governing the mechanisms, kinetics and types of the radioactive decays.

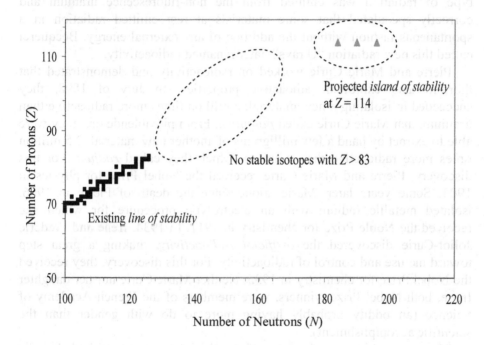

Figure 5-2. The new "island" of stable super-heavy nuclides

2. MECHANISM OF RADIOACTIVE DECAY

Radioactivity is defined as the spontaneous nuclear transformation of an unstable element resulting in the formation of a new one. The process of radioactive decay is statistical and therefore random in its nature. For example, whether a radioactive uranium atom will or will not decay at any given instant is purely a matter of probability. There is no physical difference between uranium atom that is decaying and one that is not decaying.

In 1895 Wilhelm Conrad Roentgen discovered a new phenomenon that he called the X-rays. Soon after, Henri Becquerel decided to study the correlation between newly discovered X-rays and the fluorescence phenomena of uranium salts. Once exposed to ultraviolet photons (sun light), the uranium salts radiate visible light (the fluorescence phenomenon). However, due to bad weather in Paris, Becquerel was not able to expose the samples to sun light for a few weeks, during which time he left them in a closet. Later he found that the plates were exposed and concluded that a new type of radiation was emitted from the non-fluorescence uranium and correctly speculated that some materials at rest emitted radiation in a spontaneous fashion without the addition of any external energy. Becquerel called this new radiation "U rays", later renamed radioactivity.

Pierre and Marie Curie worked on radioactivity and demonstrated that thorium also exhibited radioactive properties. In July of 1898, they succeeded in isolating a new material, a million times more radioactive than uranium, that Marie Curie called *polonium*. From pitchblende ore, they were able to extract by hand a few milligrams of another new material, 2.5 million times more radioactive than uranium, which they called *radium*. For this discovery, Pierre and Marie Curie received the Nobel Prize for physics in 1903. Some years later, Marie, alone since the death of Pierre in 1906, isolated metallic radium with an electrolytic procedure for which she received the Noble Prize for chemistry in 1911. In 1934, Irene and Frederic Joliot-Curie discovered the *artificial radioactivity*, making a great step toward the use and control of radioactivity. For this discovery, they received the Nobel Prize for chemistry in 1935. Neither Marie Curie nor her daughter Irene, both Nobel Prize winners, were members of the French Academy of Science (an oddity probably having more to do with gender than the scientific accomplishment).

The spontaneous nuclear transformations are accomplished by the emission of an alpha (α) particle, a beta (β^-) particle or a positron (β^+) as well as by orbital electron capture, neutron emission (n) or proton emission (p). Each of these reactions may or may not be accompanied by gamma radiation (γ). The α rays are the nuclei of the helium atoms that were discovered by Rutherford in 1909. The β^- particles are the same as electrons. They are emitted from the beta emitters at the speed of up to 99% of the speed of light. The γ rays emitted in the radioactive decay transitions have shorter wavelengths than the X-rays (thus are more energetic).

Radioactivity is a nuclear process that originates in the nucleus and is therefore not determined by the chemical or physical states of the atom. As discussed in Chapters 2 and 3, an isotope of a given element is an atom that contains the same number of protons and has the same electronic structure, but differs in the number of neutrons. Most elements have several isotopes;

chlorine, for example has two: 75.4% ^{35}Cl and 24.6% ^{37}Cl. A few radioisotopes arise naturally; however, most of them are created artificially. There are more than 2,930 known isotopes, but only 65 are naturally occurring and exist either as a product of cosmic radiation in the atmosphere (^{3}H, ^{7}Be and ^{14}C) or as products of radioactive decay of primordial isotopes (^{40}K and ^{238}U).

The exact mode of radioactive decay depends on two factors:

1. The particular type of nuclear instability (whether the neutron-to-proton ratio is too high or too low) and
2. The mass–energy relationship among the parent nucleus, daughter nucleus and the emitted particle.

Radioactive decay of a nucleus changes the arrangement of its nucleons. This change usually influences the entire atom by affecting the electron cloud. The change can even propagate further and affect the molecule. The decay, however, obeys a series of physical laws of conservation. The conservation laws are a direct consequence of the symmetries in nature that require certain variables to stay unchanged. The following conservation laws must be satisfied for the radioactive decay to exist:

1. *Conservation of mass–energy*: radioactive decay of a nucleus changes mass into energy. In other words, in the process of radioactive decay the total mass is reduced but the energy is increased. The difference in masses before and after the decay is emitted as energy of the emitted particles in the decay. It can be shown that conservation of mass and energy holds for the radioactive decay as a consequence of symmetry in time. Every experiment will give the same decay results for the same nucleus no matter when it was performed.
2. *Conservation of momentum*: the sum of the momentum before and after the decay must be the same. This law is a direct consequence of the symmetry in space. Namely, all points in Euclidian space are equivalent and thus the physical laws are the same for all points in space.
3. *Conservation of angular momentum*: the sum of the angular momentum and spins must remain the same before and after the decay. Angular momentum describes the degree of rotation of a system (Chapter 4). Since in space all directions are equally probable, the physical laws are independent of orientation of motion of a system in space.
4. *Conservation of charge*: the sum of charges before and after the decay remains the same. The charge can only be redistributed between the particles and cannot be lost in the process of radioactive transformations. For example, this law says that an electron cannot appear or disappear on its own. In order for an electron to disappear there should exist its counterpart; a *positron*. Interaction of these two particles will result in the annihilation of both (as described in Chapter 6). The fact that an electron

cannot change its appearance without its opposing particle (positron) assures the stability of the electron and thus the stability of all matter.

5. *Conservation of nucleons*: the total number of nucleons for any decay mode must be conserved. This law forbids neutrons (and protons) confined within a nucleus to decay into other particles and assures the stability of matter.

3.   KINETICS OF RADIOACTIVE DECAY

3.1   Decay Constant

Radioactive decay although a random process that occurs at a characteristic rate can be predicted. The length of time, the number of steps required for completing the transformation and the types of radiation released at each step of the decay are well known.

The *decay constant*, λ (also referred to as the disintegration constant), represents the probability that a radionuclide will decay in a unit time. Thus the probability that a radionuclide will decay in time dt is λdt. The characteristics of the decay constant confirmed experimentally are

• The decay constant is the same for all nuclei of a given atom. It cannot be changed by ambient pressure or temperature.
• The decay constant does not depend on the age of nuclides, i.e., it does not change with time.

3.2   Radioactive Decay

If the probability for a nucleus to decay in time dt is

$$\lambda dt \tag{5-1}$$

then from the total number of nuclei N, in time dt, the number of nuclei that will decay, dN, can be calculated as

$$-dN = N\lambda dt \tag{5-2}$$

Since the decay constant is not dependent on time (as explained in Section 3.1), the solution of the above equation is simply

$$N = N_0 e^{-\lambda t} \tag{5-3}$$

where N_0 is the starting number (amount) of nuclei and N represents the amount of nuclei that did not decay after time t. As Fig. 5-3 shows, the amount of the initial radionuclide decreases exponentially with time.

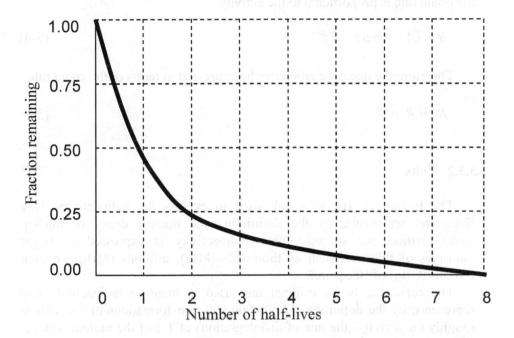

Figure 5-3. Radioactive decay

3.3 Activity

3.3.1 Definition

Activity, A, is the number of nuclei decaying per unit time. If the probability for a nucleus to decay is λ, and there are N nuclei present, the average number of decaying nuclei is $N\lambda$, and is defined as activity

$$A = N\lambda \tag{5-4}$$

Thus, from Eq. (5-3)

$$A \equiv N\lambda = N_0 \lambda e^{\lambda t} = A_0 e^{-\lambda t} \tag{5-5}$$

Radiation detectors do not usually measure the total activity, that is, the total number of decays per second, but some fraction of this called the *count rate* (see Problems 5.6.–5.8.). In any given situation, if all else is kept equal, the count rate is proportional to the activity

$$R = kA \quad \text{where} \quad k \le 1 \tag{5-6}$$

Therefore the decay equation can be expressed in terms of the count rate

$$R_1 = R_0 e^{-\lambda t} \tag{5-7}$$

3.3.2 Units

The Becquerel, Bq, is a unit used to express the radioactivity. One Becquerel represents, by the definition, one nuclear decay or nuclear transformation per second. Often radioactivity is expressed in larger multiples of this unit such as thousands (kBq), millions (MBq) or even billions (GBq) of Becquerel.

The curie, Ci, is the original unit used to measure radioactivity and represents, by the definition, 37,000,000,000 transformations in 1 s. This is roughly the activity (the rate of disintegration) of 1 g of the radium isotope, ^{226}Ra (see Problem 5.1). Radioactivity is often expressed in smaller multiples of this unit such as thousandths (mCi), millionths (µCi) or even billionths (nCi) of a curie.

As a result of having 1 Bq being equal to one transformation per second, there are 3.7×10^{10} Bq in 1 Ci.

Example 5.1 Disintegration of ^{60}Co

Determine the number of disintegrations released per one curie of ^{60}Co (Fig.5-1).

From the decay scheme shown in Fig. 5-1, it follows that each disintegration of a ^{60}Co nucleus releases one β particle and two γ rays. Therefore, the total number of radiations is: $3 \times 3.7 \times 10^{10} = 11.1 \times 10^{10}$ per second per curie ^{60}Co.

3.4   Half-Life

The *half-life*, $T_{1/2}$, of a nuclide is the time needed for half of the atoms to decay. Half-lives can range from less than a millionth of a second to millions of years. After one half-life, the level of radioactivity of a substance is halved; after two half-lives it is reduced to one quarter; after three half-lives to one-eighth, and so on (Fig. 5-3). The products of the radioactive decay are the particles emitted and the remaining nucleus called the daughter of the decaying atom.

Radioactive decay proceeds exponentially, as does the growth of the daughter product. The decay constant and the half-life of a given nuclide are related. The quantitative relationship can be found by setting

$$A = \frac{A_0}{2} \quad \Rightarrow \quad \frac{A_0}{2} = A_0 e^{-\lambda t} = A_0 e^{-\lambda T_{1/2}} \quad \Rightarrow \quad T_{1/2} = \frac{\ln 2}{\lambda} = \frac{0.693}{\lambda} \quad (5\text{-}8)$$

Example 5.2 Activity of radium

Calculate the percent of ^{226}Ra that will decay during a period of 1,000 years if the half-life is 1600 years.

$$A = A_0 e^{-\lambda t} = A_0 e^{-0.693 t / T_{1/2}} \quad \Rightarrow \quad \frac{A}{A_0} = e^{-0.693 \times 1000 / 1600} = 0.648 = 64.8\%$$

The percent that decayed away during 1000 years is $100\% - 64.8\% = 35.2\%$.

Example 5.3 Estimate of decay constant and half-life for radium

Calculate the decay constant and half-life for ^{226}Ra if one microgram emits 3.65×10^4 α particles per second.

The number of radium atoms per microgram of radium, N, is

$$N = \frac{N_a (\text{atoms/mole}) \times M(\text{g})}{A(\text{g/mole})} =$$

$$\frac{(6.02 \times 10^{23} \text{ atoms/mole}) \times (10^{-6} \text{ g})}{226 (\text{g/mole})} = 2.66 \times 10^{15} \text{ atoms}$$

The decay constant is thus obtained from the known number of radium nuclei that decayed and the number of radium nuclei that did not decay per unit time

$$-dN = N \lambda dt \quad \Rightarrow \quad \lambda = \frac{-dN / N}{dt} = -\frac{(3.65 \times 10^4 \text{ atoms}) / (2.66 \times 10^{15} \text{ atoms})}{1 \text{ s}} \Rightarrow$$

$$\lambda = 1.37 \times 10^{-11} \text{ s}^{-1} = 4.26 \times 10^{-4} \text{ yrs}^{-1}$$

This gives the half-life very close when compared with the measured value of 1,600 years (see also Example 5.2)

$$T_{1/2} = \frac{\ln 2}{\lambda} = \frac{\ln 2}{4.26 \times 10^{-4}} = 1627 \text{ yrs}$$

Example 5.4 Specific activity of a radioactive nuclide

A sample of ^{113}In has a mass of 2 µg and a physical half-life of 1.6582 h. Calculate:

a) The number of ^{113}In atoms present.
b) The number of ^{113}In atoms remaining after 4 h.
c) The activity of the sample (in Bq and Ci) after 4 h.
d) The specific activity of the ^{113}In sample.

a) Number of atoms present in the 2 µg sample is

$$N_0 = \frac{N_a(\text{atoms/mole}) \times M(\text{g})}{A(\text{g/mole})}$$

$$= \frac{(6.02 \times 10^{23} \text{ atoms/mole}) \times (2 \times 10^{-6} \text{g})}{113(\text{g/mole})} = 1.065 \times 10^{16} \text{ atoms}$$

b) Number of atoms that remain after 4 h is

$$N = N_0 e^{-\lambda t} = (1.065 \times 10^{16} \text{ atoms}) e^{-\frac{\ln 2}{1.6582} \times 4} = 2.00 \times 10^{15} \text{ atoms}$$

c) Activity of the sample after 4 h is

$$A \equiv N\lambda = (2.00 \times 10^{15} \text{ atoms}) \left(\frac{\ln 2}{1.6582 \text{h}} \right) = 8.36 \times 10^{14} \text{ atoms/h}$$

$$A = \left(8.36 \times 10^{14} \frac{\text{atoms}}{\text{h}} \right) \left(\frac{\text{h}}{3600 \text{ s}} \right) = 2.32 \times 10^{11} \text{Bq}$$

$$A = (2.32 \times 10^{11} \text{Bq}) \left(\frac{1}{3.7 \times 10^{10} \text{Ci}} \right) = 6.3 \text{ Ci}$$

d) The ratio of nuclide activity to the total mass of the element present is known as the specific activity of the sample, SA (see also Problem 5.9)

$$SA = \frac{A}{M} = \frac{6.3 \text{Ci}}{2 \times 10^{-6} \text{g}} = 3.15 \times 10^6 \text{ Ci/g}$$

3.5 Radioactive Decay Equilibrium

Most radioactive decay schemes contain more than one member, and it is therefore interesting to analyze the relation between the radioactivities and the number of nuclei that disintegrate per unit time in such a series. For example, the radioactive chain in which the parent, A, decays into a daughter nucleus, B, which is also radioactive is written as

$$A \rightarrow B \rightarrow C \rightarrow ...$$

The equilibrium of radioactive decay is attained when the ratio between the activities of the successive members in the series remains constant. Considering only the first two members in the above chain, as shown in Fig. 5-4, the rate of change is

- Rate of change of a parent nuclide

$$\frac{dN_p}{dt} = -\lambda_p N_p \tag{5-9}$$

- Rate of change of a daughter nuclide (= rate of production − rate of decay)

$$\frac{dN_d}{dt} = \lambda_p N_p - \lambda_d N_d \tag{5-10}$$

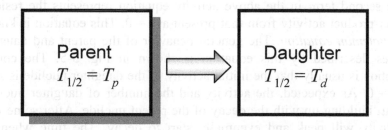

Figure 5-4. Two successive members in radioactive chain

Equations (5-9) and (5-10) are a system of the first-order linear differential equations whose solution is

$$N_d(t) = N_p(0)\frac{\lambda_p}{\lambda_d - \lambda_p}\left(e^{-\lambda_p t} - e^{-\lambda_d t}\right) + N_d(0)e^{-\lambda_d t} \tag{5-11}$$

According to the relation between the activity and number of atoms that decay, Eq. (5-5), it follows that

$$A_d(t) = A_p(0)\frac{\lambda_d}{\lambda_d - \lambda_p}\left(e^{-\lambda_p t} - e^{-\lambda_d t}\right) + A_d(0)e^{-\lambda_d t} \tag{5-12}$$

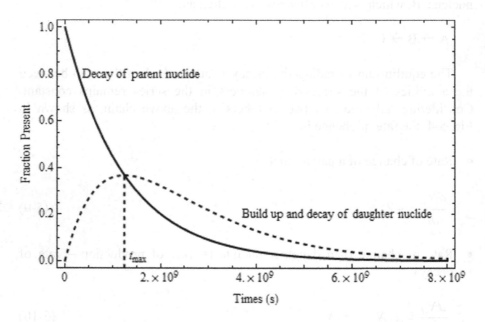

Figure 5-5. General trend of activity change with time according to the Bateman equation

The second term in the above activity equation represents the residual daughter product activity from that present at $t = 0$. This equation is known as the *Bateman equation*. The general behavior of the parent and daughter activities described by this equation is shown in Fig. 5-5. The correct assumption is usually that the initial activity of the daughter nuclide is zero, $N_d(0) = 0$. As expected, the activity and the number of daughter nuclides will start building up with the decay of the parent nuclide. After some time the activity will peak and eventually start to decay. The time when the daughter nuclide reaches its maximum activity can be estimated as follows

$$\frac{dN_d(t)}{dt} = \frac{d}{dt}\left[N_p(0)\frac{\lambda_p}{\lambda_d - \lambda_p}\left(e^{-\lambda_p t} - e^{-\lambda_d t}\right)\right] = 0 \tag{5-13}$$

which gives $\lambda_p e^{-\lambda_p t_{max}} = \lambda_d e^{-\lambda_d t_{max}}$. Solving for time (Fig. 5-5)

$$t_{max} = \frac{\ln(\lambda_d / \lambda_p)}{\lambda_d - \lambda_p} \qquad (5\text{-}14)$$

Thus, the time when the activity of the daughter nuclide reaches its maximum value depends only on the decay constants of the parent and daughter nuclides.

The Bateman equation is usually analyzed for the following cases:

1. *The daughter nuclide is stable, $\lambda_d = 0$:* assuming that $N_d(0) = 0$, Eq. (5-10) becomes

$$\frac{dN_d}{dt} = \lambda_p N_p = \lambda_p N_p(0)e^{-\lambda_p t} \quad \Rightarrow \quad N_d(t) = N_p(0)(1 - e^{-\lambda_p t}) \qquad (5\text{-}15)$$

The decay of a parent and accumulation of a stable daughter nuclide is shown in Fig. 5-6 for the decay of ^{60}Co to stable ^{60}Ni (decay scheme of ^{60}Co is given in Fig. 5-1).

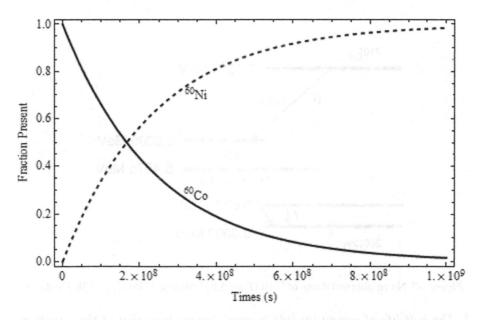

Figure 5-6. Serial decay of parent to stable daughter (^{60}Co to ^{60}Ni)

2. *The half-life of the parent nuclide is shorter than that of the daughter, $T_{1/2p} < T_{1/2d}$:* in this case the daughter nuclide builds up faster than it decays. Essentially all parent nuclei transform into daughter nuclei and the activity of the sample comes from the daughter nuclide only. This

condition is called *no equilibrium*. One example is the decay of ^{210}Bi into ^{210}Po as shown in Fig. 5-7.

3. *The half-life of the parent nuclide is longer than that of the daughter,* $T_{1/2p} > T_{1/2d}$: the change (decrease) of the activity of the parent nuclide becomes negligible. This case is called a *transient equilibrium* and is schematically depicted in Fig. 5-8. Examples include ^{132}Te (78 h) decaying to ^{132}I (2.3 h) and ^{113}Sn decaying to ^{113m}In (1.7 h). However, the best example is the ^{99}Mo (65.94 h) parent – ^{99m}Tc (6.01 h) daughter relationship. The Bateman equation reduces to the following form

$$\lambda_p < \lambda_d \ \therefore \ e^{-\lambda_d t} << e^{-\lambda_p t} \quad \rightarrow \quad N_d(t) = N_p(0)\frac{\lambda_p}{\lambda_d - \lambda_p}\left(e^{-\lambda_p t}\right) \qquad (5\text{-}16)$$

The ratio of the rate change of parent to daughter nuclides thus becomes

$$N_p(t) = N_p(0)e^{-\lambda_p t} \quad \rightarrow \quad \frac{N_p(t)}{N_d(t)} = \frac{\lambda_d - \lambda_p}{\lambda_p} \qquad (5\text{-}17)$$

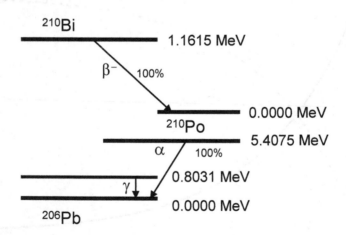

Figure 5-7. No equilibrium decay of ^{210}Bi ($T_{1/2} = 5.013$ days) to ^{210}Po ($T_{1/2} = 138.376$ days)

4. *The half-life of parent nuclide is much longer than that of the daughter,* $T_{1/2p} >> T_{1/2d}$: For example, ^{226}Ra with a half-life of 1,600 years decays into ^{226}Rn, which has a half-life of only 4.8 days. The observation period is therefore very small compared to the 1,600-year half-life of ^{226}Ra. From Eq. (5-17)

$$\frac{N_p(t)}{N_d(t)} = \frac{\lambda_d - \lambda_p}{\lambda_p} \qquad \lambda_p << \lambda_d$$

it follows

$$\frac{N_p(t)}{N_d(t)} = \frac{\lambda_d}{\lambda_p} \quad \rightarrow \quad N_p(t)\lambda_p = N_d(t)\lambda_d \quad \rightarrow \quad A_p(t) = A_d(t) \qquad (5\text{-}18)$$

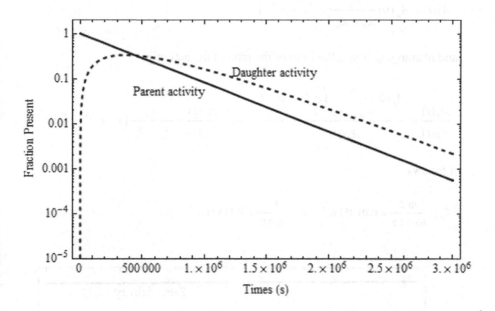

Figure 5-8. Transient equilibrium decay when $T_{1/2p} > T_{1/2d}$

The activity of the parent and daughter are the same and total activity of the sample remains effectively unchanged during the time of observation.

This is called a *secular equilibrium* and the example for ^{226}Ra $\rightarrow$ ^{226}Rn is shown in Fig. 5-9 (see also Problem 5.11). The half-life of long-lived nuclides can be estimated knowing that they are in a secular equilibrium. Knowing the atomic composition of a mixture of two radionuclides that are in a secular equilibrium, such as ^{226}Ra and ^{238}U in uranium ore, the decay constant or half-life of one nuclide can be determined given the half-life of the other using Eq. (5-18).

Example 5.5 ^{99}Mo (65.94 h) parent–^{99m}Tc (6.01 hours) daughter relationship

Sketch a diagram of the activity change in time for the transient equilibrium of these two nuclides and find the time at which the daughter reaches a maximum activity. From the decay of ^{99}Mo it is known that 87% decays into ^{99m}Tc. Assume that the activity of the parent nuclide at $t = 0$ is 1 Ci.

Starting from the Bateman equation and assuming that the activity of the daughter nuclide at $t = 0$ was zero,

$$A_d(t) = A_p(0)\frac{\lambda_d}{\lambda_d - \lambda_p}\left(e^{-\lambda_p t} - e^{-\lambda_d t}\right)$$

and rearranging it into the form of the ratio of the activities

$$\frac{A_d(t)}{A_p(t)} = \frac{A_p(0)\dfrac{\lambda_d}{\lambda_d - \lambda_p}\left(e^{-\lambda_p t} - e^{-\lambda_d t}\right)}{A_p(0)e^{-\lambda_p t}} \quad \rightarrow \quad \frac{A_d(t)}{A_p(t)} = \frac{\lambda_d}{\lambda_d - \lambda_p}\left(1 - e^{-(\lambda_d - \lambda_p)t}\right)$$

it follows

$$\lambda_p = \frac{\ln 2}{65.94\text{ h}} = 0.01051\text{ h}^{-1} \quad \lambda_d = \frac{\ln 2}{6.01\text{h}} = 0.11531\text{ h}^{-1}$$

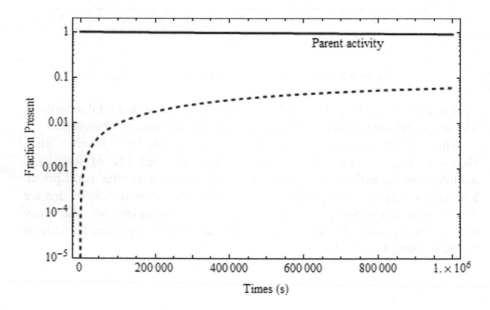

Figure 5-9. Secular equilibrium decay: buildup of daughter activity when $T_{1/2p} \gg T_{1/2d}$

$$A_d(t) = 0.87 \times A_p(t)\frac{0.11531}{0.11531 - 0.01051}\left(1 - e^{-0.1048t}\right)$$

$$= 0.957 A_p(t)\left(1 - e^{-0.1048t}\right)$$

$$= 0.957 A_p(0)e^{-0.01051t}\left(1 - e^{-0.1048t}\right)$$

The time when the daughter reaches its maximum activity is obtained by differentiating the above equation (Fig. 5-10):

$$\frac{dA_d(t)}{dt} = 0 \quad \Rightarrow \quad t_{max} = 22.8\,\text{h} \quad A_d(22.8\text{h}) = 0.684\,\text{Ci}$$

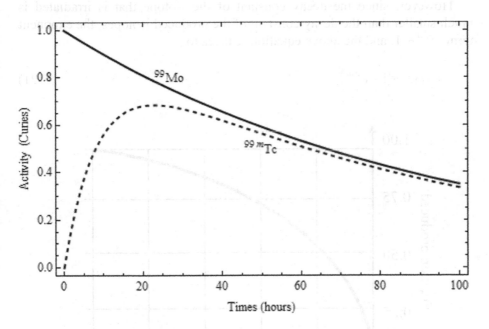

Figure 5-10. Activity change for ^{99}Mo and ^{99m}Tc

3.6 Production of Radioisotopes

The activity of isotopes irradiated in nuclear reactors or accelerators changes according to secular equilibrium of radioactive decay. If a nuclear reaction produces an isotope with concentration N_2 from N_1 atoms at a rate $R = \lambda_1 N_1(0)$, then assuming the activity of the isotope that is produced by this reaction at $t = 0$ is zero:

$$N_2(t) = N_1(0)\frac{\lambda_1}{\lambda_2 - \lambda_1}\left(e^{-\lambda_1 t} - e^{-\lambda_2 t}\right) \qquad (5\text{-}19)$$

The production of radioisotopes is constant (similar to secular equilibrium in which the half-life of a parent is much longer than the half-life of a daughter):

$$\lambda_1 \ll \lambda_2 \quad \Rightarrow \quad N_2(t) \cong N_1(0)\frac{\lambda_1}{\lambda_2}\left(e^{-\lambda_1 t} - e^{-\lambda_2 t}\right) \quad \Rightarrow$$

$$A_2(t) \cong R\left(e^{-\lambda_1 t} - e^{\lambda_2 t}\right) \qquad (5\text{-}20)$$

However, since the decay constant of the isotope that is irradiated is much smaller than the decay constant of the produced isotopes, the exponent term, $e^{-\lambda_1 t} \sim 1$, and the above equation reduces to

$$A_2(t) \cong R\left(1 - e^{\lambda_2 t}\right) \qquad (5\text{-}21)$$

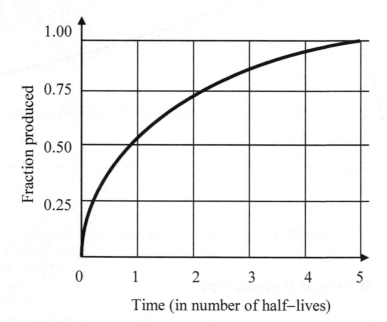

Figure 5-11. Activation curve – production of radioisotopes

This equation is called the *activation equation* (Fig. 5-11). Initially, when λt is small, the activity of the produced radioisotope increases almost linearly due to the behavior of $\left(1-e^{-\lambda_2 t}\right)$. After some time the activity reaches its saturated value. At an irradiation time equal to one half-life of the radioisotope, half of the maximum activity is formed. It is easy to realize that the activity of the produced isotope will saturate and therefore irradiation times that exceed twice the half-life are usually not worthwhile.

Radioisotopes are produced at large scale for various applications in medicine (for imaging and cancer treatment), agriculture, hydrology, radiography and for scientific research.

4. ALPHA DECAY

4.1 Mechanism of Alpha Decay

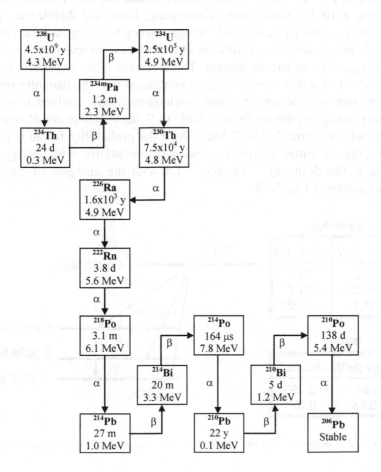

Figure 5-12. Uranium series

   An alpha (α) particle is a highly energetic helium nucleus consisting of two protons and two neutrons. Whenever a nucleus has too many protons, causing excessive repulsion, it is unstable and has a tendency to decay by emitting an α particle, which reduces the repulsive force. Most α emitters are toward the end of the periodic table.

   The products of the decay are called the radioactive series, and there are four natural α radioactive series: (1) uranium series, (2) thorium series, (3) actinium series and (4) neptunium series. The uranium series starts with the ^{238}U isotope, which has a half-life of 4.5×10^9 years, and is shown in Fig. 5-12. Because of the very long half-life of the parent nuclide, this chain is still present today. The thorium and actinium series are also present today, whereas the parent nucleus of the neptunium series, $^{237}_{93}$Np, has a half-life of 2.2×10^6 years and has already disappeared. However, it is possible to produce this element artificially and thus determine the half-lives of the series. Alpha particles do not exist as such inside the heavy nuclei. Instead, they form, exist for some time, disintegrate, form and disintegrate again. Occasionally, some of them have enough energy to overcome the potential barrier of the nucleus; this results in a net decrease in mass of the nucleus and consequently an energy release. The emission of an α particle leads to the formation of a more stable nuclear configuration. The daughter product, however, may also be unstable and continue to decay. Emitted α particles may have energies ranging from 4 MeV to 7 MeV. There are almost no α particles with energies below 2 MeV since the probability for an α particle to cross the potential barrier decreases exponentially with energy. One example is the decay of ^{232}Th into ^{228}Ra with the energies of emitted α particles shown in Fig. 5-13.

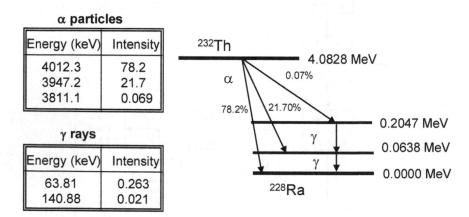

α particles

Energy (keV)	Intensity
4012.3	78.2
3947.2	21.7
3811.1	0.069

γ rays

Energy (keV)	Intensity
63.81	0.263
140.88	0.021

Figure 5-13. α decay of ^{232}Th → ^{228}Ra and energies of the emitted α particles and γ rays

The half-lives of α emitters range from microseconds to 10^{17} years. The half-life of an α emitter is directly dependent on the energy of the emitted α particle. For example, the half-life of ^{213}Po is 4.2 μs and the energy of α particle emitted in the decay is 8.37 MeV. Thorium-232 emits the α particles of energy around 4 MeV with a half-life of 1.405×10^{10} years, while ^{218}Th emits α particles of energy of about 10 MeV and has a half-life of only 0.11 μs. To further illustrate this point, the energy and half-lives of the ^{238}U decay series (shown in Fig. 5-12) are plotted in Fig. 5-14 indicating similar correlation between energy of the α particle and half-lives of the nuclides.

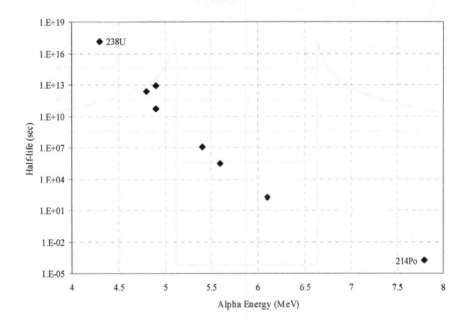

Figure 5-14. Geiger–Nuttall plot of the energy of emitted α particles versus the half-lives of nuclides in ^{238}U decay series

The mechanism of α decay as well as the observed relation is explained by considering the α particle as being bound in the potential of the nucleus (Fig. 5-15). Alpha particles in a level with negative energy cannot penetrate the Coulomb barrier, but those in a positive energy level may have enough energy to overcome the barrier and exit the nucleus. The region between $-a$ and $+a$ in Fig. 5-15 represents the inner part of the nucleus where α particles are bound by a strong nuclear potential. The regions left and right of this are governed by the Coulomb repulsive potential between the charge of the α particle, $+2e$, and the charge of the remaining nucleus, $+(Z-2)e$. The three energy levels shown in Fig. 5-15 are not to scale and are shown only to illustrate the mechanism of α decay: if an α particle has energy corresponding to energy level one, the energy is described as negative in order to indicate that it cannot exit the nucleus, and thus no α decay is expected to take place; however, if an α particle has an energy corresponding to levels two or three, then it may penetrate the potential barrier via the tunnel effect and α decay will happen. The following example illustrates how the potential energy can be estimated (refer also to Chapter 4, Section 3.7).

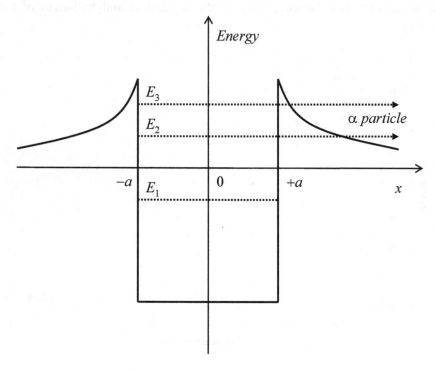

Figure 5-15. Coulomb potential of an α particle in a nucleus
(simplified one-dimensional representation)

Example 5.6 Coulomb potential barrier in α decay

Calculate the Coulomb potential at the nuclear surface felt by an α particle emitted by the parent nucleus ^{212}Po, and compare with the decay energy of 8.954 MeV. Approximate the daughter nucleus as well as the α particle as uniformly charged spheres and plot energy versus center-to-center separation. Also, estimate the velocity of the particle inside the nucleus and the frequency of hitting the wall of the Coulomb potential. See Chapter 4, Section 3.7 for definition of tunneling probability.

The Coulomb repulsion potential energy, U_0 (also known as the height of the potential barrier), when two spheres just touch is given by (Fig. 5-16)

$$U_0 = \frac{2(Z-2)e^2 k}{R} \tag{5-22}$$

$$= \frac{2(82)(1.6 \times 10^{-19}\,\text{C})^2 (8.99 \times 10^9\,\text{Nm}^2/\text{C}^2)}{8 \times 10^{-15}\,\text{m}}$$

$$= 4.72 \times 10^{-12}\,\text{J}\,\frac{1\,\text{eV}}{1.6 \times 10^{-19}\,\text{J}} = 29\,\text{MeV}$$

where the radius, R, is estimated by the Fermi model (see Chapter 3, Section 2.1)

$$R = 1.07 A^{1/3} = 1.07(4^{1/3} + 208^{1/3}) = 8\,\text{fm}$$

$$f = \frac{\upsilon}{2R} = \frac{2.08 \times 10^7\,\text{m/s}}{2(8 \times 10^{-15}\,\text{m})} = 1.3 \times 10^{21}\,/\text{s}$$

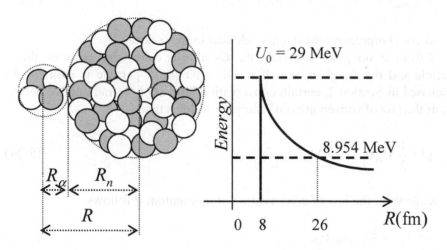

Figure 5-16. Coulomb potential barrier in ^{212}Po α decay

The distance at which the Coulomb potential becomes equal to the energy of the observed decay of ^{212}Po is

$$8.954\,\text{MeV} = \frac{2(Z-2)e^2 k}{R'} \quad \Rightarrow \quad R' = 26\,\text{fm}$$

Thus the width of the ^{212}Po Coulomb barrier seen by the α particle is

$$R' - R = 26\,\text{fm} - 8\,\text{fm} = 18\,\text{fm}$$

For the known α particle energy, its velocity, υ, moving inside the nucleus, and the frequency, f, of hitting the wall of the potential barrier may be estimated as

$$T = 8.954 \text{ MeV} = \frac{m\upsilon^2}{2} = \frac{3727 \text{ MeV}}{2} \frac{\upsilon^2}{c^2} \quad \Rightarrow \quad \upsilon/c = 0.0693$$

$$\upsilon = 2.08 \times 10^7 \text{ m/s}$$

Applying the derivation describe in Chapter 4, Section 3.7 the probability of this decay can be estimated (see Problem 5.36).

4.2   Kinetics of Alpha Decay

Generally, for α particle emission to happen the following conservation equation must be satisfied

$$M_{parent}c^2 = M_{daughter}c^2 + M_\alpha c^2 + Q \tag{5-23}$$

where Q represents the energy released in the α decay.

If there is no γ ray emission, the Q-value is distributed between the α particle and the daughter, which recoils after the α particle is emitted. As discussed in Section 2, certain conservation laws apply to radioactive decay. From the law of conservation of energy it follows that

$$Q = \frac{1}{2}M_\alpha \upsilon_\alpha^2 + \frac{1}{2}M_{daughter}\upsilon_{daughter}^2 \tag{5-24}$$

while from the law of conservation of momentum it follows

$$M_\alpha \upsilon_\alpha = M_{daughter}\upsilon_{daughter} \tag{5-25}$$

When combined, Eqs. (5-24) and (5-25) lead to

$$Q = \frac{1}{2}M_\alpha \upsilon_\alpha^2 + \frac{1}{2}M_{daughter}\frac{M_\alpha^2}{M_{daughter}^2}\upsilon_\alpha^2$$

$$= \frac{1}{2}M_\alpha \upsilon_\alpha^2\left(1 + \frac{M_\alpha}{M_{daughter}}\right) = E\left(1 + \frac{M_\alpha}{M_{daughter}}\right) \tag{5-26}$$

where E represents the kinetic energy of the α particle. From this equation describing the kinetics of α decay, it is understood that the spontaneity condition for the decay is $Q > 0$.

Example 5.7 Kinetics of α decay

Find the energy released in the decay of ^{238}U

$$^{238}_{92}U \rightarrow {}^{234}_{90}Th + \alpha$$

Calculate the energy of the emitted α particle and the recoil nucleus, if m_α = 4.002603 amu, m_{Th} = 234.043583 amu and m_U = 238.050786 amu.

The energy released in the reaction obtained from the mass difference between the nuclei and particle involved (see Chapter 3, Section 2.3) is

$$Q = \left[m_U - (m_{Th} + m_\alpha) \right] c^2 = 4.28 \text{ MeV}$$

This energy is also equal to

$$Q = E\left(1 + \frac{m_\alpha}{m_{Th}}\right)$$

which gives the energy of the emitted α particle

$$E = Q \left/ \left(1 + \frac{m_\alpha}{m_{Th}}\right)\right. = 4.208 \text{ MeV}$$

The energy released in the reaction is also equal to the energy of emitted α particle plus the energy of recoil nucleus.

It follows

$$Q = E + E_{Th} \quad \Rightarrow \quad E_{Th} = 4.28 - 4.208 = 0.072 \text{ MeV}$$

5. BETA DECAY

5.1 Mechanism of Beta Decay

Beta decay is a process that involves nucleon transformation and therefore is a unique decay mode. Beta unstable nuclei can decay in one of the following three modes:

- β^- particle (electron) emerges in a weak decay process where one of the neutrons inside the nucleus decays to a proton, an electron and an anti-electron-type neutrino: $n \rightarrow p + e^- + \nu_e$

- β^+ particle (positron) emerges in a process where a proton decays into a neutron, a positron and an electron-type neutrino: $p \rightarrow n + e^+ + \nu_e$
- Nuclei having an excess number of protons may capture an electron from one of the inner orbits which immediately combines with a proton in the nucleus to form a neutron and an electron-type-neutrino: $p + e^- \rightarrow n + \nu_e$

All of these reactions are a result of restructuring the nucleus within an unstable nuclide in order to approach the region of stability as discussed in Chapter 3. In all of these decay modes, the laws of conservation as described in Section 2 must be satisfied. In order to determine which nuclei are β emitters, it is useful to compare the masses of the isobars, $_Z^A M$:

$$_Z^A M > _{Z+1}^A M \qquad \text{possible } \beta^- \text{ decay} \qquad Z \rightarrow Z+1$$

$$_Z^A M < _{Z+1}^A M + E_b \qquad \text{possible orbital electron capture} \qquad Z+1 \rightarrow Z$$

$$_Z^A M + 2m_0 c^2 > _{Z+1}^A M \quad \text{possible } \beta^+ \text{ decay} \qquad Z+1 \rightarrow Z$$

Table 5-1. Decay modes of six isobars with $A = 90$

Nuclide	Mass (amu)	BE (MeV)	Decay mode	Half-life
$_{36}^{90}$Kr	89.9195238	773.217	β^-	32.32 s
$_{37}^{90}$Rb	89.9148089	776.826	β^-	158 s
$_{38}^{90}$Sr	89.9077376	782.631	β^-	28.79 years
$_{39}^{90}$Y	89.9071514	782.395	β^-	64.00 h
$_{40}^{90}$Zr	89.9047037	783.892	Stable	Stable
$_{41}^{90}$Nb	89.9112641	776.999	β^+	14.60 h
$_{42}^{90}$Mo	89.9139362	773.728	β^+	5.56 h

Beta-minus decay will occur if an atom of higher Z has a smaller mass. Orbital electron capture requires the mass difference to be greater than the binding energy of the electron to be captured. Beta-plus decay is possible only in the case when the mass difference is greater than two electron rest masses. An example that illustrates the conditions for various beta decay modes is shown in Table 5-1 for six isobars with atomic mass number $A = 90$ (Zr). It can be seen that the even–even nucleus, $_{40}^{90}$Zr, has the smallest mass and the highest binding energy. This is the only stable element among the six listed isobars. Masses of all other nuclei, up and down from the $_{40}^{90}$Zr, are increasing. This is illustrated in Fig. 5-17. Nuclei left of $_{40}^{90}$Zr decay by β^- forming a decay chain whereas those to the right decay either through β^+ or electron capture and also form a decay chain. Since β decay is caused by the difference in the masses of the two isobars, there are no two stable neighboring isobars.

The energy spectrum of every β emitter is continuous up to a maximum finite value. Every emitted electron or positron particle is accompanied by the emission of an anti-neutrino or neutrino, whose energies are equal to the difference between the kinetic energy of the β particle and the maximum energy of the spectral distribution for the β decay of that nuclide. The anti-neutrino and neutrino have no electrical charge and have a small mass that is usually neglected in analyzing the kinetics of the decay.

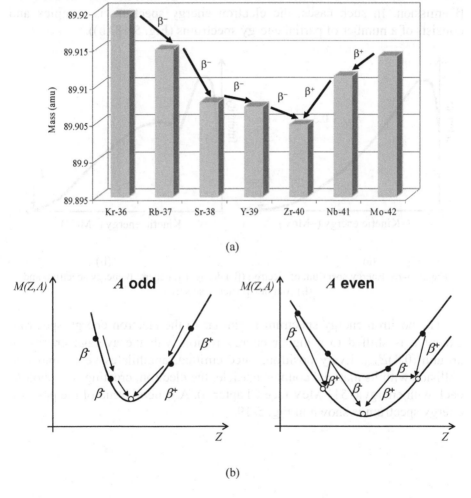

(a)

(b)

Figure 5-17. (**a**) Isobars with $A = 90$; (**b**) mass of nuclei with fixed A (see also Section 3.1)

The energy released in β⁻ decay is distributed between the emitted particles: electron, neutrino and the recoil nucleus, which usually has negligibly small energy and is not taken into account. Therefore

$$E_{max} = E_\beta + E_\nu \qquad (5-27)$$

The electron spectrum is asymmetric, with a higher population of emitted electrons at lower energies and the average energy of around $(0.3)E_{max}$. A general β energy spectrum is shown in Fig. 5-18 (a). Beta-minus decay does not often lead to only one element, but a series of nuclides that all decay by β emission. In such cases, the electron energy spectrum is complex and consists of a number of partial energy spectrums (Fig. 5-18 (b)).

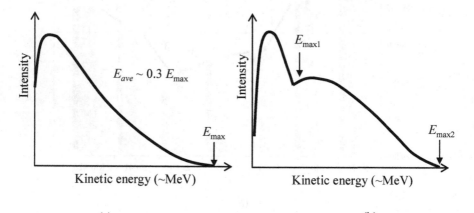

(a)   (b)

Figure 5-18. Energy spectrum of electron (β⁻) decay: (**a**) general β energy spectrum and (**b**) complex β energy spectrum

The positron energy spectrum is similar to the electron energy spectrum except it is shifted to a higher energy region with the average energy of around $(0.4)E_{max}$. Every positron, once emitted, annihilates very rapidly in collision with its material counter-particle, the electron, creating two photons each with energy 0.511 MeV (see Chapter 6). A general trend of the positron energy spectrum is shown in Fig. 5-19.

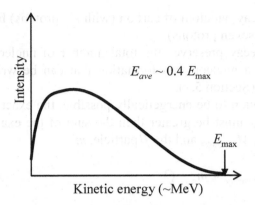

Figure 5-19. Energy spectrum of a positron (β^+) decay

When β decay leaves the residual nucleus in an excited state, the nucleus, in order to achieve stability, may either emit a γ ray or transfer the excitation energy to an electron. The latter is called internal conversion and is an alternative mechanism for an excited nucleus to relax into the ground state. It is an interaction in which a tightly bound electron interacts with the nucleus, absorbs the excitation energy and is ejected from the atom (Section 5.5). A list of the most commonly used β emitters (sources) is shown in Table 5-2.

Table 5-2. Most commonly used β emitters

Nuclide	Half-life	E_{max} (MeV)	Production
^{3}H	12.33 years	0.019	^{9}Be(d, 2α)
			^{6}Li(n, α)
^{14}C	5,730 years	0.156	^{13}C(d, p)
			^{14}N(n, p)
^{32}P	14.262 days	1.711	^{31}P(n, γ)
			^{32}S(n, p)
^{45}Ca	162.61 days	0.257	^{44}Ca(n, γ)
^{90}Sr	28.79 years	0.546	Fission
^{90}Y	64.00 h	2.280	^{90}Sr$\rightarrow$^{90}Y, fission
^{210}Bi	5.013 days	1.163	^{209}Bi(n, γ)

5.2 Kinetics of Beta-Minus Decay

Beta-minus decay produces an electron, an anti-neutrino that always accompanies the decay and the daughter atom that is left in an ionized state. The electron and anti-neutrino move away from the nucleus and the residual nucleus has one more proton than the parent. Since an atom gains a proton during β^- decay, it changes from one element to another. For example, after undergoing β^- decay, an atom of carbon (with six protons) becomes an atom of nitrogen (with seven protons).

Beta-minus decay preserves the total number of nucleons, A. The beta decay process is a nucleon transformation that can be written as follows: $n \rightarrow p + e^- + \bar{\nu}_e$ (Section 5.1).

For beta emission to be energetically possible, the exact nuclear mass of the parent, M_{parent}, must be greater than the sum of the exact masses of the daughter nucleus, $M_{daughter}$, and the β particle, m_e

$$M_{parent} = M_{daughter} + m_e + Q \qquad (5-28)$$

If atomic masses are used the above equation reduces to

$$M_{parent} = M_{daughter} + Q \qquad (5-29)$$

The energy of β decay, Q, appears as kinetic energy of the β particle and is equivalent to the difference in mass between the parent nucleus and the sum of daughter nucleus and β particle. An extremely small part of the released energy is dissipated by the recoil nucleus, since the ratio of beta particle mass to that of the recoil nucleus is very small. The following example illustrates the energy conservation of β^- decay and explains how atomic masses may be used in a calculation instead of nuclear masses (masses of nuclei only).

Example 5.8 Kinetics of beta-minus decay

Calculate the energy released in the β decay of ^{32}P. The atomic mass of the parent nucleus is 31.9739072 amu. The daughter product is ^{32}S with an atomic mass of 31.9720707 amu.

The reaction is

$$^{32}_{15}P \rightarrow {}^{32}_{16}S + e^- + \bar{\nu}_e$$

Although it was explained in Chapter 3 (Section 2.3) that the atomic masses can be used as nuclear masses, here again is a summary of this explanation based on the example of phosphorus decay. From the reaction equation it follows that the mass balance equation (neglecting the mass of anti-neutrino) is

$$M_{nucleus}\left({}^{32}P\right) = M_{nucleus}\left({}^{32}S\right) + m_e$$

However, the tabulated masses are the atomic masses, thus

$$M_{atom} = M_{nucleus} + Zm_e$$

Adding the same number of electron masses to both sides of the reaction equation

$$M_{nucleus}\left(^{32}P\right) + 15m_e = M_{nucleus}\left(^{32}S\right) + m_e + 15m_e$$

It can now be seen that the masses obtained are equal to the atomic masses. Thus, for β^- decay the energy equation becomes

$$M_{parent} = M_{daughter} + Q \implies$$
$$Q = \left[M_{atom}\left(^{32}P\right) - M_{atom}\left(^{32}S\right)\right]c^2 = 1.7107\,\text{MeV}$$

This energy corresponds to the maximum energy of the ^{32}P β spectrum and is exactly equal to the measured value as shown in Table 4-2. For example, if the energy of the emitted electron is 650 keV, the energy of the anti-neutrino will be 1.06 MeV. As discussed, an extremely small part of the β decay energy is observed in the recoil nucleus, because the ratio of β particle mass to the recoil nucleus mass is very small. In this example it is 0.00055/31.9720707 = 0.000017.

5.3   Kinetics of Beta-Plus Decay

When the neutron-to-proton ratio is too low and α emission is not energetically possible, the nucleus may reach stability by emitting a positron. During β^+ decay, a proton in the nucleus transforms into a neutron, emitting a positron and neutrino. The positron and neutrino move away from the nucleus, and the residual nucleus has one less proton than the parent nucleus. Since the atom loses a proton during β^+ decay, it changes from one element to another. For example, after undergoing β^+ decay, an atom of carbon (with six protons) becomes an atom of boron (with five protons). The beta-plus decay preserves the total number of nucleons, A. The decay process is a nucleon transformation and as discussed in Section 5.1 can be written as follows: $p \rightarrow n + e^+ + \nu_e$. The reaction equation in the form of nuclear masses is identical to that for β^- decay given in Eq. (5-28)

$$M_{parent} = M_{daughter} + m_e + Q$$

If the atomic masses are used the above equation becomes

$$M_{parent} = M_{daughter} + 2m_e + Q \tag{5-30}$$

The following example illustrates the energy conservation of β^+ decay and again illustrates that atomic masses can be used instead of nuclear masses.

Example 5.9 Kinetics of beta-plus decay

Calculate the energy released in the β^+ of $^{13}_{7}N$ given the following masses: $M(^{13}_{7}N) = 13.0057386$ amu, $M(^{13}_{6}C) = 13.0033548$ amu, $m_e = 0.00055$ amu.

Starting from the reaction

$$^{13}_{7}N \rightarrow {}^{13}_{6}C + e^+ + \nu_e$$

From the above reaction equation it follows that the mass balance equation (neglecting the mass of neutrino) is

$$M_{nucleus}\left(^{13}N\right) = M_{nucleus}\left(^{13}C\right) + m_{e^+}$$

However, the tabulated masses are atomic masses, thus

$$M_{atom} = M_{nucleus} + Zm_e$$

Adding the same number of electron masses to both sides for the reaction equation

$$M_{nucleus}\left(^{13}N\right) + 7m_e = M_{nucleus}\left(^{13}C\right) + m_{e^+} + 7m_e$$

It can now be seen that the masses obtained are equal to the atomic masses if (mass of an electron is obviously equal to the mass of a positron)

$$M_{atom}\left(^{13}N\right) = M_{atom}\left(^{13}C\right) + 2m_e$$

Thus

$$M_{parent} = M_{daughter} + 2m_e + Q \quad \Rightarrow$$
$$Q = \left[M_{atom}\left(^{13}N\right) - M_{atom}\left(^{13}C\right) - 2m_e\right]c^2 = 1.196\,\text{MeV}$$

The positive Q-value indicates that the decay is possible, i.e., $^{13}_{7}N$ is unstable and decays by positron emission.

5.4 Kinetics of Orbital Electron Capture

The energy balance equation for positron emission

$$M_{parent} = M_{daughter} + 2m_e + Q \qquad (5\text{-}31)$$

indicates that if a neutron-deficient atom is to attain stability by positron emission, it must exceed the weight of its daughter by at least two electron rest masses. If this requirement cannot be met, then the neutron deficiency is overcome by a process called electron capture or *K*-capture. In this radioactive transformation, one of the lowest positioned (*K*-shell) electrons is captured by the nucleus and combines with a proton to produce a neutron and a neutrino (see Section 5.1): $p + e^- \rightarrow n + v_e$. A schematic of this decay mode is shown in Fig. 5-20.

The energy conservation for *K*-capture is therefore

$$M_{parent} + m_e = M_{daughter} + E_b + Q \qquad (5\text{-}32)$$

where E_b is the binding energy of the captured electron and the masses are nuclear masses. The following example illustrates the kinetics of the *K*-capture decay mode.

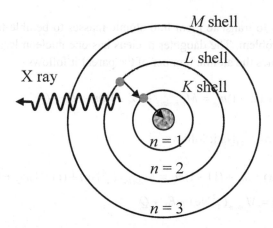

Figure 5-20. Orbital electron capture (K-capture)

Example 5.10 Kinetics of the orbital electron capture (*K*-capture)

For the decay shown in Fig. 5-21, determine the energy of the orbital electron capture decay mode. The *K*-shell binding energy of $^{22}_{11}$Na is 1.0721 keV, and the atomic masses of $^{22}_{11}$Na and $^{22}_{10}$Ne are 21.9944368 amu and 21.9913855 amu, respectively.

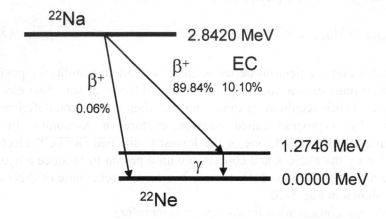

Figure 5-21. $^{22}_{11}$Na decay scheme

In order to calculate the decay energy from the equation that is expressed in nuclear masses

$$M_{parent} + m_e = M_{daughter} + E_b + Q$$

it is necessary to translate them into atomic masses to be able to use the values as given in the problem. The daughter nucleus has one nucleon less than the parent and thus if Z denotes the atomic number of the parent it follows

$$M_{parent} + m_e + (Z-1)m_e = M_{daughter} + (Z-1)m_e + E_b + Q$$

For the decay of $^{22}_{11}$Na it follows

$$M_{nucleus}(^{22}_{11}Na) + m_e + (11-1)m_e = M_{nucleus}(^{22}_{10}Ne) + (11-1)m_e + E_b + Q$$
$$M_{atom}(^{22}_{11}Na) = M_{atom}(^{22}_{10}Ne) + E_b + Q$$

The Q-value is

$$Q = M_{atom}(^{22}_{11}Na) - M_{atom}(^{22}_{10}Ne) - E_b = 2.841MeV$$

From the decay shown in Fig. 5-21 it can be seen that a photon is emitted with energy of 1.2746 MeV. Thus, the excess energy is equal to 2.841 MeV

– 1.2746 MeV = 1.567 MeV. The recoil energy associated with the emission of a gamma ray photon is insignificantly small; therefore the excess energy is carried away by the neutrino. In order to conserve energy, whenever radioactive decay involves the capture or emission of an electron, a neutrino must be emitted. In the middle of the periodic table, the isotopes that are lighter than the most stable isotopes tend to decay by electron capture, and the heavier ones decay by beta-minus emission. One example is silver, which has two stable isotopes: one of lower mass which decays by electron capture and one of greater mass which decays by beta emission.

5.5 Kinetics of Internal Conversion

Internal conversion is an alternative mechanism to γ decay for an excited nucleus to release excess energy and return to its ground state. It is a process in which a tightly bound electron interacts with the nucleus, absorbs the excitation energy (which concurrently can be emitted as a γ ray) and is then ejected from the atom. Internally converted electrons called conversion electrons are monoenergetic, and the kinetic energy of the converted electron, E_e, is equal to the difference in excitation energy between the initial and final state, E_γ, and the binding energy of the converted electron of the daughter element, E_b

$$E_e = E_\gamma - E_b \tag{5-33}$$

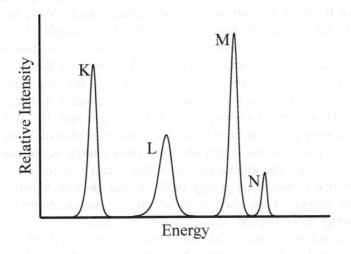

Figure 5-22. Internal conversion electron spectrum (not to scale)

Notice that in this process, the emitted electron was previously one of the orbital electrons, whereas the electron in β decay is produced by the decay of a neutron from the nucleus.

Since the internal conversion process can interact with any of the orbital electrons (from *K*, *L*, *M*, *N*... shells), which all have different binding energies, the energy spectrum consists of many lines (see Fig. 5-22).

Example 5.11 Kinetics of the internal conversion

^{141}Ce decays into ^{141}Pr that is left in its excited state in 70.2% of the cases at the energy of 145.4405 keV. ^{141}Pr returns to its ground state by either emitting a γ ray or emitting a conversion electron. Assuming that the return to a ground state is through internal conversion of the *K*-electron (having a binding energy of 42 keV) what is the kinetic energy of the *K*-electron? What is the energy of the γ ray that would be emitted instead of the internal conversion?

According to Eq. (5-33), the kinetic energy of the *K*-electron is 145.4405 − 42 = 103.4405 keV. The energy of the γ ray is equal to the energy of the excited state, i.e., 145.4405 keV.

5.6 Auger Electrons

In the process of internal conversion and orbital electron capture, an electron leaves its atomic orbit and the vacancy is soon filled. There are two competing processes: emission of an X-ray due to the transition of an electron from an outer shell to the vacancy in a shell closer to the nucleus and another process that is similar to internal conversion in which the energy difference between the two orbits is not released as an X-ray but rather is used to knock another electron from its orbit. For example, if a vacancy in the *K*-shell is filled in with an electron from the *L*-shell, then the energy difference is enough to remove another electron from the *L*- or *M*-shell which causes another vacancy to form. This process is called Auger electron emission and can consist of a number of vacancies and thus emitted electrons. However, the entire process is not longer than 10^{-9} s. The Auger process is more probable in light nuclei, while emission of X-rays is more probable in heavy nuclei. Auger electrons have energy in the range of 100 eV to a few keV. The kinetic energy of the Auger electron corresponds to the difference between the energy of the initial electron transition and the ionization energy of the shell from which the Auger electron is ejected. These energy levels depend on the properties of the atom.

This effect was discovered independently by both Lise Meitner and Pierre Auger. The discovery made by Meitner was published in 1923 in the *Journal Zeitschrift fur Physik*, 2 years before Auger discovered the same effect. However, the English-speaking scientific community adapted Auger's name for this phenomenon.

6. GAMMA DECAY

6.1 Mechanics of Gamma Decay

Gamma (γ) decay follows α and β decay. The most common γ sources are β radioactive isotopes, because they are easy to produce and have higher γ ray intensity than α emitters. Very penetrating γ rays were discovered in 1900 by Paul Villard, a French physicist. They are similar to X-rays, but are emitted from the nucleus and generally have much shorter wavelengths. Gamma rays are the most energetic forms of electromagnetic radiation, with more than 10,000 times the energy of visible light photons.

In γ decay, a nucleus rearranges its constituent protons and neutrons in order to transition from a higher energy state to a lower energy state through the emission of electromagnetic radiation. The number of protons and neutrons in the nucleus does not change in this process, so the parent and daughter atoms are the same chemical element. The emitted γ ray is monoenergetic having energy equal to the energy level difference less than the small fraction transferred to the recoil nucleus (Fig. 5-1). Gamma decay is a fast process with half-lives that are usually in the range of 10^{-13} s to days. The most common sources of γ radiation are listed in Table 5-3. In the case when lower energy photon radiation is required, the isotopes that decay by electron capture are used to produce X-rays. A list of some of the isotopes that can be produced in a reactor and are X-ray emitters is shown in Table 5-4.

Table 5-3. Most commonly used γ ray emitters

Isotope	Half-life	$E_{max} > 5\%$ (MeV)
^{24}Na	14.9590 h	2.75
^{72}Ga	14.10 h	2.20
^{140}La	1.6781 days	2.52
^{110m}Ag	249.79 days	1.52
^{152,154}Eu	13.537 years, 8.593 years	1.40
^{60}Co	1925.1 days	1.33
^{187}W	23.72 h	0.78

Table 5-4. Most commonly used X-ray emitters

Isotope	Half-life	E(K-shell) (keV)
^{55}Fe	2.73 years	6.404
^{65}Zn	244.26 days	8.639
^{75}Se	119.79 days	11.222
^{170}Tm	128.6 days	50.741
^{204}Tl	3.78 years	72.872

6.2   Kinetics of Gamma Decay

In the γ decay of a nucleus, a γ ray is produced by a transition between nuclear levels:

$$E_{initial} = E_{final} + E^* = E_{final} + E_{nucleus} + E_\gamma \tag{5-34}$$

where the excitation energy, E^*, is shared between the γ, E_γ and the kinetic energy of the recoiling nucleus, $E_{nucleus}$. In general, the transition energy and γ energy may be considered equal, because the energy of the recoil nucleus is much smaller than the energy of the emitted γ ray. Thus, the mass–energy equation for the γ decay reduces to

$$E_{initial} = E_{final} + E^* = E_{final} + E_\gamma \tag{5-35}$$

Example 5.12 Fine structure of the α decay spectrum

^{236}U decays into ^{232}Th as shown in Fig. 5-23. What does the fine structure of the α decay spectrum indicate about the nucleus? [The decay of ^{232}Th is shown in Fig. 5-13.]

From the decay it can be seen that there are three different α particles that could be emitted in the decay with different energies and different probabilities. Only one of them leads to the ground state of the daughter nucleus, and this α particle has the energy of 4.494 MeV. The other two decay pathways leave the daughter nucleus in its excited states. The excited nucleus reaches its ground state through the emission of γ rays. Therefore, the fine structure of the α decay spectrum indicates the energy levels in the daughter nucleus. The existence of different α particle emissions and different γ ray emissions prove the existence of discrete energy levels in a nucleus. The challenging experiments in nuclear physics are those to determine these energy levels of a nucleus excited states and the transitions during its process of de-excitation. These types of experiments are usually based on nuclear spectroscopy used to establish relationships between the excited states transitions and often based on the coincidence technique. The decay scheme shown in Fig. 5-23 and in similar figures in this textbook are actually obtained from nuclear spectroscopy.

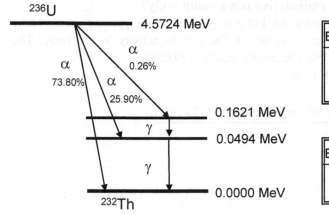

Figure 5-23. ^{236}U decay scheme

7. NATURAL RADIOACTIVITY

Since its creation, the world has been naturally radioactive. The level of radioactivity differs from area to area depending on the naturally occurring radioisotope concentration and their half-lives. Background radiation describes the total exposure to existing, natural radionuclides, which are found not only in air, water and soil, but also in human bodies and are divided into three general categories:

a) Primordial radioisotopes that have existed since before the creation of the Earth,
b) Cosmogenic radioisotopes that are formed as a result of cosmic ray interactions, and
c) Those produced due to human actions.

The primordial radionuclides are typically long lived, with half-lives often on the order of hundreds of millions of years:

- ^{235}U (703800000 years): 0.72% of all natural uranium
- ^{238}U (4.468×10^9 years): 99.2745% of all natural uranium
- ^{232}Th (1.405×10^{10} years)
- ^{226}Ra (1,600 years)
- ^{222}Rn (3.8235 days)
- ^{40}K (1.277×10^9 years).

One example of a primordial nuclide that is a constituent of every living creature is ^{40}K. The following example demonstrates the radiation level due to potassium decay in an average human.

Example 5.13 How radioactive is a human body?

Considering that there are about 140 g of potassium in a typical person's body, determine the total number of atoms of ^{40}K and its activity in the body. The abundance of ^{40}K is 0.0117%. The atomic weight is 39.0983.

The total number of ^{40}K atoms

$$N_{^{40}K} = \frac{(140g)(6.02 \times 10^{23} \text{ atoms/mole})(0.0117 \times 10^{-2})}{39.0983 \text{ g/mole}}$$

$$= 2.52 \times 10^{20} \text{ atoms}$$

The activity is thus

$$A_{^{40}K} = \lambda_{^{40}K} N_{^{40}K} = \frac{\ln 2}{1.277 \times 10^{9} \times 365 \times 24 \times 3600 \text{ s}} \times \left(2.52 \times 10^{20} \text{ atoms}\right)$$

$$= 4.3 \text{ kBq} = 116 \text{ nCi}$$

Cosmic radiation can exist in many forms, from high-speed heavy particles to high-energy photons and muons. Cosmic radiation interacts with the upper atmosphere and produces radioactive nuclides. Although they can have long half-lives, the majority have shorter half-lives than the primordial nuclides. Three of the main cosmogenic radionuclides are

- ^{14}C (5730 years): created by ^{14}N(n, p)^{14}C
- ^{3}H (12.33 years): created through cosmic radiation interactions with N and O, or ^{6}Li(n, α)^{3}H
- ^{7}Be (53.29 days): created through cosmic radiation interactions with N and O

The most interesting is the cycle of radiocarbon in nature (Fig. 5-24) and is explained in more details as follows. From the known content (activity) of radiocarbon (^{14}C) in organic matter it is possible to determine its age. This method is called the *carbon dating* and was developed after World War II by Willard F. Libby. It is used to determine the age of specimens (for example wood, charcoal, marine and freshwater shells) in archeology, geology, geophysics and other branches of science. Carbon has many unique properties which are essential for life on earth. As sketched in Fig. 5-24, ^{14}C is created in a series of events in the atmosphere that starts with cosmic radiation interactions in the upper atmosphere by removing neutrons from nuclei. These neutrons interact with ordinary nitrogen (^{14}N) at lower altitudes, producing ^{14}C. Unlike common carbon (^{12}C), ^{14}C is unstable and decays to nitrogen. Ordinary carbon is a constituent of carbon dioxide (CO_2) in air and is consumed by plants. Since plants are consumed by animals (including humans), carbon enters the food chain. ^{14}C also combines with oxygen to form carbon dioxide ($^{14}CO_2$) that follows the same cycle as the non-radioactive CO_2. The ratio of these two molecules of CO_2 can be determined by measuring a sample of the air. The ratio $^{14}C/^{12}C$ is fairly constant in air, leaves or the human body because ^{14}C is intermixed with ^{12}C. In living things the ^{14}C atoms decay into ^{14}N, and at the same time living things exchange carbon with the environment, so that the ratio remains approximately the same as in the atmosphere. However, as soon as a plant or animal dies, the ^{14}C atoms continue to decay but are no longer replaced. Thus, the amount of ^{14}C decreases with time; in other words, the $^{14}C/^{12}C$ ratio becomes smaller. The "clock" of radiocarbon dating thus starts ticking at the moment a living organism dies. Because the half-life of ^{14}C is 5,730 years anything over approximately 50,000 years should theoretically have no detectable ^{14}C. That is why radiocarbon dating cannot approximate an age of millions of years.

Humans have used radioactivity for 100 years, and through its use, have added to the natural inventories. However, the amount of radionuclides created by humans is small compared to the natural amounts discussed above. In addition, the majority of created radionuclides have short half-lives. Major radionuclides produced by humans that are included into the food chain are fission products produced from weapons testing: ^{131}I (8.0207 days), ^{137}Cs (30.07 years) and ^{90}Sr (28.79 years).

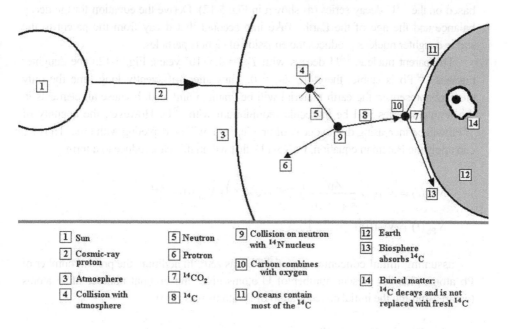

1 Sun	5 Neutron	9 Collision on neutron with ^{14}N nucleus	12 Earth
2 Cosmic-ray proton	6 Protron	10 Carbon combines with oxygen	13 Biosphere absorbs ^{14}C
3 Atmosphere	7 $^{14}CO_2$	11 Oceans contain most of the ^{14}C	14 Buried matter: ^{14}C decays and is not replaced with fresh ^{14}C
4 Collision with atmosphere	8 ^{14}C		

Figure 5-24. Carbon cycle

Example 5.14 Radiocarbon dating

The ^{14}C content in living things decreases after death with a half-life of 5,730 years. If the ^{14}C content of an animal bone is found to be 22.5% of that of an equivalent present-day sample, determine its age. Calculate the activity of the bone assuming the initial activity to be 15 dis/min/g.

The age of the specimen is determined as follows:

$$N(t) = N(0)e^{-\lambda t} \quad \lambda = \frac{\ln 2}{T_{1/2}} = 0.000121 \text{ yrs}^{-1}$$

$$\frac{N(t)}{N(0)} = 0.225 \quad \rightarrow \quad t = 12,331 \text{ years}$$

The activity of the specimen knowing that the rate of disintegration is constant, $A(0) = 15$ dis/min/g, is

$$\frac{\lambda N(t)}{\lambda N(0)} = \frac{A(t)}{A(0)} = 0.225 \quad \rightarrow \quad A(t) = 15 \times 0.225 = 3.4 \text{ dis/min/g}$$

Example 5.15 The age of the Earth

Determine the geological age of the Earth using the lead method. The lead method is based on the ^{238}U decay series (as shown in Fig. 5-12). Derive the equation for the decay balance and the age of the Earth. Take into account that decay from the parent to the stable daughter nucleus produces the emission of eight α particles.

The parent nucleus ^{238}U decays with $T_{1/2} = 4.5 \times 10^9$ years, Fig. 5-12. The daughter nucleus ^{206}Pb is stable, therefore $\lambda_{Pb} = 0$. Thus after sufficiently long time the only elements present in the earth material will be uranium and lead, because all elements in the uranium series will be in secular equilibrium with ^{238}U. However, the quantity of ^{206}Pb will be increasing due to decay of uranium that will be depleting with time. Thus we can apply the Bateman equation, Eq. (5-11), that will in this case reduce to a form

$$N_{Pb}(t) = N_U(0)\frac{\lambda_U}{\lambda_{Pb} - \lambda_U}\left(e^{-\lambda_U t} - e^{-\lambda_{Pb} t}\right) + N_{Pb}(0)e^{-\lambda_{Pb} t}$$

$$N_{Pb}(t) = N_U(0)\left(1 - e^{-\lambda_U t}\right)$$

assuming initial concentration of ^{206}Pb was zero. In addition, the present number of Pb atoms plus the present number of U atoms gives the original number of U atoms (again assuming the initial concentration of Pb atoms was zero)

$$N_{Pb}(t) + N_U(t) = N_U(0)$$

The solution of last two equations gives the time or the age of the Earth

$$N_{Pb}(t) = [N_{Pb}(t) + N_U(t)](1 - e^{-\lambda_U t}) \Rightarrow \quad t = \frac{1}{\lambda_U} \ln\left(\frac{N_{Pb}(t) + N_U(t)}{N_U(t)}\right)$$

The spectro-chemical analysis of the earth sample will give the concentration of uranium and lead, and by using the equation just derived, the age of the sample is determined. The age of the Earth is estimated to be 4.5×10^9 years based on the oldest sample found on Earth.

8. NUCLEAR ISOMERISM

A nuclear isomerism refers to a metastable state of a nucleus corresponding to an excited state of one or more of the nucleons. The nuclear isomers release the extra energy and decay into the ground states through different mechanisms, by releasing the γ ray or conversion electrons, or by decaying through other modes of the decay such as for example β⁻ decay. One example is a doubly β⁻ decay of a ^{234}Pa with two different half-lives of 1.17 min and 6.7 h corresponding to two different excited states. These excited states are called the isomeric or metastable states. Their half-lives range from a fraction of a second to several months or even million of years as it is in the case of the only naturally occurring nearly stable nuclear isomer ^{180m}Ta with the half-life of 10^{15} years; it is found in natural tantalum at about 1 part in 8,300.

The nuclear isomerism is related to a combination of large angular momentum change between the final and initial states of a nucleus with a low energy of transition to a ground state thus producing a long half-life for that energy state. All states with the half-lives longer than 1 ps are considered as isomeric or metastable states.

Some isomers are produced by nuclear reaction or fusion. Their application is found in medicine and industry. Most widely used are ^{99m}Tc with a half-life of 6.01 h) and ^{95m}Tc (with a half-life of 61 days).

NUMERICAL EXAMPLE

Solution of the Bateman Equation

The ^{20}O decays by beta-minus according to the following decay scheme:

$$^{20}\text{O} \xrightarrow[13.51s]{\beta^-} {}^{20}\text{F} \xrightarrow[11.163s]{\beta^-} {}^{20}\text{Ne (stable)}$$

Calculate the decay as a function of time for ^{20}O as well as its daughter product ^{20}F by numerical solution of the Bateman equations. Also compute the time of maximum ^{20}F concentration (Fig. 5-25).

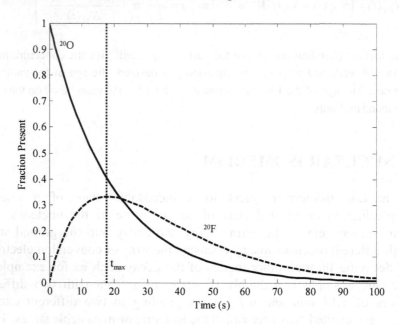

Figure 5-25. Numeric solution of the Bateman equations for ^{20}O decay

Solution in MATLAB:

```
clear all
global lambda_p lambda_d
Thalf_p = 13.51; %s half life of O-20
Thalf_d = 11.163; %s half life of F-20
lambda_p = log(2)/Thalf_p; % s^-1
lambda_d = log(2)/Thalf_d; % s^-1
% Numerical Solution
[t,N] = ode45(@Bateman, [0 100], [1 0]);
% Tmax calculation
tmax = log(lambda_d/lambda_p) / (lambda_d - lambda_p)
plot(t,N(:,1),'k')
hold on
plot(t,N(:,2),'k--')
plot([tmax tmax],[0 1],'k:')
xlabel('Time (s)')
ylabel('Fraction Present')
function dN = Bateman(t,N)
```

```
global lambda_p lambda_d
dN = zeros(2,1);
dN(1) = -lambda_p*N(1);
dN(2) = lambda_p*N(1) - lambda_d*N(2);
```

PROBLEMS

5.1. Write the decay of ^{226}Ra and show that the activity of 1 gram of pure radium is equal to 1 Ci. Then calculate the activity of this sample 100 and 1,000 years later. The half-life of ^{226}Ra is 1,600 years.

5.2. A solution with radioactive sodium of activity of 12,000 disintegrations/min was injected into the blood stream of a patient. After 30 h the activity of 1 cm^3 of the blood was 0.5 disintegrations/min. If the half-life of the sodium isotope is 15 h calculate the volume of blood in the body.

5.3. A sample contains an isotope of magnesium, ^{27}Mg, which undergoes β decay with a half-life of 9.46 min. A Geiger-counter measured the activity of the sample to be 1.69×10^{21} Bq. Write the decay of this isotope. Calculate the decay constant. How many moles of the isotope are present in the sample? How many radioactive isotopes are present after 1 h? What is the activity of the sample after 10 h?

5.4. Use the nuclide table from http://atom.kaeri.re.kr/ton/nuc7.html and find the decay of ^{27}Mg. Sketch the decay scheme and find the energy of γ rays emitted with the probabilities of 29% and 71%.

5.5. A canister was found in a laboratory to contain 1,000,000 atoms of a certain isotope in 2004. The label on the canister showed that the number of nuclei in 1984 was 2,000,000. Calculate the decay constant and half-life of this isotope.

5.6. The disintegrations of radioactive nuclides are detected by an appropriate counting apparatus. The efficiency of such equipment is determined as the ratio of counts per unit time (usually a minute) to the number of disintegrations per same unit of time. If dpm represents the number of disintegrations per minute and cpm number of counts per minute then the efficiency of measurement is given with

Efficiency of measurement = [cpm / dpm] × 100%

Calculate the efficiency of measurement if a sample had 1,000 disintegrations per minute while the counter recorded 800 counts per minute.

5.7. Every measurement of sample radioactivity includes the background radioactivity caused by cosmic rays, natural radioactivity, radioactive fallout and electronic noise in the circuitry of the equipment. Therefore the true value of the cpm of a sample must be reduced for the background value: sample cpm – background cpm. The efficiency of measurement is

Efficiency of measurement = [(sample cpm – background cpm)/dpm]×100%

Calculate the efficiency for the counts from Problem 4.7 if the background radiation is 15 cpm.

5.8. A 0.01 µCi (1 µCi = 2.22×10^6 dpm) source of ^{35}S ($T_{1/2}$ = 87.51 days) was counted in a liquid scintillation counter after 200 days and was found to contain 2,600 cpm. Calculate the efficiency of the counting apparatus.

5.9. Calculate the specific activity of ^{60}Co. The half-life is 1925.1 days.

5.10. Consider a decay chain $C_1 \rightarrow C_2 \rightarrow C_3 \rightarrow \ldots \rightarrow C_n$. Write the coupled system of decay equations.

5.11. Use the nuclide table from http://atom.kaeri.re.kr/ton/nuc7.html to find the decay of ^{238}U and its half-life. Compare the half-lives of ^{238}U and its daughter nuclide and define the condition for secular equilibrium. Calculate the molar concentration of the daughter nuclide at secular equilibrium if the activity of ^{238}U is 2.3 dpm/kg.

5.12. Use the nuclide table from http://atom.kaeri.re.kr/ton/nuc7.html to find the decay of ^{220}Ra and read the atomic masses of the parent and daughter nuclide. Calculate the Q-value for the decay, kinetic energy of the α particle and the Coulomb barrier potential.

5.13. Use the nuclide table from http://atom.kaeri.re.kr/ton/nuc7.html to find decay of ^{51}Ti. Calculate the decay energy, the maximum kinetic energy of the emitted β particle and the maximum kinetic energy of the anti-neutrino. The decay emits one γ ray of energy 0.32 MeV.

5.14. Show that the atomic mass of ^{252}Cf is 252.0816196 amu knowing that it decays by emission of α particle of energy 6.118 MeV. Show also that the decay energy is 6.217 MeV. Use the nuclide table from http://atom.kaeri.re.kr/ton/nuc7.html to find the necessary data.

5.15. A sample consists of mixture containing ^{239}Pu and ^{240}Pu. If the specific activity is 3.42×10^8 dpm/mg, calculate the proportion of plutonium in the sample.

5.16. The half-life of ^{22}Na is 2.6019 years. It decays 89% by positive electron emission and 11% by electron capture. Calculate the partial decay constants.

5.17. Complete the decay schemes

$$^{13}O \xrightarrow{???} {}^{13}N \xrightarrow{???} {}^{13}C$$
$$^{20}O \xrightarrow{???} {}^{20}F \xrightarrow{???} {}^{20}Ne$$

5.18. Calculate the time when the rock is solidified if the ratio of ^{40}K to ^{40}Ar was found to be 0.1.

5.19. A rock sample of 200 g was found to contain 25 g of ^{40}K. Determine the age of the rock sample.

5.20. Estimate the age of the ore sample that contained 10.67 mg of ^{238}U and 2.81 g of ^{206}Pb.

5.21. Calculate the age of a sample containing 25 g of carbon with ^{14}C activity measured to be 4 Bq. Assume that $^{14}C/^{12}C = 1.3 \times 10^{-12}$.

5.22. What is the age of a bone sample that is found to contain 1 mg of ^{14}C.

5.23. Write the equation of ^{14}C decay.

5.24. Define the unit of Bq and Ci. Explain the relation between these two units and correlate to the disintegration per minute.

5.25. Explain the decay of tritium.

5.26. There are a number of sources containing α-emitting radionuclides. One such case is an americium–beryllium source. Use the nuclide table from http://atom.kaeri.re.kr/ton/nuc7.html to find decay of ^{241}Am. What is the most probable energy of the emitted α particle? What is the α particle energy spectrum for this decay?

5.27. What is the activity of 1 mCi of ^{14}C after 10 weeks?

5.28. What is the activity of a sample of 250 μCi of ^{32}P after 10 weeks?

5.29. Calculate the density of water and standard gas in unit of molecules per liter and comment on the density of ionization interactions in these two media.

5.30. Explain what causes atoms to be radioactive.

5.31. Explain how does the radioactivity take place and how stable atoms can become radioactive? What are the isotopes and what are the isotones?

5.32. Calculate the activity of 2 g-mole of ^{40}K in 2005 and million years later? What is the number of atoms present in 2 g-mole in 2005 and million years later?

5.33. The biological removal of radioisotopes from the human body is taken into account through so-called biological half-life. Very often, the radioactive half-life and the biological half-life are evaluated through the effective half-life

$$\lambda_{eff} = \lambda + \lambda_{bio}$$

Knowing that radioactive half-life for ^{131}I is 8 days and its biological removal half-life is 120 days, calculate the effective half-life of ^{131}I.

5.34. Knowing that after 500 years the activity of a sample containing ^{226}Ra was reduced to 80.4% of its original value, determine the half-life of ^{226}Ra. Compare the value you can find in the table of elements provided at http://atom.kaeri.re.kr/ton/nuc7.html.

5.35. Calculate the maximum kinetic energy of a positron emitted in the decay of vanadium-48.

5.36. Use the barrier penetration model (the tunneling formalism explained in Chapter 4, Section 3.7) to determine the probability of α decay of ^{212}Po.

5.37. The mean lifetime of a nuclide is a measure of the average life expectancy of that nuclide; it is defined as the expected amount of time before the decay takes place. Show that the mean lifetime of a radionuclide, τ, is 1.443 times its half-life. [*Hint*: $\tau = 1/\lambda = T_{1/2}/0.693 = 1.443\ T_{1/2}$]

5.38. Find the decay of ^{40}K. What is the total decay constant and the total half-life? How would you determine the branching ratios (*BR*) for different decay modes?

5.39. When the ^{60}Co source activity used in medical applications drops to 40% of its initial value the source need to be replaced. Calculate the time when this shall take place.

Chapter 6

INTERACTIONS OF RADIATION WITH MATTER
Basic Principles, Evidence and Examples

> The social system of science begins with the apprenticeship of the graduate student with a group of his peers and elders in the laboratory of a senior scientist; it continues to collaboration at the bench or the blackboard, and on to formal publication – which is a formal invitation to criticism. The most fundamental function of the social system of science is to enlarge the interplay between imagination and judgment from a private into a public activity. The oceanic feeling of well-being, the true touchstone of the artist, is for the scientist, even the most fortunate and gifted, only the midpoint of the process of doing science.
> *Horace Freeland Judson* (b. 1931)

1. INTRODUCTION

Radiation interaction with matter is generally analyzed by considering charged particles and electromagnetic radiation separately. As discussed in Chapter 5 the two types of charged particles emitted are α and β particles (electrons or positrons). The mass of these particles differs by many orders of magnitude and the types of nuclear interactions they undergo are thus dramatically different. Other important heavy charged particles that need to be considered are protons, deuterons and helium. The characteristics of electromagnetic radiation (γ rays and X-rays) interactions are quite different (photons are massless and travel at the speed of light) to that of charged particles and are considered separately. This chapter discusses the mechanisms of interaction for both charged particles (α, protons and

T. Jevremovic, *Nuclear Principles in Engineering*,
DOI 10.1007/978-0-387-85608-7_6, © Springer Science+Business Media, LLC 2009

electrons) and electromagnetic radiation (γ rays and X-rays). The characteristics and interactions of neutrons with matter are described in Chapter 7.

2.   INTERACTIONS OF CHARGED PARTICLES

2.1   Types of Interactions

An incoming charged particle may interact either with the atom's electron cloud or directly with the nucleus. The difference in size, mass and binding energy of the nucleus and electrons determines the type of interaction the incoming particle will undergo. In every collision energy is exchanged between the target and the incoming particle but the energy before and after the collision must be preserved (conservation of energy law). There are two types of collisions: elastic and inelastic scatterings which differ in energy distribution after the collision. Incoming particle brings kinetic energy into the system and the collision is elastic if none of this energy is transferred to the target. These collisions are conceptually similar to collisions between the billiard balls. However, if a portion of the incoming kinetic energy is imparted on the target atom then the collision is considered to be inelastic.

2.1.1 Elastic Scattering of Charged Particles

When a charged particle passes through matter there is a significant probability that an elastic scattering event will take place. In principle, there are no elastic interactions with the bound electrons in an atom because the electron subsequently transfers the energy to the nucleus and the collision is analyzed as a collision with the whole atom. Since the mass and charge of an atom are dominated by that of its nucleus, such a collision is generally considered as a direct collision between the charged particle and the nucleus. As described in Chapter 2, Rutherford's gold foil experiment uncovered many aspects of the charged particle collisions with matter. For example, if a charged particle passes very close to the nucleus, the electron cloud distribution is nearly symmetric with respect to the incoming particle, and the electronic Coulomb forces are neglected. When the charged particle passes further away from the nucleus through the electron cloud, the electrons reduce the effect. Due to this screening effect, the analyses of such collisions require a correction to the total charge seen by the incoming particle, called the effective charge that is always less than the charge of the nucleus, $Z_{eff} < Z$. This allows the elastic scattering of a charged particle with an atom to be simplified as a direct interaction with a nucleus of charge Z_{eff}.

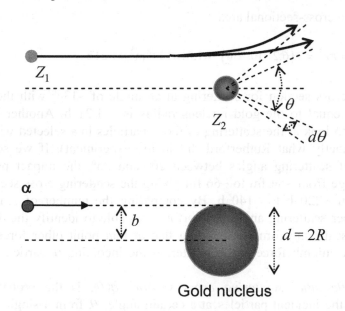

Figure 6-1. Rutherford scattering of charged particle with the nucleus

In the elastic collisions, also known as *Rutherford scattering*, the direction and the energy of the incoming charged particle may be changed. Since the interaction is strongly dependent on Coulomb forces it is also referred to as *Coulomb scattering*.

The Rutherford scattering formula gives angular deflection probability of the incident particle (Fig. 6-1) usually discussed in terms of the cross section, i.e., the effective target area seen by the incident particle (see Chapter 7) as shown in Fig. 6-1. For a gold nucleus ($A = 197$), the radius can be estimated according to the Fermi model given in Eq. (3-1) (assume $R_0 = 1.07$ fm):

$$R = 1.07 A^{1/3} = 6.2 \text{ fm}$$

For example, if an α particle with kinetic energy, $E = 7.7$ MeV, approaches the gold nucleus, with the impact parameter, b, equal to the radius of a gold nucleus, it will be scattered at an angle of $\sim 134°$ according to Eq. (2-3):

$$b = \frac{k(Ze)(ze)}{2E \tan(\theta/2)} = \frac{k(79e)(2e)}{2E \tan(\theta/2)} \equiv \frac{(79e)(2e)}{8\pi\varepsilon_0 E} \cot(\theta/2)$$

$$b = \frac{k(Ze)(ze)}{E\tan(\theta/2)} = \frac{(1.44\,\text{eVnm})(2.79)}{7.7\times10^6\,\text{eV}}\frac{1}{\tan(\theta/2)} \Rightarrow \theta = 134°$$

with the cross-sectional area

$$Area: \quad \sigma = \pi r^2 = \pi\left(6.2\times10^{-15}\,\text{m}\right)^2 = 1.21\times10^{-28}\,\text{m}^2 = 1.21\,\text{b}$$

Thus the cross section for scattering at an angle of ~134° with the impact parameter equal to the gold nucleus radius is ~ 1.21 b. Another example would be to look at the scattering of the α particles in a selected window of angles (exactly what Rutherford did in his experiment). If we select the window of scattering angles between 20° and 25°, the impact parameter would range from ~84 fm to ~66 fm giving the scattering cross section in a range from ~220 b to ~140 b. By comparing the number of scattered α particles per scattering angle, Rutherford was able to identify the departure from his scattering formula knowing that at that point other forces rather than just Coulomb force act in scattering the incoming α particles off the gold foil.

The *differential scattering cross section*, σ(θ), is the probability of scattering the incident particles at a certain angle, θ, from a single nucleus, and is given by the Rutherford scattering formula

$$\sigma(\theta) = \frac{k^2 Z_1^2 e^2 Z_2^2 e^2}{m_1^2 \upsilon^4}\frac{1}{\sin^4(\theta/2)} = \frac{k^2 Z_1^2 Z_2^2 e^4}{4E^2}\frac{1}{\sin^4(\theta/2)} \tag{6-1}$$

where $E = m_1\upsilon^2/2$ is the kinetic energy of an incoming charged particle with charge Z_1e and Z_2e is the charge of the target nucleus. The above relation indicates the following:

- The probability of deflection is proportional to the square of the product of the charges of the incident particle and the target nucleus. That is, a larger deflection is obtained for incoming particles of a greater charge or for heavier target nuclei.
- The incoming charged particle deflection is smaller if its energy is larger, since the probability of the angular distribution is inversely proportional to the square of the kinetic energy of the incoming charged particle.
- Smaller angles have a higher probability since the differential scattering cross section is inversely proportional to the fourth power of half the scattering angle, θ/2.

The probability that an incoming charged particle will transfer all or part of its energy to a target nucleus (which is assumed to be at rest) depends on a number of factors and may be written as

$$\frac{d\sigma(E)}{dE} \propto \frac{Z_1^2 Z_2^2 e^4}{E^2 M \upsilon^2} \tag{6-2}$$

- The probability of energy transfer is directly proportional to the charges of the incoming particle and the target nucleus squared.
- The probability of smaller energy transfer is inversely proportional to the energy of the incoming particle (energy that is to be transferred) squared.
- If the target is lighter, the transfer of energy is more probable since it is inversely proportional to the mass of the target nucleus (M).
- If the incident particle velocity (υ) is small, it will deflect more and transfer more energy since the probability of energy transfer is inversely proportional to the square of the incoming particle velocity.

The maximum energy transfer, governed by the energy and momentum conservation laws, occurs during a head-on collision. It should, however, be mentioned that the head-on collision is a very rare event. The energy exchange directly depends on the masses of the incoming particle, m, and the target, M; for example, when the mass of the incoming particle is smaller than the mass of the target nucleus the incoming particle is repelled:

$$\frac{E^{'}}{E} = \left(\frac{M-m}{M+m}\right)^2 \tag{6-3}$$

where E is the particle energy before the collision and E' is the particle energy after the collision. From this relation it can be seen that when two masses are equal, the energy difference of the incoming particle, before and after the collision, is zero. The incoming particle is thus stopped and all of its energy is transferred to the target nucleus. Conversely, if $M \gg m$, the right hand side of Eq. (6-3) approaches unity and the energy of the incoming particle remains unchanged after the collision. Equation (6-3) suggests that in a collision of an α particle with a gold nucleus, as in the Rutherford's experiment, the maximum energy that an α particle will lose is ~10%. The maximum energy that the target nucleus may receive (head-on collision) from an incoming charged particle of energy E is

$$E_{t\,arget} = E\frac{4Mm}{(M+m)^2} \tag{6-4}$$

Since the incoming particles are usually much lighter than the target nuclei, the collision leads to a change in their directions while the change in their energies can almost be neglected.

2.1.2 Inelastic Scattering of Charged Particles with Electrons

In an inelastic scattering, the incoming particle may transfer all or part of its energy to the electrons in an atom. The energy transferred to the electrons may cause excitation of the electron or ionization of the atom. The excitation and ionization processes are described in Chapter 2. If the energy of the incoming charged particle is larger than the binding energy of the electron it interacts with, then the collision is similar to elastic scattering. In this case, the probability of the angular deflection is proportional to the square of the target's charge. In the collision with an electron, the incoming particle is deflected at an angle which is a factor of $1/Z^2$ smaller than in a direct collision with a nucleus of charge Z.

2.1.3 Inelastic Scattering of Charged Particles with a Nucleus

When passing close to the nucleus, charged particles are attracted or repelled by the Coulomb force. The acceleration of a particle, a, is proportional to the charges of the nucleus and of the particle itself:

$$a \propto \frac{Z_1 Z_2 e^2}{m} \tag{6-5}$$

This acceleration causes the particle to deflect from the original trajectory (see Fig. 6-2). An electron will, for example, deflect toward the nucleus, while an α particle will deflect away from the nucleus because of the opposite signs of their respective charges. According to classical electrodynamics, every charged particle that accelerates emits electromagnetic radiation with an intensity that is proportional to the square of the acceleration. Thus

- The intensity of electromagnetic radiation decreases for heavier particles. For example, the intensity of this radiation for the α particles is a million times smaller than for the electrons.
- The intensity of radiation is greater for heavier target materials.

This type of electromagnetic radiation is also called the radiative loss, "bremsstrahlung" (German for *braking* radiation), or continuous X radiation. Quantum mechanics gives a correct interpretation of the bremsstrahlung since it defines it as a quantum process in which an electron emits a photon. Bremsstrahlung photons have a continuous energy distribution that ranges from zero to a theoretical maximum that is equal to the kinetic energy of the β particle (electron). The emitted energy spectrum is in the range of X-rays and the energy is higher for the materials of higher Z. For example, since the use of light materials reduces bremsstrahlung, Plexiglas® is often used to shield against β radiation. The greatest bremsstrahlung occurs when high-energy β particles interact with high-density materials such as lead. In general, the probability of bremsstrahlung production increases with the energy of the incoming charged particle and the mass (charge) of the target material.

The bremsstrahlung hazard due to β particles of maximum energy E that interact with a target material with atomic number Z may be estimated from the following approximation:

$$f = 3.5 \times 10^{-4} ZE \tag{6-6}$$

where f represents the fraction of incident β particle energy converted into bremsstrahlung photons.

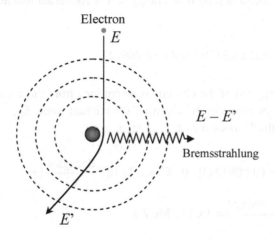

Figure 6-2. Bremsstrahlung radiation

For electrons and positrons, one of the approximate relations describing the ratio of energy loss due to bremsstrahlung and the ionization is as follows:

$$\frac{\text{Bremsstrahlung energy loss}}{\text{Ionization energy loss}} \approx \frac{E(Z+1.2)}{700} \tag{6-7}$$

where E is kinetic energy in MeV and Z is the atomic number of the medium (see also Eq. (6-26)). It is usual practice to define a critical energy of electron above which the bremsstrahlung (or radiation loss of energy) is dominant. This energy is approximately defined as $700/Z$ expressed in MeV where Z represents the charge number of the target material. The electron loses its energy due to the emission of bremsstrahlung photon over certain distance. The distance over which its energy is reduced by a factor e is usually called the radiation (or the attenuation) distance (given in g/cm^2 or in cm), x_0, and is determined as follows (see Section 2.2):

$$-\left(\frac{dE}{dx}\right)_{rad} \approx \frac{E}{x_0} \quad \Rightarrow \quad E = E_0 e^{-x/x_0} \tag{6-8}$$

Example 6.1 Bremsstrahlung radiation

A 1 mCi ^{60}Co source is encapsulated in a spherical lead (atomic number 82) shield that has a thickness that is sufficient to stop all the β particles emitted by the source. Calculate the bremsstrahlung radiation flux as a function of distance outside the spherical shield. The maximum energy of the emitted β particle is 0.3179 MeV.

The fraction of incident β particle energy that is converted into an X-ray in each decay of ^{60}Co is

$$f = (3.5 \times 10^{-4})(82)(0.3179 \text{ MeV}) = 0.009124$$

The source energy (S) of the photons is determined from the assumption that the average β energy is one-third of the spectrum's maximum value, $E_{\beta max} = 0.3179$ MeV. For the activity of the ^{60}Co source of $A = 1$ mCi

$$S = fA\frac{E_{\beta max}}{3} = (0.009124)\left(10^{-3}\text{Ci} \times 3.7 \times 10^{10}\frac{\text{decays/s}}{\text{Ci}}\right) \times$$

$$\left(\frac{0.3179 \text{ MeV/decay}}{3 \text{ Ci}}\right) = 35,773 \text{ MeV/s}$$

It is important to mention that the maximum β energy spectrum value should be assumed for the photons whenever bremsstrahlung is considered in the radiation exposure of humans. In other cases it is sufficient to assume that each photon receives the average β energy. Thus, the photon generation rate for the maximum β energy is expressed as

$$\frac{35,773 \text{ MeV/s}}{0.3179 \text{ MeV/photon}} = 1.125 \times 10^5 \text{ photon/s}$$

Assuming a point ^{60}Co source, the bremsstrahlung flux as a function of distance from the spherical shield is

$$\phi(r) = \frac{1.125 \times 10^5 \text{ photons/s}}{4\pi r^2}$$

2.2   Loss of Energy

The mechanisms by which charged particles transfer their energy in inelastic collisions with matter are expressed in one or more of the following forms: stopping power, relative stopping power, specific ionizations and loss of energy per ionization.

2.2.1 Stopping Power ($- dE/dx$)

Stopping power is defined as the amount of energy, dE, which a charged particle loses along the length of its path through matter, dx. This quantity always represents the average energy loss per number of interacting particles. It is proportional to the square of the charge of the incoming particle, Z_1^2, and it is inversely proportional to its velocity; thus the stopping power increases as the particle velocity is decreased. The value of $-dE/dx$ along a particle track is also called a *specific energy loss, S,* the rate of energy loss or the linear energy transfer (LET). The classical formula that describes the specific energy loss is known as the *Bethe—Bloch formula*, which is valid for all types of heavy charged particles with velocities that are large compared to the orbital electron velocities. The *Bethe—Bloch* formula has different forms for heavy and light charged particles:

- For heavy charged particles (α particles and protons):

$$-\frac{dE}{dx} = \frac{4\pi Z^2 e^4}{m_e \upsilon^2} NB$$

$$B = Z\left[\ln\frac{2m_e\upsilon^2}{I} - \ln\left[1 - \frac{\upsilon^2}{c^2}\right] - \frac{\upsilon^2}{c^2}\right]$$

$$(6\text{-}9)$$

υ: velocity of the charged particle
Ze: charge of the charged particle
N: number density of the target
m_e: rest mass of the electron
I: experimentally evaluated average excitation and ionization potential (see Chapter 2)
B: stopping number
- For light charged particles (electrons and positrons):

$$-\frac{dE}{dx} = \frac{2\pi e^4}{m_e v^2} NB$$

$$B = \left[\begin{array}{l} \ln \dfrac{m_e v^2 E}{2I^2\left(1-\beta^2\right)} - \left(\ln 2\right)\left(\sqrt{1-\beta^2}-1+\beta^2\right) + \\[2mm] 1-\beta^2 + \dfrac{1}{8}\left(1-\sqrt{1-\beta^2}\right)^2 \end{array} \right]$$

(6-10)

$$\beta = v/c$$

For the charged particles with $v \ll c$ (non-relativistic particles) only the first term in the stopping number (B) equation is necessary. Equations (6-9) and (6-10) show that B varies slowly with particle energy and is proportional to the atomic number (Z) of the absorber material. Thus the stopping power varies as $1/v^2$, or inversely with particle energy.

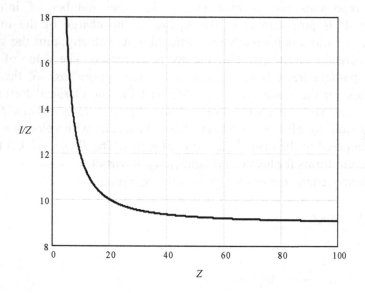

Figure 6-3. Average ionization and excitation potential as a function of Z

The Bethe—Bloch formula also shows that the higher-Z materials have greater stopping powers. The ionization/excitation parameter I is an experimentally determined value (see Section 2.2.4) and the ratio I/Z is approximately constant for absorbers with $Z > 13$. This ratio ranges from 10 eV for heavy elements to 15 eV for light elements (see Fig. 6-3).

The loss of energy due to ionization and excitation shows a general trend for all charged particles (see Fig. 6-4) summarized as follows:

- For each charged particle, the maximum energy loss occurs at a characteristic incoming particle velocity.
- The stopping power then decreases to a minimum value on the order of 1 MeV for electrons and higher for heavier charged particles.

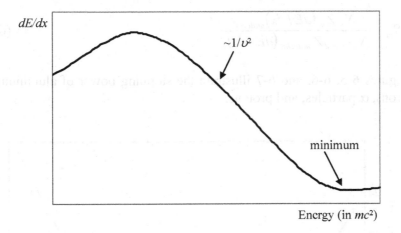

Figure 6-4. General trend of *dE/dx* as a function of particle energy *E*

2.2.2 Relative Stopping Power

The relative stopping power for a charged particle that interacts with a given material is defined as the ratio of a particle energy loss per atom of the given material to the energy loss per atom that will be experienced by the same particle in standard material (which is the air for β particles and the aluminum for α particles).

- Relative linear stopping power (MeV/cm)

$$S_{linear} = \frac{(dE/dx)_{material}}{(dE/dx)_{st}}$$ (6-11)

- Relative mass stopping power (MeV/g cm^2)

$$S_m = \frac{\rho_{st}(dE/dx)_{material}}{\rho_{material}(dE/dx)_{st}}$$ (6-12)

- Relative stopping power per atom (MeV/atom cm)

$$S_m = \frac{N_{st}(dE/dx)_{material}}{N_{material}(dE/dx)_{st}} \qquad (6\text{-}13)$$

- Relative stopping power per electron (MeV/electron cm)

$$S_m = \frac{N_{st}Z_{st}(dE/dx)_{material}}{N_{material}Z_{material}(dE/dx)_{st}} \qquad (6\text{-}14)$$

Figures 6-5, 6-6, and 6-7 illustrate the stopping power of aluminum for electrons, α particles, and protons.

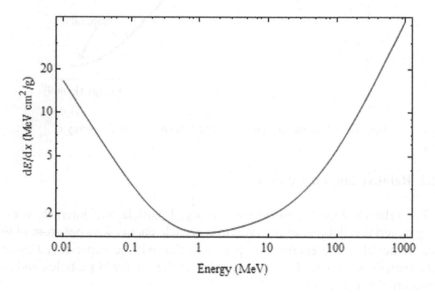

Figure 6-5. Total mass stopping power of electron in aluminum (National Institute of Standards and Technology tables)

2.2.3 Secondary Electrons

As a charged particle passes through a medium it ionizes some of the atoms by ejecting the electrons. This involves a transfer of energy from the charged particle to the electrons, which may receive sufficient kinetic energy to cause further ionizations. This process is known as secondary ionization. The total ionization is thus the sum of both primary and secondary processes.

The energy of secondary electrons ranges from zero to a theoretical maximum which depends on the mass and energy of the primary charged particle. For example, in a collision between an α particle with energy E and mass m_α and an electron of mass m_e, the maximum energy that the α particle can transfer to the electron, E_{target}, is given by Eq. (6-4)

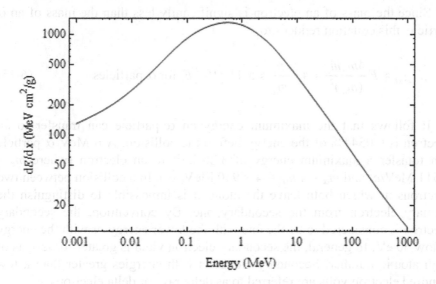

Figure 6-6. Total mass stopping power of α particle in aluminum (National Institute of Standards and Technology tables)

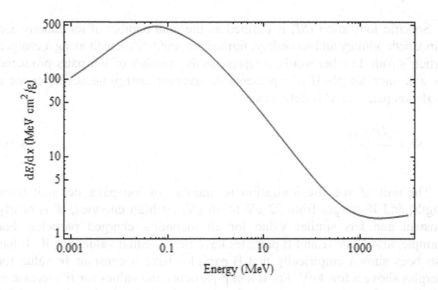

Figure 6-7. Total mass stopping power of proton in aluminum (National Institute of Standards and Technology tables)

$$E_{target} = E \frac{4m_\alpha m_e}{(m_\alpha + m_e)^2}$$

Since the mass of an electron is significantly less than the mass of an α particle, this equation reduces to

$$E_{target} \approx E \frac{4m_\alpha m_e}{(m_\alpha)^2} \approx 4E \frac{m_e}{m_\alpha} \approx 5.44 \times 10^{-4} E \text{ for } \alpha \text{ particles} \tag{6-15}$$

It follows that the maximum energy an α particle can transfer to an electron is 0.0544% of the energy before the collision. A 6 MeV α particle can transfer a maximum energy of 3.26 keV to an electron (where $m_e = 0.511$ MeV/c^2 and $m_\alpha = 4\ m_H = 4 \times 940$ MeV/ c^2). In a collision between two electrons in which both leave the atom, it is impossible to distinguish the primary electron from the secondary one. By convention, the secondary electron is considered to be the one with the lower energy or with the energy below 50 eV. In general, the secondary electron yield is greater for targets of high atomic number. Secondary electrons with energies greater than a few hundred electron volts are referred to as delta rays or delta electrons, δ.

2.2.4 Specific Ionization and Ion Pairs

Specific ionization (*SI*) is defined as the total number of ionizations (ion pairs), both primary and secondary, formed per unit track length along a charged particle's path. In other words, it represents the number of ion pairs produced per unit track length. If *W* represents the average energy needed to create a single ion pair, the *SI* is defined as

$$SI = \frac{dE/dx}{W} \tag{6-16}$$

The unit of specific ionization is [number of ion pairs per unit track length] and *W* ranges from 22 eV to 46 eV. At high energies, *W* is nearly constant and has similar value for all incoming charged particles. For example, at 4 MeV α and β particles have nearly equal values for *W*. It has also been shown empirically that β particles have a constant *W* value for energies above a few keV. For α and β particles, the values for *W* increase at lower energies, since the probability for ionization is reduced. Table 6-1 shows some measured *W* values. The ionization density produced by a single charged particle depends on its charge and velocity. For example, a slower moving particle spends more time in the vicinity of an atom or molecule, thereby increasing the chance of ionization events. Thus, an α particle creates thousands of ion pairs per centimeter than an electron (β particle) of the same energy (approximately 100 ion pairs per cm).

Table 6-1. Ion-pair generation energy for different materials

Material	Ion-pair generation energy, W (eV)
Air	33.9
Silicon (Si)	3.6
Germanium (Ge)	2.8
Silicon oxide (SiO$_2$)	17
Hydrogen (H)	37
Helium (He)	46
Nitrogen (N)	36
Oxygen (O)	32
Neon (Ne)	37

Example 6.2 Ion-pair production from α particle interaction

An α particle loses about 35.5 eV for each ion pair formed. Calculate the number of ion pairs produced by an α particle with a kinetic energy of 5.5 MeV.

The number of ion pairs produced is

$$\frac{5.5 \times 10^6 \, eV}{35.5 eV / i.p.} = 154,930 \text{ ion pairs}$$

2.2.5 Range of Interactions

Heavy particles such as protons and α particles will deposit all their energies along a definite depth of penetration in a medium. This depth or distance is called the *range* of the particle and it depends on the energy and mass of the particle. The range is longer for the particles of higher energy and shorter for the heavier particles. For example, consider two particles with the same kinetic energy; the heavier particle has a shorter range.

The range may either be defined as linear range (units of length) or the mass range (units of mass per area). Theoretically, the range of a charged particle in a medium may be obtained from the integration of the inverse of a particle's energy loss per unit length, i.e.,

$$R(E_{in}) = \int_0^{E_{in}} \frac{dE}{dE / dx} \tag{6-17}$$

where dE/dx represents the total stopping power and the final particle velocity is assumed to be zero.

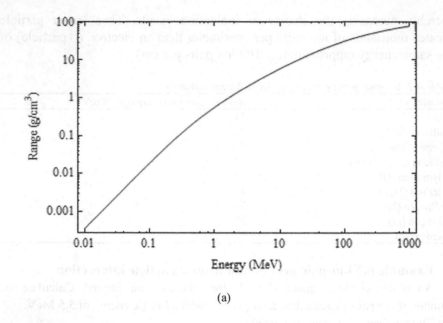

(a)

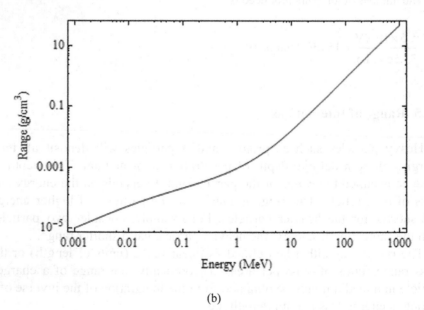

(b)

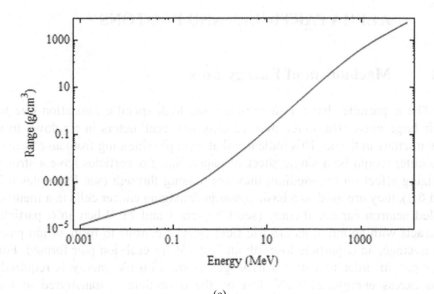

(c)

Figure 6-8. CSDA range of (**a**) electron, (**b**) α particle and (**c**) proton in aluminum (National Institute of Standards and Technology tables)

The evaluation of this term is complicated, especially for light charged particles (see Section 4.1), and its reciprocal is commonly assumed to become zero at zero particle energy and increases linearly to the known value of least energy. The use of these assumptions in the calculation of range is referred to as the continuous slowing down approximation (CSDA). Figure 6.8 shows the ranges of a proton, electron and an α particle passing through aluminum. This illustrates the tendency of the CSDA range and its difference for light and heavy charged particles.

Example 6.3 Ion-pair density from α particle interaction

If the range of an α particle is 10 cm, determine the average ion-pair density using the data from Example 6.2.

The ion-pair density is obtained as the ratio of the number of ion pairs produced to the length of track of the ionizing particle:

$$\frac{154,930 \text{ i.p.}}{10 \text{ cm}} = 15,493 \text{ ion pairs per unit length (cm)}$$

3.   ALPHA PARTICLES AND PROTONS

3.1   Mechanism of Energy Loss

The α particles have a short range and high specific ionization due to their large mass. The range may be only few centimeters in air down to a few microns in tissue. This indicates that a simple shielding from an external α emitter would be a single sheet of paper. Since α particles have a strong ionizing effect on the medium they are passing through (see Examples 6.2 and 6.3), they are used as a basic *agent* in damaging cancer cells in a method called neutron capture therapy (see Chapters 1 and 7). When an α particle interacts with an atom, its electric field ejects electrons to form an ion pair. On average, an α particle loses about 35.5 eV for each ion pair formed. For example, in order to ionize a hydrogen atom, 13.6 eV energy is required. The excess energy, 21.9 eV, lost by the α particle is transferred to the electron as its kinetic energy. Thus, the ejected electron is set into a motion and can produce another ion pair or the secondary electrons. A fast-moving α particle may lose energy without causing ionization as it passes through a medium. In such cases, the electrons do not receive sufficient energy to be ejected and they simply change orbits (moving to higher energy levels). Thus, the α particle causes only excitation of the medium but not ionization. The fast-moving α particle has less time for interactions and the specific ionization consequently decreases at higher energies (Fig. 6-9).

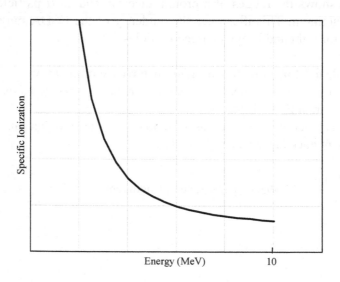

Figure 6-9. Specific ionization of α particle versus its energy

In other words, the specific ionization of a high-energy α particle will increase as the velocity decreases. This tendency of increasing ionization probability with the continuous slowing down of the α particle toward the end of its range is known as the Bragg curve and is illustrated in Fig. 6-10. As soon as the energy of the α particle drops below the energy required for ionization of the atoms in a medium, its ionization efficiency abruptly reduces to zero. The highest localized ionization energy deposition is expected around the Bragg peak.

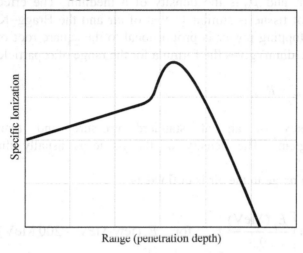

Figure 6-10. Specific ionization of α particle versus its range

3.2   Range–Energy Relationship

The general expression for the linear stopping power (linear energy loss) for a charged particle that is slowing down in a medium is the Bethe–Bloch formula (as discussed in Section 2.2.1). Since the energy loss is proportional to the square of the charge of the incoming particle, an α particle is expected to stop much faster than a proton in a given medium. Semi-empirical formulas express the range of charged particles as a function of kinetic energy. For α particles, the range in air at a temperature of 15°C and 760 mm pressure is given by the equations

$$R_{air}(\text{cm}) = \begin{cases} 0.56\left(\dfrac{\text{cm}}{\text{MeV}}\right)E(\text{MeV}) & E < 4\,\text{MeV} \\[3mm] 1.24\left(\dfrac{\text{cm}}{\text{MeV}}\right)E(\text{MeV}) - 2.62(\text{cm}) & 4\,\text{MeV} < E < 8\,\text{MeV} \end{cases} \qquad (6\text{-}18)$$

The range (expressed as density thickness) of an α particle in any other medium, R_m, is given by

$$R_m \left(\text{cm}\right) = \frac{0.00056 A^{1/3}}{\rho_m} R_{air} \qquad (6\text{-}19)$$

where A is the atomic mass number of a medium, R_{air} is the range of an α particle in air, and ρ_m is the density of a medium. The effective atomic composition of tissue is similar to that of air and the Bragg–Kleeman rule (the atomic stopping power is proportional to the square root of the atomic weight of a medium) gives the formula for the range of α particles in tissue

$$R_{tissue} \rho_{tissue} = R_{air} \rho_{air} \qquad (6\text{-}20)$$

The density of air at standard pressure and temperature is 1.293×10^{-3} g/cm^3. The density of the tissue is usually that of water 1.0 g/cm^3.

The proton range in the air is defined as

$$R_{air} \left(\text{m}\right) = \left[\frac{E_p \left(\text{MeV}\right)}{9.3}\right]^{1.8} \quad \text{for} \quad E_p \left(\text{few MeV} \sim 200 \text{ MeV}\right) \qquad (6\text{-}21)$$

The range of protons in aluminum is given by the semi-empirical formula

$$R_{Al} \left(\mu\text{m}\right) = \begin{cases} \dfrac{10.5 E_p^2}{0.68 + 0.434 \ln\left(E_p\right)} & 2.7 \text{ MeV} \le E_p \le 20 \text{ MeV} \\ 14.21 E_p^{1.5874} & 1 \text{ MeV} < E_p \le 2.7 \text{ MeV} \end{cases} \qquad (6\text{-}22)$$

The rate of the energy loss for α particles and protons and their ranges are shown in Tables 6-2, 6-3, 6-4, and 6-5 for different materials and particle energies. Data were taken from the National Institute of Standards and Technology ASTAR and PSTAR tables.

The charged particle range is affected by the following factors:

- *Energy*: The range is approximately linear with energy since the Bethe–Bloch formula for stopping power is inversely proportional to E.
- *Mass*: With the same kinetic energy, an electron is much faster than an α particle because of its smaller mass; therefore, incoming electrons spend less time near the orbital electrons, thus reducing the effect of Coulomb interactions (consequently stopping power) and increasing the range.

- *Charge*: Stopping power increases with charge while the range decreases. Range is inversely proportional to the square of the charge of the incoming particle. For example, a tritium particle with $Z=1$ will have $\frac{1}{4}$ the stopping power of a ^{3}He particle with $Z=2$.
- *Density*: Stopping power increases with density. The range is inversely proportional to the density of the absorbing medium.

Table 6-2. Total stopping power (MeV cm^2/g) of α particles in different materials

α particle energy (MeV)	Hydrogen	Air	Water	Tissue	Aluminum
0.001	1.264E+03	2.215E+02	3.271E+02	3.688E+02	1.305E+02
0.005	1.136E+03	2.937E+02	3.667E+02	4.382E+02	2.095E+02
0.01	1.292E+03	3.625E+02	4.304E+02	5.227E+02	2.790E+02
0.05	2.746E+03	7.310E+02	8.230E+02	9.924E+02	6.444E+02
0.1	4.123E+03	1.031E+03	1.151E+03	1.375E+03	9.056E+02
0.5	8.220E+03	1.964E+03	2.184E+03	2.698E+03	1.300E+03
1	7.167E+03	1.924E+03	2.193E+03	2.522E+03	1.226E+03
1.5	5.654E+03	1.626E+03	1.898E+03	2.062E+03	1.100E+03
2	4.593E+03	1.383E+03	1.625E+03	1.729E+03	9.859E+02
3	3.354E+03	1.072E+03	1.257E+03	1.324E+03	8.217E+02
4	2.678E+03	8.865E+02	1.035E+03	1.086E+03	6.991E+02
5	2.244E+03	7.612E+02	8.855E+02	9.267E+02	6.053E+02
10	1.284E+03	4.637E+02	5.344E+02	5.556E+02	3.762E+02
15	9.200E+02	3.425E+02	3.930E+02	4.070E+02	2.809E+02
20	7.245E+02	2.748E+02	3.146E+02	3.250E+02	2.272E+02

Table 6-3. CSDA range (g/cm^2) of α particles in different materials

α particle energy (MeV)	Hydrogen	Air	Water	Tissue	Aluminum
0.001	7.480E–07	5.377E–06	3.273E–06	2.985E–06	9.964E–06
0.005	4.230E–06	2.079E–05	1.489E–05	1.292E–05	3.329E–05
0.01	8.380E–06	3.605E–05	2.746E–05	2.334E–05	5.381E–05
0.05	2.882E–05	1.097E–04	9.179E–05	7.632E–05	1.418E–04
0.1	4.344E–05	1.665E–04	1.425E–04	1.186E–04	2.059E–04
0.5	1.034E–04	4.188E–04	3.699E–04	3.063E–04	5.421E–04
1	1.671E–04	6.698E–04	5.931E–04	4.922E–04	9.343E–04
1.5	2.459E–04	9.520E–04	8.374E–04	7.118E–04	1.365E–03
2	3.445E–04	1.287E–03	1.123E–03	9.777E–04	1.845E–03
3	6.030E–04	2.116E–03	1.829E–03	1.646E–03	2.961E–03
4	9.391E–04	3.147E–03	2.711E–03	2.485E–03	4.283E–03
5	1.349E–03	4.368E–03	3.759E–03	3.485E–03	5.825E–03
10	4.424E–03	1.309E–02	1.130E–02	1.072E–02	1.666E–02
15	9.097E–03	2.581E–02	2.236E–02	2.138E–02	3.224E–02
20	1.527E–02	4.222E–02	3.668E–02	3.523E–02	5.216E–02

Example 6.4 Range of an α particle and shielding

Estimate the range of a 3 MeV α particle in air and tissue. Calculate the linear aluminum thickness required to totally stop these α particles. Investigate the accuracy by comparing with Table 6-3. The density of aluminum is 2.7 g/cm^3.

The range of a 3 MeV α particle in air is

$$R_{air}(\text{cm}) = 0.56(\text{cm/MeV})E(\text{MeV}) = 1.68 \text{ cm}$$

From Table 6-3 the range of 3 MeV α particle in the air is 2.116×10^{-3} g/cm^2 giving a linear range of 1.64 cm.

The range of a 3 MeV α particle in a tissue is

$$R_{tissue} = \frac{R_{air}\rho_{air}}{\rho_{tissue}} = 1.293 \times 10^{-3} R_{air} = (1.293 \times 10^{-3})(1.68 \text{ cm}) = 0.0022 \text{ cm}$$

Table 6-4. Total stopping power (MeV cm^2/g) of protons in different materials

Proton energy (MeV)	Hydrogen	Air	Water	Tissue	Aluminum
0.001	9.730E+02	1.414E+02	1.769E+02	2.180E+02	1.043E+02
0.005	1.741E+03	2.776E+02	3.153E+02	4.067E+02	2.131E+02
0.01	2.402E+03	3.850E+02	4.329E+02	5.620E+02	2.966E+02
0.05	3.818E+03	6.897E+02	7.768E+02	9.887E+02	4.749E+02
0.1	3.493E+03	7.301E+02	8.161E+02	1.004E+03	4.477E+02
0.5	1.160E+03	3.501E+02	4.132E+02	4.395E+02	2.550E+02
1	6.771E+02	2.229E+02	2.608E+02	2.737E+02	1.720E+02
1.5	4.902E+02	1.683E+02	1.957E+02	2.045E+02	1.328E+02
2	3.885E+02	1.371E+02	1.586E+02	1.653E+02	1.095E+02
3	2.788E+02	1.018E+02	1.172E+02	1.217E+02	8.250E+01
4	2.197E+02	8.197E+01	9.404E+01	9.738E+01	6.707E+01
5	1.825E+02	6.909E+01	7.911E+01	8.174E+01	5.695E+01
10	1.019E+02	4.006E+01	4.567E+01	4.692E+01	3.376E+01
15	7.239E+01	2.894E+01	3.292E+01	3.373E+01	2.466E+01
20	5.679E+01	2.294E+01	2.607E+01	2.667E+01	1.969E+01

Table 6-5. CSDA range (g/cm^2) of protons in different materials

Proton energy (MeV)	Hydrogen	Air	Water	Tissue	Aluminum
0.001	1.091E–06	9.857E–06	6.319E–06	5.418E–06	1.471E–05
0.005	4.058E–06	2.891E–05	2.262E–05	1.825E–05	3.981E–05
0.01	6.473E–06	4.400E–05	3.599E–05	2.857E–05	5.943E–05
0.05	1.849E–05	1.152E–04	9.935E–05	7.769E–05	1.560E–04
0.1	3.194E–05	1.842E–04	1.607E–04	1.268E–04	2.632E–04
0.5	2.598E–04	1.021E–03	8.869E–04	7.801E–04	1.503E–03
1	8.476E–04	2.867E–03	2.458E–03	2.270E–03	3.945E–03
1.5	1.728E–03	5.479E–03	4.698E–03	4.410E–03	7.287E–03
2	2.883E–03	8.792E–03	7.555E–03	7.147E–03	1.146E–02
3	5.968E–03	1.737E–02	1.499E–02	1.429E–02	2.210E–02
4	1.004E–02	2.839E–02	2.458E–02	2.355E–02	3.563E–02
5	1.506E–02	4.173E–02	3.623E–02	3.481E–02	5.188E–02
10	5.346E–02	1.408E–01	1.230E–01	1.191E–01	1.705E–01
15	1.126E–01	2.899E–01	2.539E–01	2.467E–01	3.462E–01
20	1.913E–01	4.855E–01	4.260E–01	4.147E–01	5.748E–01

The tabulated value shown in Table 6-3 is 0.0018 cm. The aluminum thickness required to totally stop a 3 MeV α particle is

$$R_{Al}\,(cm) = \frac{0.00056A^{1/3}}{\rho_m}\,R_{air} = \frac{0.00056A^{1/3}}{2.7\ g/cm^3}(1.68\ cm) = 0.00105\ cm$$

This is in good agreement with the value given in Table 6-3, 0.002961/2.7=0.001 cm.

Example 6.5 Range of a proton

Estimate the range of a 3 MeV proton in air and aluminum and compare with the value given in Table 6-5.

The range of a 3 MeV proton in air is

$$R_{air}\,(m) = \left[\frac{E_p(MeV)}{9.3}\right]^{1.8} = \left[\frac{3}{9.3}\right]^{1.8} = 0.130\ m$$

The range of a 3 MeV proton in aluminum is

$$R_{Al}\,(\mu m) = \frac{10.5E_p^2}{0.68 + 0.434\ln(E_p)} = \frac{10.5(3)^2}{0.68 + 0.434\ln(3)} = 0.0082\ cm$$

Table 6-5 gives a range of 13.4 cm in air and 0.0082 cm in aluminum for a 3 MeV proton.

4. BETA PARTICLES (ELECTRONS AND POSITRONS)

4.1 Mechanism of Energy Loss

The mechanism of energy loss and the type of interactions for β particles in matter are more complex than for α particles due to the smaller mass and higher speed. Beta particles are emitted during the decay of radionuclides with a continuous energy spectrum with a maximum energy that is characteristic of the radionuclide (see Chapter 5). This maximum value is taken as the total transition energy. The difference between this maximum value and the emitted β particle energy is carried off by an electrically neutral particle. The maximum energy for β radiation from the majority of radionuclides is in the range of 0.5 MeV to 3.5 MeV. When passing through a medium, the β particles interact with atomic nuclei and electrons; the β particle range is not as well defined as for α particles. This is due to the combined effect of the continuous energy spectrum and the scattering characteristics. The characteristics of the range are described in Section 4.2.

The loss of β particle energy in a medium consists of two components. The total stopping power (energy loss) is expressed as a summation of the two terms:

- The collision term, $(dE/dx)_{coll}$, represents the energy loss due to Coulomb interactions (ionization and excitation).
- The radiative term, $(dE/dx)_{rad}$, accounts for the energy loss due to bremsstrahlung, Cerenkov radiation or nuclear interactions.

Therefore, the total stopping power (as illustrated in Fig. 6-11) is

$$\frac{dE}{dx} = \left(\frac{dE}{dx}\right)_{coll} + \left(\frac{dE}{dx}\right)_{rad} \tag{6-23}$$

where the collision term is referred to as the linear energy transfer (LET), i.e., the linear rate of energy loss of a β particle due to ionization and excitation (see Section 2.2.1). The LET is related to the local energy deposition, while the radiative stopping power takes into account the total energy loss due to bremsstrahlung radiation and the formation of secondary and δ electrons. The collision term and total stopping power are nearly equal for heavy charged particles (see Figs. 6-12 and 6-13). Notice that in the case of heavy charged particles, the collision stopping power is called *electronic stopping power*. It is also important to note that the nuclear stopping power is only significant for heavy charged particles, and it represents an average rate of energy loss per unit path length due to the transfer of energy to recoiling atoms in elastic collisions. Except for highly relativistic electrons, ionization and excitation are the main forms of energy loss, which may therefore be calculated with the Bethe–Bloch equation.

The general form of this equation is described in Section 2 for different types of particles. The condensed form of this equation for the ionization and excitation energy loss of electrons can be written as

$$-\left(\frac{dE}{dx}\right)_{coll} = \rho\frac{Z}{A}f_{coll}(I,\beta) \tag{6-24}$$

The equation illustrates the following:
- The collision energy loss is proportional to the electron density in the medium, ρ, and the ratio Z/A, where Z is the atomic number of a medium and A is its atomic weight. This ratio varies slowly with increasing Z (for example the ratio is 0.5 for low-Z materials and reduces to ~0.39 for uranium).

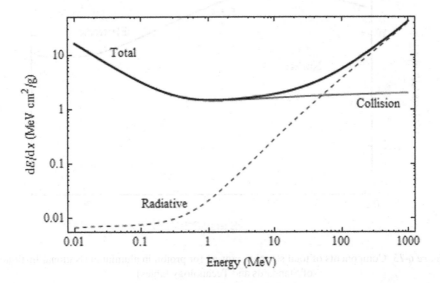

Figure 6-11. Components of total stopping powers for electrons in aluminum (National Institute of Standards and Technology tables)

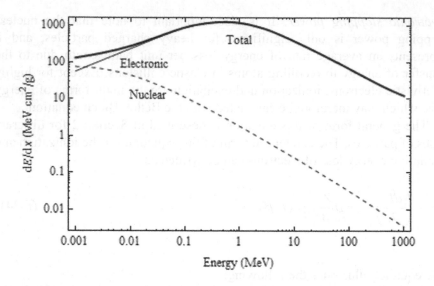

Figure 6-12. Components of total stopping powers for α particle in aluminum (National Institute of Standards and Technology tables)

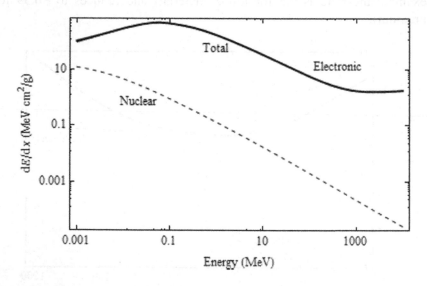

Figure 6-13. Components of total stopping powers for proton in aluminum (National Institute of Standards and Technology tables)

- For electron energies up to 1 MeV, the collision term of the total stopping power decreases due to the increase of the $\beta = \upsilon/c$ term. The collisional stopping power decreases as $1/\beta^2$ for increasing velocity until it reaches a minimum value at electron energy of about 1.5 MeV (Fig. 6-14).

For higher electron energies where $\beta \sim 1$, the energy loss due to ionization and excitation increases logarithmically (relativistic rise) until it reaches a constant value (Fermi plateau).

- As discussed in Chapter 2, the ionization potential (I) increases with Z, but loses significance due to the logarithmic dependence in the collision stopping power equation. However, as illustrated in Fig. 6-14, the loss of energy decreases with increasing Z of a medium.

For electron energies above a few MeV, an additional density–effect correction is required. This accounts for the reduction in the collision stopping power due to the incident electron's polarization of the medium. Descriptions of the average radiative loss of electron energy in the form of bremsstrahlung radiation are only approximate. Although there are no adequate equations to express the radiative stopping power over a wide range of electron energies, a general equation can be used

$$-\left(\frac{dE}{dx}\right)_{rad} = \frac{Z^2 \rho}{A}\left(E + m_e c^2\right) f_{rad}(E, Z) \tag{6-25}$$

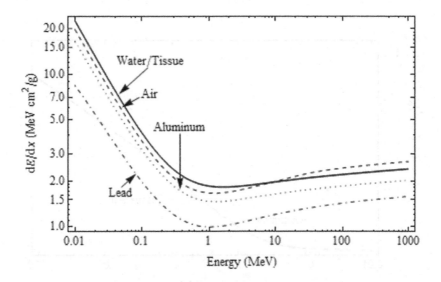

Figure 6-14. Energy loss of electrons due to ionization and excitation (collision term) in water, air, aluminum, lead and tissue (National Institute of Standards and Technology ESTAR tables)

From this equation and the trends illustrated in Fig. 6-15 (radiative stopping power curves for various materials), the following is understood about the radiative stopping power:

- It is proportional to Z^2 and as a result the radiative stopping power becomes comparable to the collision term for the higher Z.
- It is proportional to the electron energy, E, and as a result it becomes comparable to the energy loss in ionization and excitation at specific electron energy values. At even higher energy values it begins to exceed these competing energy loss contributors.

For the relativistic electron energies the ratio of the radiative and the collision stopping power becomes

$$\frac{-(dE/dx)_{rad}}{-(dE/dx)_{coll}} \approx \frac{EZ}{F} \tag{6-26}$$

where E is in MeV and F has a value of 700 for lighter elements and 800 for higher-Z materials (see Example 6.6). The above relations can be generalized for any charged particle of rest mass M and energy $E \gg Mc^2$

$$\frac{-(dE/dx)_{rad}}{-(dE/dx)_{coll}} \approx \frac{EZ}{F} \left(\frac{m_e}{M}\right)^2 \tag{6-27}$$

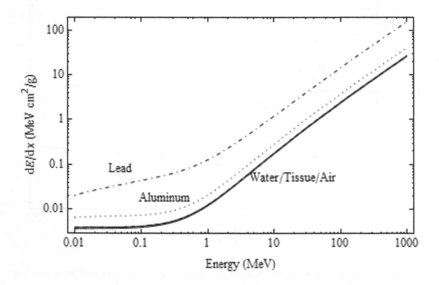

Figure 6-15. Energy loss of electrons due to radiative processes (radiative term) in water, air, aluminum, lead and tissue (National Institute of Standards and Technology ESTAR tables)

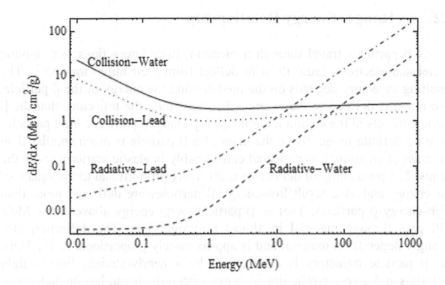

Figure 6-16. Collision and radiative energy losses of electrons (National Institute of Standards and Technology ESTAR tables)

For example, this ratio is equal to one, that is, the bremsstrahlung and ionization/excitation energy losses become equal, at the electron energy of 47 MeV in aluminum and at 7 MeV in lead (Fig. 6-16).

Example 6.6 Energy loss of an electron to bremsstrahlung and ionization and excitation

Estimate the energy at which an electron will start losing its energy equally in both bremsstrahlung and ionization/excitation while moving through lead ($Z = 82$) and water. Compare the results with Fig. 6-16.

- Lead

$$\frac{-\left(dE/dx\right)_{rad}}{-\left(dE/dx\right)_{coll}} = 1 \approx \frac{EZ}{800} \quad \Rightarrow \quad E = 9.76 \text{ MeV}$$

which is in good agreement with ~ 10 MeV shown in Fig. 6-16.
- Water

$$\frac{-\left(dE/dx\right)_{rad}}{-\left(dE/dx\right)_{coll}} = 1 \approx \frac{EZ}{700} \quad \Rightarrow \quad E = 53.8 \text{ MeV}$$

This is in good agreement with the value of ~ 52 MeV shown in Fig. 6-16.

4.2   Range–Energy Relationship

As β particles travel through a medium, their interactions with atomic nuclei and electrons cause them to deflect from their initial trajectory. The resulting trajectory depends on the medium and the energy of the β particle. The range is defined as the average distance along the trajectory that the β particle travels in the medium. Unlike heavy charged particles, light particles have no definite range. Since the mass of a β particle is much smaller than the mass of an atom, it is deflected considerably in elastic scattering with the atoms. The probability of deflection is inversely proportional to the square of the energy and as a result low-energy β particles are deflected more than high-energy β particles. That is, β particles with energy above a few MeV will pass through material in almost a straight line. The deflection also strongly depends on material and is approximately proportional to Z^2. Thus, the β particle trajectory is expected to be a nearly-straight line in light materials and a very erratic line in heavy materials. It can be concluded that the trajectory of β particles

- is nearly a straight line for energies above 1 MeV. With decreased energy of β particles the deflection is more pronounced and particles start to diffuse in a medium.
- fluctuates along the trajectory more than heavy particles.
- the depth of penetration for most of the β particles is smaller than the length of range.

The range of β particles in material is a complex function of their energy, the type and the atomic number of the absorber material. From the macroscopic point of view, the absorption of β particles is a function of distance traveled and the density of material. It has been observed that the absorption (attenuation of β particle beam as sketched in Fig. 6-17) in the absorber is approximately an exponential function of the density of the absorber (ρ), distance through the absorber (x), and the absorbing property of the material (μ). Thus the absorption is approximately similar to that of photon beam: $\ln(I/I_o) = (x\rho)(\mu/\rho)$. However due to a multicollision nature of β particle interactions with absorber the absorption curve does not follow exactly the exponential decline in beam intensity.

- The ionization caused by β particles falls off exponentially with distance.
- For thick absorbers the absorption curve (curve of activity versus the absorber thickness as sketched in Fig. 6-17) becomes almost horizontal, indicating a nearly constant absorption, i.e., ionization occurs. Thick absorbers are expected to stop all incoming β particles while only a part of the β particle energy is lost in thin absorbers.

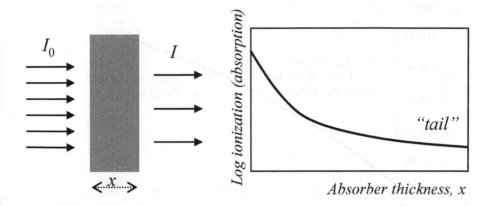

Figure 6-17. Beta particles attenuation and the absorption curve

- The absorption depth (range) depends on β particle energy as well; for example, very high-energy β particles can penetrate to a depth of about one centimeter in tissue.
- The "tail" at the end of the absorption curve indicates the presence of a bremsstrahlung radiation. This is especially prevalent in materials with the high atomic number such as lead. Even though high atomic number materials are the most effective in stopping high-energy β particles, the presence of bremsstrahlung makes lighter materials such as Lucite or plywood a better choice as absorber materials.
- Although β particles do not have definite range, it is possible to specify absorber thickness that will reduce the ionization to a *zero* level.
- Given the complexity of β interactions and the β spectrum, it is difficult to develop theoretical range energy dependence. As a result, experimental energy–range measurements are used to approximate the range of β particles. Figure 6-18 shows the range–energy curve and equations that may be used to compute the range. The range is often expressed in terms of the mass thickness rather than the linear distance, because mass thickness is independent of material density. The mass thickness is given in mg/cm². The linear thickness (range) in cm is obtained by dividing the mass thickness with the density of absorber.

The total stopping power and the CSDA range for electrons in various materials are shown in Tables 6-6 and 6-7, respectively (data were from the ESTAR tables of the National Institute for Standards and Technology).

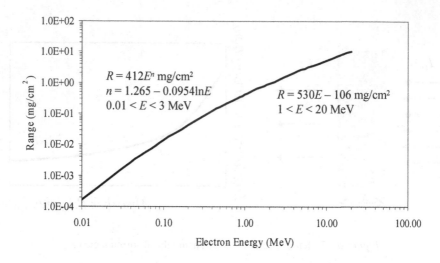

Figure 6-18. Range–energy curve for β particle transport

Example 6.7 Range of β particles

Calculate the distance a β particle will travel in aluminum ($\rho = 2.7$ g/cm^3) as it slows down from 15 MeV to 0.5 MeV. What is the average linear stopping power?

Table 6-7 gives the CSDA mass range values and the resulting linear distances are

$$R(15 \text{ MeV}) = (8.328 \text{ g/cm}^2)/(2.7 \text{ g/cm}^3) = 3.084 \text{ cm}$$

$$R(0.5 \text{ MeV}) = (0.226 \text{ g/cm}^2)/(2.7 \text{ g/cm}^3) = 0.084 \text{ cm}$$

Thus the β particle travels $3.084 - 0.084 = 3.0$ cm in aluminum while it deposits an energy of 14.5 MeV. The average stopping power is obtained as

$$\frac{dE}{dx} = \frac{14.5 \text{ MeV}}{3.0 \text{ cm}} = 4.833 \text{MeV/cm}$$

Table 6-6. Total stopping power (MeV cm^2/g) for β particles in different materials

Electron energy (MeV)	Hydrogen	Air	Water	Tissue	Aluminum
0.01	5.124E+01	1.976E+01	2.256E+01	2.257E+01	1.650E+01
0.05	1.424E+01	5.822E+00	6.607E+00	6.597E+00	5.046E+00
0.1	8.738E+00	3.637E+00	4.119E+00	4.111E+00	3.185E+00
0.5	4.196E+00	1.809E+00	2.041E+00	2.034E+00	1.604E+00
1.0	3.821E+00	1.674E+00	1.862E+00	1.851E+00	1.486E+00

Table 6-6. (Continued)

Electron energy (MeV)	Hydrogen	Air	Water	Tissue	Aluminum
1.5	3.796E+00	1.680E+00	1.841E+00	1.829E+00	1.491E+00
2	3.835E+00	1.711E+00	1.850E+00	1.838E+00	1.518E+00
3	3.943E+00	1.783E+00	1.889E+00	1.876E+00	1.580E+00
4	4.047E+00	1.850E+00	1.931E+00	1.917E+00	1.637E+00
5	4.140E+00	1.911E+00	1.971E+00	1.957E+00	1.690E+00
10	4.479E+00	2.159E+00	2.149E+00	2.131E+00	1.921E+00
15	4.714E+00	2.359E+00	2.306E+00	2.283E+00	2.134E+00
20	4.903E+00	2.539E+00	2.454E+00	2.425E+00	2.340E+00

Table 6-7. CSDA range (g/cm^2) for β particles in different materials

Electron energy (MeV)	Hydrogen	Air	Water	Tissue	Aluminum
0.01	1.076E–04	2.884E–04	2.515E–04	2.512E–04	3.539E–04
0.05	1.970E–03	4.913E–03	4.320E–03	4.324E–03	5.738E–03
0.1	6.650E–03	1.623E–02	1.431E–02	1.433E–02	1.872E–02
0.5	8.480E–02	1.995E–01	1.766E–01	1.770E–01	2.260E–01
1.0	2.117E–01	4.912E–01	4.367E–01	4.385E–01	5.546E–01
1.5	3.433E–01	7.901E–01	7.075E–01	7.110E–01	8.913E–01
2	4.744E–01	1.085E+00	9.785E–01	9.839E–01	1.224E+00
3	7.316E–01	1.658E+00	1.514E+00	1.523E+00	1.869E+00
4	9.819E–01	2.208E+00	2.037E+00	2.050E+00	2.491E+00
5	1.226E+00	2.740E+00	2.550E+00	2.566E+00	3.092E+00
10	2.383E+00	5.192E+00	4.975E+00	5.011E+00	5.861E+00
15	3.470E+00	7.405E+00	7.219E+00	7.276E+00	8.328E+00
20	4.510E+00	9.447E+00	9.320E+00	9.401E+00	1.056E+01

Example 6.8 Energy deposition in tissue cell

Using the data provided for the CSDA range for 1.5 MeV α and β particles in Tables 6-3 and 6-7, respectively, comment on the linear ranges in a human tissue cell. Assume the radius of a human cell is 15 μm with a density of 1 g/cm^3.

The ranges are

- α particle

$$R = (7.840 \times 10^{-4}\,\text{g/cm}^2)/(1\,\text{g/cm}^3) = 7.84 \times 10^{-4}\,\text{cm} = 7.84\,\mu\text{m}$$

- β particle

$$R = (0.711\,\text{g/cm}^2)/(1\,\text{g/cm}^3) = 0.711\,\text{cm}$$

It can be seen that a 1.5 MeV α particle travels only 7.84 µm and thus deposits all of its energy inside the cell volume, while the β particle leaves the cell and will deposit its energy over a much longer range. In boron–neutron capture therapy for brain cancer treatments, the α particle emitted in boron–neutron interaction has the energy of nearly 1.5 MeV and because of its short range represents the key agent in killing the cancer cells.

Example 6.9 Summation of ranges in different materials

The range of an unknown β particle is measured to be 0.111 mm in aluminum. Calculate the energy of the β particle if the β emitter is placed in air at 1 cm from the aluminum sheet and with a 1.7 mg/cm^2 mica absorber between the counter and the aluminum sheet.

Summation of ranges is allowed if the ranges are expressed as density thicknesses:

- Air

$$R_{air} = 1.293 \text{ mg/cm}^3 \times 1 \text{ cm} = 1.293 \text{ mg/cm}^2$$

- Aluminum

$$R_{Al} = 2.7 \text{ g/cm}^3 \times 0.0111 \text{ cm} = 29.97 \text{ mg/cm}^2$$

- Mica

$$R_{mica} = 1.7 \text{ mg/cm}^2$$

Thus, the total range is

$$R_{total} = 1.293 \text{ mg/cm}^2 + 29.97 \text{ mg/cm}^2 + 1.7 \text{ mg/cm}^2 = 32.96 \text{ mg/cm}^2$$

Fig.6-18 shows that this range corresponds to an energy of 0.17 MeV.

5.   PHOTONS (GAMMA AND X - RAYS)

5.1   Exponential Absorption Law

Many nuclear reactions, radioactive decays and particle interactions result in the emission of gamma (γ) rays, the highest energy electromagnetic waves (or photons). Their energies range from thousands of electron volts (keV) to millions of electron volts (MeV) and thus their wavelengths are very short (10^{-11} m to 10^{-13} m). These high-energy particles have found application in the medical profession, especially in cancer treatments.

As explained in previous sections both α and β radiations can be completely absorbed by properly selected materials and their thicknesses. Gamma radiation, however, can only be reduced in intensity. This intensity reduction or attenuation is governed by the exponential absorption law

$$I = I_0 e^{-\mu_l x_l} \tag{6-28}$$

where
I: γ ray intensity transmitted through an absorber of thickness x
I_0: γ ray intensity at zero absorber thickness
x_l: linear absorber thickness
μ_l: linear absorption coefficient.

The linear absorption coefficient is related to the mass absorption coefficient, μ_m, through the density of the absorber materials, ρ

$$\mu_l (\text{cm}^{-1}) = \mu_m (\text{cm}^2/\text{g}) \times \rho(\text{g/cm}^3) \tag{6-29}$$

If the mass thickness, x_m, is defined as the mass per unit area obtained by multiplying the linear thickness x_l by the density ($x_m = \rho \, x_l$), then the exponential absorption law can be written in the following way

$$I = I_0 e^{-\mu_l x_l} = I_0 e^{-(\mu_l / \rho) x_m} \tag{6-30}$$

and μ_l / ρ can be obtained from the empirical measurements of I_0, I and x. These values are tabulated for different materials and photon energies. Figure 6-19 (a) and (b) shows the mass absorption coefficient for aluminum and lead, respectively.

The *total absorption coefficient* or *attenuation coefficient* represents the fraction of the γ ray beam attenuated per unit thickness of the absorber.

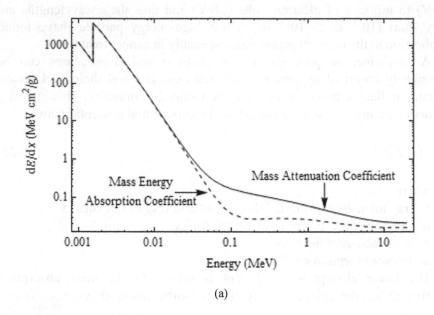

(a)

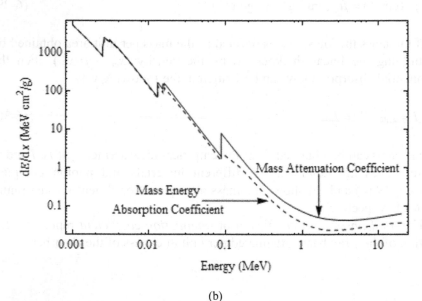

(b)

Figure 6-19. Mass absorption coefficient for γ rays in (**a**) aluminum and (**b**) lead (National Institute of Standards and Technology tables)

The *atomic absorption coefficient,* μ_a, is the fraction of an incident γ ray beam that is absorbed by a single atom, i.e., the probability that an absorber atom will interact with the γ rays in the incoming beam. If N is the number of absorber atoms per cm^3, the atomic absorption coefficient is

$$\mu_a = \frac{\mu_l}{N} \equiv \sigma \left[\frac{cm^2}{atom} \right] \tag{6-31}$$

The atomic absorption coefficient has units of area and is thus referred to as the "cross section" of the absorber (for the discussion on the cross section see Chapter 7). The atomic absorption coefficient is called the microscopic cross section (σ), while the linear absorption coefficient is called the macroscopic cross section (Σ). The microscopic cross section may be expressed in barns (1 b = 10^{-24} cm^2) while the unit of the macroscopic cross section is cm^{-1}. These two cross sections are related as

$$\Sigma(cm^{-1}) = \sigma \left[\frac{cm^2}{atom} \right] \times N \left[\frac{atoms}{cm^3} \right] \tag{6-32}$$

Thus the attenuation of γ rays can be expressed in terms of cross sections

$$I = I_0 e^{-\mu_a N x_l} = I_0 e^{-\Sigma x_l} \tag{6-33}$$

Gamma ray interaction data are usually expressed as mass attenuation coefficients (examples shown graphically in Fig. 6-19 and values given in Table 6-8). Neutron interaction data are usually expressed as cross sections.

Table 6-8. γ rays mass attenuation (absorption) coefficients (cm^2/g) from the NIST X-ray attenuation database

Photon energy (MeV)	Hydrogen	Air	Water	Tissue	Aluminum
0.01	3.854E–01	5.120E+00	5.329E+00	4.937E+00	2.623E+01
0.05	3.355E–01	2.080E–01	2.269E–01	2.223E–01	3.681E–01
0.1	2.944E–01	1.541E–01	1.707E–01	1.688E–01	1.704E–01
0.5	1.729E–01	8.712E–02	9.687E–02	9.593E–02	8.445E–02
1.0	1.263E–01	6.358E–02	7.072E–02	7.003E–02	6.146E–02
1.5	1.027E–01	5.175E–02	5.754E–02	5.699E–02	5.006E–02
2	8.769E–02	4.447E–02	4.942E–02	4.893E–02	4.324E–02
3	6.921E–02	3.581E–02	3.969E–02	3.929E–02	3.541E–02
4	5.806E–02	3.079E–02	3.403E–02	3.367E–02	3.106E–02
5	5.049E–02	2.751E–02	3.031E–02	2.998E–02	2.836E–02
10	3.254E–02	2.045E–02	2.219E–02	2.191E–02	2.318E–02
15	2.539E–02	1.810E–02	1.941E–02	1.913E–02	2.195E–02
20	2.153E–02	1.705E–02	1.813E–02	1.785E–02	2.168E–02

The energy absorption coefficients shown in Fig. 6-19 are the total absorption coefficients and they account for both primary and secondary radiations. Primary radiation considers the local energy deposition during the photon interactions with matter, while secondary radiation considers the energy deposited elsewhere via secondary radiation such as Compton scattered photons, bremsstrahlung, fluorescence, and annihilation photons.

For a mixture or composite materials, the mass attenuation coefficient, $\mu_{m\text{-}mixture}$, is the weighted average of the individual mass coefficients, μ_{mi}

$$\mu_{m-mixture} = w_1\mu_{m1} + w_2\mu_{m2} + ...$$ (6-34)

Example 6.10 Attenuation of γ rays

Calculate the linear and density thicknesses of aluminum and lead needed to transmit not more than 5 % of a 0.60 MeV γ ray beam and compare the density thicknesses. The density of aluminum is 2.7 g/cm^3 and the density of lead is 11.35 g/cm^3. The mass absorption coefficients are 7.802×10^{-2} cm^2/g for aluminum and 0.1248 cm^2/g for lead.

- Aluminum linear thickness

$$\mu_l\left(cm^{-1}\right) = \mu_m\left(cm^2/g\right)\times\rho\left(g/cm^3\right) = 0.07802\times2.7 = 0.2107 \text{ cm}^{-1}$$

$$\frac{I}{I_0} = \frac{5}{100} = e^{-\mu_l x_l} \quad\Rightarrow\quad x_l = 14.22 \text{ cm}$$

Density thickness

$$x_d = x_l \times \rho = 14.22 \times 2.7 = 38.4 \text{ g/cm}^2$$

- Lead linear thickness

$$\mu_l\left(cm^{-1}\right) = \mu_m\left(cm^2/g\right)\times\rho\left(g/cm^3\right) = 0.1248\times11.35 = 1.416 \text{ cm}^{-1}$$

$$\frac{I}{I_0} = \frac{5}{100} = e^{-\mu_l x_l} \quad\Rightarrow\quad x_l = 2.115 \text{ cm}$$

Density thickness

$$x_d = x_l \times \rho = 2.115 \times 11.35 = 24 \text{ g/cm}^2$$

Example 6.11 Attenuation coefficient and cross section for γ ray interactions

Knowing that the linear absorption coefficient represents the macroscopic cross section for γ ray interactions that are predominantly with the electrons of an atom, calculate the microscopic cross section for lead if the mass attenuation coefficient for 0.6 MeV γ ray is 0.1248 cm^2/g (density of lead is 11.35 g/cm^3). The atomic weight of the lead is 207.2.

$$\mu_i\left(\text{cm}^{-1}\right) = \mu_m\left(\text{cm}^2/g\right) \times \rho\left(g/\text{cm}^3\right) = 0.1248 \times 11.35 = 1.416 \text{ cm}^{-1}$$

$$\mu_i = N\sigma \equiv \Sigma \Rightarrow$$

$$\sigma = \frac{\mu_i}{N} = \frac{\mu_i}{N_a Z \rho / A} = \frac{1.416}{6.02 \times 10^{23} \times 82 \times 11.35 / 207.2} = 0.524b$$

Example 6.12 Attenuation of γ rays in a composite material

The soft tissue can be approximated as a mixture of four elements: 10.1174% of hydrogen, 11.1% of carbon, 2.6% of nitrogen and 76.1826% of oxygen. This composition gives a soft tissue density of 1 g/cm^3. Determine the linear attenuation coefficient for 0.6 MeV γ rays for which the mass attenuation coefficients in these four elements are 0.1599 cm^2/g in hydrogen, 0.08058 cm^2/g in carbon, 0.08063 cm^2/g in nitrogen and 0.08070 cm^2/g in oxygen.

The general definition for the total mass attenuation coefficient for the mixture is as follows:

$$\mu_{m-tissue} = w_H \mu_{mH} + w_C \mu_{mC} + w_N \mu_{mN} + w_O \mu_{mO}$$

$$\mu_{m-tissue} = 0.101174 \times 0.1599 + 0.111 \times 0.08058 + 0.026 \times 0.08063 +$$

$$0.761826 \times 0.08070 = 0.088698 \text{ cm}^2/g$$

$$\mu_{l-tissue} = \mu_{m-tissue} \times \rho_{tissue} = 0.088698 \times 1.0 = 0.088698 \text{ cm}^{-1}$$

5.2 Mechanism of Energy Loss

Photons are energy quanta of electromagnetic nature and interact with particles that have electrical charge or, with smaller probability, with particles that behave as small magnets (possess magnetic momentum). The main interactions of photons with matter are with the electrons and nuclei through the following:

- *Absorption of photons*: in this interaction the initial photon disappears as it transfers all of its energy to an electron or nucleus.
- *Scattering of photons*: photon can be scattered through an elastic or inelastic interaction. In elastic scattering, the wavelength of the scattered photon is almost the same as that of the initial photon. If the interaction leads to interference, it is referred to as coherent scattering. In inelastic, incoherent scattering the initial photon transfers its energy to the matter and scatters with a longer wavelength.

The types of photon interactions are summarized in Table 6-9 and show that photoelectric absorption and Compton scattering are interactions that are limited to the orbital electrons of the absorber. These interactions are probable for incident photon energies less than or not significantly higher than the energy equivalent of the rest mass for two electrons (1.022 MeV). Pair production dominates in the energy range above this threshold.

Table 6-9. Types of photon interactions

Interaction with	*Absorption*	*Elastic scattering*	*Inelastic scattering*
Electrons in atoms	Photoelectric effect	Rayleigh scattering	Compton scattering
Electromagnetic field of a nucleus or electron	Pair production		

5.2.1 Photoelectric Effect (γ + atom → e^- + ion)

In 1886, Heinrich Hertz discovered that photons in the ultraviolet region of the spectrum (wavelengths of 200–400 nm) could eject electrons from a metal surface (see Fig. 6-20). The experiment showed that the emission of electrons and the incoming light had certain dependencies.

- The number of electrons emitted by the metal was found to directly depend on the intensity of the light, i.e., the number of emitted electrons increased with increasing light intensity.
- The emitted electrons moved faster if the light had a higher frequency.
- There was a cut-off frequency, f_c, for the incident photons, below which no electrons were emitted.

According to classical mechanics and the wave theory of light that was valid at the time, it was expected that the intensity of the emitted light would determine the kinetic energy of the ejected electrons. The experiments, however, showed that the kinetic energy of the ejected electrons depended on the incoming photon frequency instead of its intensity. The photon intensity thus only affected the number of ejected electrons but not their kinetic energies. This was the discovery of a new phenomenon called the photoelectric effect and it was defined as the emission of electrons from a metal surface exposed to photon radiation. The full physical explanation of the phenomenon was given in 1905 by Albert Einstein who applied Planck's idea of energy quanta and additionally assumed that the light had also the particle properties. He proved that the incoming photon could be represented as the discrete quanta of energy, hf, where f is the photon frequency and h is Planck's constant (see Chapter 4, Section 2). It thus follows that every photon carries a specific energy that is related to its frequency or its wavelength, such that photons of short wavelength (for example the blue light) transmit more energy than long-wavelength (for example the red light) photons. Einstein's equation that explained the photoelectric effect based on the experimental observations is

$$hf = W + \frac{1}{2}m_e\upsilon^2 \qquad (6\text{-}39)$$

where W is called the work function and represents the minimum energy required to remove an electron from the metal surface and $m_e\upsilon^2/2$ is the maximum kinetic energy of the emitted photoelectron. The work function for most metals is around 4.5 eV (Table 6-10).

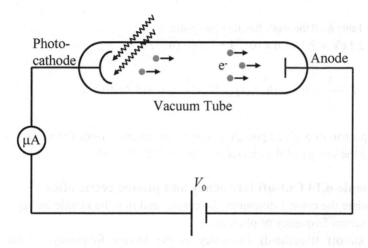

Figure 6-20. Schematics of the experiment for the photoelectric effect

Table 6-10. Work function, W (eV), for some metals

Element	W (eV)
Aluminum	4.08
Beryllium	5.0
Cadmium	4.07
Calcium	2.9
Carbon	4.81
Cesium	2.1
Cobalt	5.0
Copper	4.7
Gold	5.1
Iron	4.5
Lead	4.14
Magnesium	3.68
Mercury	4.5
Nickel	5.01
Niobium	4.3
Potassium	2.3
Platinum	6.35
Silver	4.73
Uranium	3.6
Zink	4.3

Example 6.13 Work function and photoelectric effect

Using the data shown in Table 6-10, determine if green light with $\lambda = 505$ nm can cause electrons to be ejected from cesium.

The energy of the incoming photon:

$$hf = \frac{hc}{\lambda} = \frac{(6.63 \times 10^{-34}\,\text{Js})(3 \times 10^{8}\,\text{m/s})}{505 \times 10^{-9}\,\text{m}} = 3.94 \times 10^{-19}\,\text{J}$$

From Table 6-10 the work function for cesium is
$$W = 2.1\,\text{eV} = 2.1 \times 1.6 \times 10^{-19}\,\text{J} = 3.36 \times 10^{-19}\,\text{J}$$

$$hf = W + \frac{1}{2}m_e\upsilon^2 \quad \rightarrow \quad hf - W = \frac{1}{2}m_e\upsilon^2 = 5.8 \times 10^{-20}\,\text{J}$$

The photon of a given energy will eject an electron from the surface of cesium metal and the energy of the ejected electron will be 0.36 eV.

Example 6.14 Cut-off frequency and photoelectric effect

Calculate the cut-off frequency for cesium and plot the kinetic energy of ejected electron versus frequency of photons.

The cut-off (threshold) frequency is the lowest frequency, or the longest wavelength, that permits photoelectrons to be ejected from the surface of a metal. At this frequency the photoelectrons have zero kinetic energy:

$$hf = W + \frac{1}{2}m_e\upsilon^2 \quad \Rightarrow \quad hf_c = W \quad \Rightarrow \quad f_c = 5.07 \times 10^{14}\,\text{Hz}$$

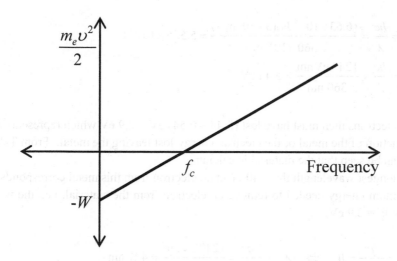

Figure 6-21. Kinetic energy of the ejected electron versus frequency of photons (the slope of the curve is always equal to the Planck's constant, *h*)

Since the kinetic energy of electrons is equal to the stopping potential, eV_0, in the experimental measurements (Fig. 6-21), Einstein's equation for the photoelectric effect can be written in the following alternative form:

$$eV_0 = hf - W \qquad (6\text{-}40)$$

Example 6.15 Stopping potential and photoelectric effect

Assume that a potential of 0.54 V is required to stop all the electrons in a photoelectric experiment. Calculate the maximum electron kinetic energy and determine the material (comparing the work function with the values listed in Table 6-10) if the incident photons have the wavelength of 360 nm. What is the longest wavelength that will eject any electron from this metal?

A potential of 0.54 V stops all of the electrons and thus the maximum kinetic energy of the electrons must be equal to the kinetic energy equivalent of a potential of 0.54 V, that is

$$\left(m_e\upsilon^2/2\right)_{max} = eV_0 = (1.6 \times 10^{-19}\,\text{C}) \times (0.54\text{V}) = 0.864 \times 10^{-19}\,\text{J} =$$

$$= (0.864 \times 10^{-19}\,\text{J})\frac{\text{eV}}{1.6 \times 10^{-19}\,\text{J}} = 0.54\,\text{eV}$$

The incident photons thus have an energy of

$$hf = \frac{hc}{\lambda} = \frac{(6.63 \times 10^{-34}\,\text{Js})(3 \times 10^{8}\,\text{m/s})}{360 \times 10^{-9}\,\text{m}} = 5.525 \times 10^{-19}\,\text{J}$$

$$hf = \frac{hc}{\lambda} = \frac{1240\,\text{eVnm}}{360\,\text{nm}} = 3.44\,\text{eV}$$

The electrons then must have lost $(3.44 - 0.54)$ eV = 2.9 eV which represents the work function of the metal or the electron energy lost leaving the metal. From Table 6-10 it can be seen that the material is calcium.

The longest wavelength that will eject an electron from this metal corresponds to the minimum energy needed to remove an electron from the material, i.e., the work function, $W = 2.9$ eV:

$$hf_c = \frac{hc}{\lambda_{max}} = W \quad \Rightarrow \quad \lambda_{max} = \frac{hc}{W} = \frac{1240\,\text{eVnm}}{2.9\,\text{eV}} = 427\,\text{nm}$$

In the photoelectric effect, the incoming photon is absorbed through interaction with an orbital electron (for example the K-shell electron, see Chapter 2). The process can be sketched as shown in Fig. 6-22. If the photon energy is above the work function, the orbital electron will be ejected from an atom. The vacancy is then filled by an electron from an outer shell and this produces either fluorescence X-rays (as indicated in Fig. 6-22) or the Auger electrons. The probability for X-ray emission is given by the fluorescence yield and, for K-shell electrons, varies from 0.005 for $Z = 8$ to 0.965 for $Z = 90$. During the photoelectric absorption of light by an atom, one quantum (photon) is absorbed by one of the orbital electrons. The orbital electron is ejected such that the incoming photon energy, hf, and the binding energy of electron, E_b, are distributed between the recoil atom and ejected electron:

$$E_{pe} = hf - E_b \qquad (6\text{-}41)$$

Virtually all of this energy is carried away by the ejected electron (also called the *photoelectron*), E_{pe}, because the electron has much smaller mass than the recoil atom.

The electron binding energy, E_b, depends on its orbit (shell) and assumes discrete values (see Chapter 2)

$$E_b = E_K, E_{L1}, E_{L2}, E_{L3}, E_{M1}, \ldots \qquad (6\text{-}42)$$

Nearly all photoelectric events in light nuclei involve K-shell electrons. Binding energy in the K-shell varies from 13.6 eV for hydrogen to 7.11 keV for iron and 88 keV in lead. The cross section for the photoelectric effect thus depends on the binding energy of the electrons in different materials. Figure 6-23 illustrates the cross sections for various photon interactions in aluminum and lead.

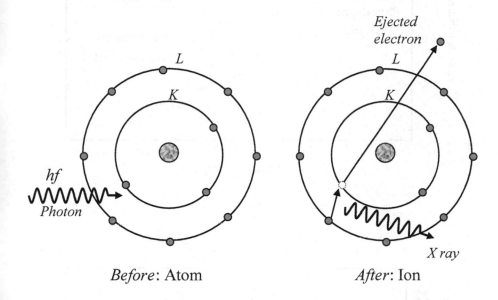

Before: Atom   *After*: Ion

Figure 6-22. Photoelectric effect

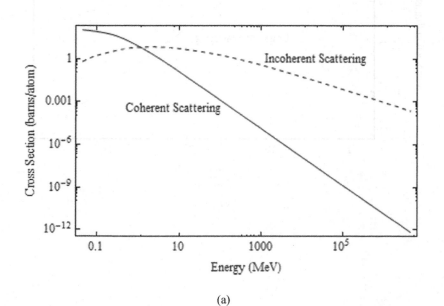

(a)

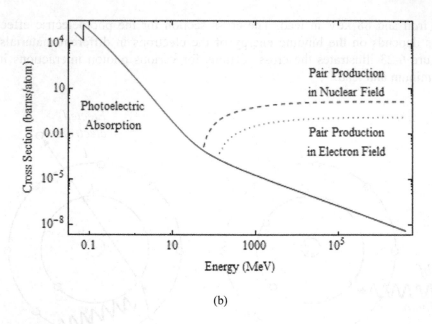

(b)

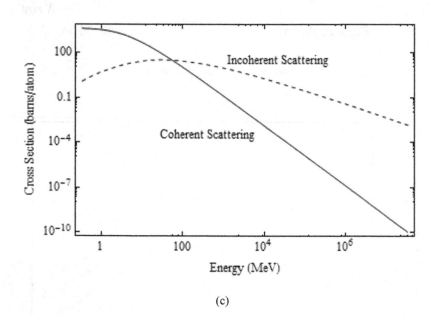

(c)

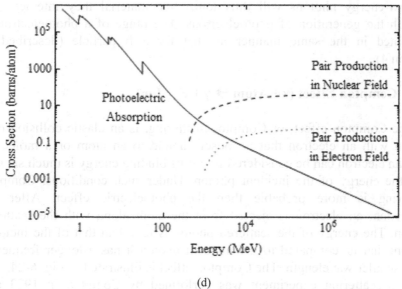

(d)
Figure 6-23. (**a**) Scattering cross sections of photon interactions with aluminum;
(**b**) photoelectric effect and pair production cross sections of photon interactions with
aluminum; (**c**) scattering cross sections of photon interactions with lead; (**d**) photoelectric
effect and pair production cross sections of photon interactions with lead

The probability for the photoelectric effect for a given orbital electron is maximum if $hf = E_b$ and it is zero, that is, the photoelectric effect can not occur, when $hf < E_b$. As the incident photon energy increases above E_b, the probability for the photoelectric effect decreases. This trend can be observed in Fig. 6-23 for any of the indicated edges (peaks). The edges correspond to the electron shells, K, L, M, etc. For energies below 150 keV, the cross section varies as $(hf)^{-3}$. Above 150 keV but below 5 MeV, it varies as $(hf)^{-2}$ and at energies above 5 MeV it becomes proportional to $(hf)^{-1}$.

The atomic cross section for the photoelectric effect is proportional to Z^m, where m depends on the incident photon energy. For a 100 keV photon, $m = 4$ and for a photon energy of 3 MeV, $m = 4.6$. Thus, the cross section for the photoelectric effect is strongly dependent on the photon energy as well as on Z, the atomic number of the material:

$$\sigma_{ph} \propto \begin{cases} \dfrac{Z^5}{(hf)^{7/2}} & \text{for low photon energies} \\[3mm] \dfrac{Z^{4.5}}{(hf)} & \text{for } 0.1 \text{ MeV} < hf < 5 \text{ MeV} \end{cases}$$

(6-43)

Low-energy photons will thus ionize the material they interact with through the generation of photoelectrons. The range of a photoelectron is calculated in the same manner as that for a β particle (described in Section 4).

5.2.2 Compton Effect (γ + Atom → γ + e⁻ + Ion)

The Compton effect, or Compton scattering, is an elastic collision of a photon with an electron that is loosely bound to an atom or a molecule. Such an electron can be considered free if its binding energy is much smaller than the energy of the incident photon. Under such conditions, Compton scattering is more probable than the photoelectric effect. After the interaction, an electron is ejected from the atom along with the scattered photon. The energy of the scattered photon is less than that of the incident photon; that is, compared to the incident photon it has a longer frequency and a smaller wavelength. The Compton effect is illustrated in Fig. 6-24.

This scattering experiment was performed by Compton in 1923 and showed that light had a corpuscular nature as well as wave-like characteristics. This conclusion was mainly due to the difference in wavelength between the incident and the scattered photons. This change in wavelengths could not be explained by the wave theory of light alone. Compton analyzed the experimental results by adopting Planck's hypothesis of considering light as an energy quanta and assigned energy values of $E = hf$ to the photons. Accordingly, the momentum of a massless particle is given by $p = h/\lambda$. Consequently, Compton assumed the incident photon to be equivalent to a particle with mass $m = hf/c^2$. By the conservation of energy (see Fig. 6-24),

$$hf + m_e c^2 = E + hf'$$
(6-44)

and by the conservation of momentum

$$\vec{p} = \vec{p'} + \vec{p_e}$$
(6-45)

where $p = E / c$ for photons and $p_e = m_e \upsilon$ for the electron. Squaring this equation and using the scalar product (see Fig. 6-25) gives

$$p_e^2 = \left(\vec{p} - \vec{p'}\right)\left(\vec{p} - \vec{p'}\right) = p^2 + p'^2 - 2pp'\cos\theta$$
(6-46)

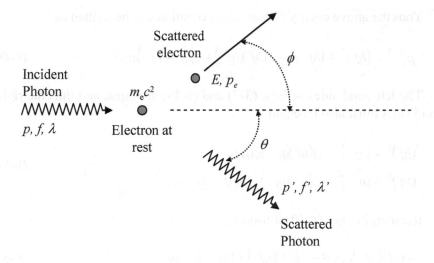

Figure 6-24. Compton scattering

$$\vec{a} \cdot \vec{b} = ab\cos\theta$$

Figure 6-25. Scalar product of two vectors

Multiplying the above equation by c^2 and replacing the momentum of photons with $p = E/c = hf/c$, the momentum conservation equation becomes

$$p_e^2 c^2 = (hf)^2 + (hf')^2 - 2(hf)(hf')\cos\theta \qquad (6\text{-}47)$$

The energy conservation equation can be squared and rewritten in the following way:

$$hf - hf' + m_e c^2 = E$$
$$(hf - hf')^2 + 2(hf - hf')m_e c^2 = E^2 - m_e^2 c^4 \qquad (6\text{-}48)$$

Recall from Chapter 3, Eq. (3-7), the energy–momentum relation for a relativistic particle is

$$E^2 = (p_e c)^2 + (m_e c^2)^2$$

Thus the above energy conservation equation can be written as

$$p_e^2 c^2 = (hf)^2 + (hf')^2 - 2(hf)(hf') + 2(hf - hf')m_e c^2 \qquad (6\text{-}49)$$

The left-hand sides of Eqs. (3-7) and (6-49) are equal and thus the right-hand sides must also be equal:

$$(hf)^2 + (hf')^2 - 2(hf)(hf')\cos\theta =$$
$$(hf)^2 + (hf')^2 - 2(hf)(hf') + 2(hf - hf')m_e c^2 \qquad (6\text{-}50)$$

Rearranging Eq. (6-50) it follows

$$-(hf)(hf')\cos\theta = -(hf)(hf') + (hf - hf')m_e c^2 \qquad (6\text{-}51)$$

$$\cos\theta = 1 - \frac{(hf - hf')}{(hf)(hf')} m_e c^2 \qquad (6\text{-}52)$$

Finally, the Compton scattering formula is

$$\frac{1}{hf'} - \frac{1}{hf} = \frac{1 - \cos\theta}{m_e c^2} \quad \text{or} \quad \lambda' - \lambda = \Delta\lambda = \frac{h}{m_e c}(1 - \cos\theta) \qquad (6\text{-}53)$$

This shows that the wavelength change of the incoming photon in a Compton scattering event depends only on the scattering angle for a given target particle. The constant in the Compton formula above can be calculated explicitly as

$$\frac{h}{m_e c} = \frac{hc}{m_e c^2} = \frac{1240 \text{ eV nm}}{0.511 \times 10^6 \text{ eV}} = 0.00243 \text{ nm} \qquad (6\text{-}54)$$

and is called the *Compton wavelength for the electron*. It corresponds to the wavelength of a photon that has energy equal to the rest mass of an electron. With this taken into account, the Compton scattering formula for an electron can be written as

$$\lambda' = \lambda + \Delta\lambda = \lambda + \frac{h}{m_e c}(1 - \cos\theta) = \lambda + 0.00243 \text{ nm}(1 - \cos\theta) \qquad (6\text{-}55)$$

From this equation it can be concluded that

- the change in wavelength, $\Delta\lambda$, does not depend on the wavelength of the incident photon
- for higher energy photons, the wavelength decreases such that the same change in wavelength corresponds to a larger difference in energies
- for low photon energies, the energy difference, $(hf) - (hf')$, is small, while for high photon energies (for example order of MeV) the electron may receive over 75% of energy of the incoming photon
- the change in wavelength, $\Delta\lambda$, depends only on the electron scattering angle
- the change in wavelength, $\Delta\lambda$, is independent of the medium.

The following relation between the scattering angles of the photon and the recoil electron may be determined from Fig. 6-24:

$$\cot\frac{\theta}{2} = \left(1 + \frac{hf}{m_e c^2}\right)\tan\phi \qquad (6\text{-}56)$$

Example 6.16 Energy of Compton scattered photons

For a photon of energy $hf = 200$ keV, that is Compton scattered on electron through an angle of 45°, calculate the energy and frequency of the scattered photon as well as the energy and the momentum of the recoil electron.

Applying the Compton scattering formula it follows that

$$\lambda' = \lambda + \Delta\lambda = \lambda + \frac{h}{m_e c}(1 - \cos\theta) = \lambda + 0.00243 \text{ nm}(1 - \cos\theta)$$

$$\frac{\lambda'}{hc} = \frac{\lambda}{hc} + \frac{(1 - \cos\theta)}{m_e c^2} \quad \rightarrow \quad \frac{1}{E'} = \frac{1}{E} + \frac{(1 - \cos\theta)}{m_e c^2}$$

$$\frac{1}{E'} = \frac{1}{200 \text{ keV}} + \frac{(1 - \cos 45°)}{511 \text{ keV}} \quad \rightarrow \quad E' = (hf') = 179 \text{ keV}$$

The frequency of the scattered photon is

$$f' = \frac{179 \text{ keV}}{h} = \frac{(179 \times 10^3 \text{ eV})(1.6 \times 10^{-19} \text{ J/eV})}{6.63 \times 10^{-34} \text{ Js}} = 4.32 \times 10^{19} \text{ Hz}$$

The energy given to a recoil electron is

$$\Delta E = hf - hf' = 200 - 179 = 21 \text{ keV}$$

The total energy and the momentum of the recoil electron are

$$E = \Delta E + m_e c^2 = 21 \text{ keV} + 511 \text{ keV} = 532 \text{ keV}$$

$$E^2 = \left(p_e c \right)^2 + \left(m_e c^2 \right)^2 \quad \rightarrow$$

$$p_e = \frac{\sqrt{E^2 - m_e^2 c^4}}{c} = \frac{144.94 \times 10^3 \text{ eV}}{3 \times 10^8 \text{ m/s}} \times 1.6 \times 10^{-19} \text{ J/eV} \quad \rightarrow$$

$$p_e = 7.73 \times 10^{-23} \text{ kg m/s}$$

The following is an analysis of some aspects of the Compton scattering:

1. **The dependence of the scattered photon energy on incident photon energy and photon scattering angle (θ):**
   The energy of the scattered photon ($E' = hf'$) depends on the energy of incident photon ($E = hf$) and the scattering angle, (θ) as

$$\frac{1}{hf'} - \frac{1}{hf} = \frac{1 - \cos\theta}{m_e c^2} \quad \rightarrow \quad E' = \frac{E\left(m_e c^2 \right)}{m_e c^2 + E\left(1 - \cos\theta \right)} \tag{6-57}$$

If the ratio of energies, E'/E, is plotted against the incident photon energy for various scattering angles (see Fig. 6-26), the following can be observed:

- For incoming photon energies smaller than ~ 50 keV, the energy of the scattered photon is nearly equal to that of the incident photon. At these low-incident photon energies, Compton scattering is similar to Rayleigh (coherent) scattering in which the energy of the scattered photon remains unchanged (the scattered photon is only deflected).
- In the case of complete forward photon scattering ($\theta = 0$), the energy of the incident photon is unchanged.
- Compton scattering is an efficient interaction type for reducing photon energy at large scattering angles. The maximum reduction in photon energy is obtained for backscattering, at $\theta = 180°$.

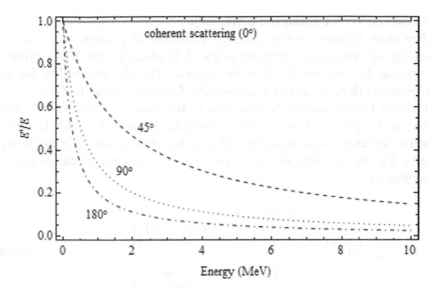

Figure 6-26. Ratio of scattered to incident photon energy versus incident photon energy and photon scattering angle in the Compton Effect

2. The angular distribution of the scattered photons and recoil electrons:

The scattering angle for the recoil electron varies from 0° to 90°. The maximum energy that a recoil electron can obtain in Compton scattering is in a head-on collision in which the electron scatters at nearly zero angle (it continues its trajectory in straight line of the impact photon) and the photon is scattered backward at the angle of 180°. The minimum energy that a recoil electron receives is during the collision in which the photon trajectory is constant (scattering angle of a photon is zero) while the electron scatters at nearly 90°. The probability for an electron to be scattered at an angle of zero increases with incident photon energy as shown in Table 6-11. This table further shows that for a given photon scattering angle, the recoil electron scattering angle decreases with increasing incident photon energy.

Table 6-11. Angular distribution of recoil electron (ϕ) as a function of incident photon energy (hf) and photon scattering (θ)

hf (MeV)	$\theta = 1°$	$\theta = 5°$	$\theta = 10°$	$\theta = 30°$	$\theta = 60°$	$\theta = 90°$	$\theta = 120°$	$\theta = 150°$
0.01	89.49	87.45	84.90	74.72	59.52	44.44	29.52	14.72
0.1	89.40	87.01	84.03	72.24	55.38	39.91	25.77	12.63
1.0	88.52	82.64	75.50	51.61	30.36	18.68	11.05	5.18
10	79.82	48.07	29.06	10.28	4.81	2.78	1.61	0.75
100	30.22	6.64	3.33	1.09	0.50	0.29	0.17	0.08

3. **Cross section for Compton scattering:**

The cross section for the Compton scattering of photons with incident energy hf, through a scattering angle θ, is given by the Klein–Nishina formula. In general, the formula suggests that the probability for the Compton effect to occur for an element Z is proportional to Z/hf. If the incident photon energy is measured in the units of electron rest mass, i.e., $\alpha = hf/m_e c^2$, then the Klein–Nishina formula for total Compton scattering cross section (integrated over the photon scattering angle θ to give the energy dependence of the cross section per electron) can be written as

$$\sigma_{Compton} = 2\pi r_e^2 \left\{ \begin{array}{l} \dfrac{1+\alpha}{\alpha^2}\left[\dfrac{2(1+\alpha)}{1+2\alpha} - \dfrac{1}{\alpha}\ln(1+2\alpha)\right] \\[3mm] + \dfrac{1}{2\alpha}\ln(1+2\alpha) - \dfrac{1+3\alpha}{(1+2\alpha)^2} \end{array} \right\} \tag{6-58}$$

where r_e is the classical electron radius, also called the Compton radius, and is defined as the radius, r_e, of a sphere which has charge e and electrostatic potential energy, U, equal to the rest mass energy of the electron. That is

$$U = \frac{ke^2}{r_e} = m_e c^2 \tag{6-59}$$

Solving for the electron radius

$$r_e = \frac{ke^2}{m_e c^2} = \frac{(8.987\times10^9\,\text{Nm}^2/\text{C}^2)(1.6\times10^{-19}\,\text{C})^2}{(0.511\times10^6\,\text{eV})(1.6\times10^{-19}\,\text{Nm/eV})} \tag{6-60}$$

$$= 2.8\times10^{-15}\,\text{m}$$

The total Compton scattering cross section plotted against incident photon energy is shown in Fig. 6-27. This shows that the probability of Compton scattering decreases with increasing incident photon energy.

5.2.3 Correction for Bound Electrons and Coherent (Rayleigh) Scattering

Compton scattering is valid under the assumption that the electron is free. This assumption is only applicable when the binding energy of the electron is much smaller than the energy of the incident photon. When the incident photon energy is comparable to the electron binding energy, a more complicated, semi-empirical relation must be used to evaluate the incoherent scattering. Such scattering interactions occur for low-incident photon energies, small photon scattering angles and highly bound electrons (electrons in the inner shells of an atom). Thus, the scattering cross section for bound electrons decreases at low photon energies. The dominant interaction at low photon energies is the photoelectric effect. The effect of electron binding energy thus becomes negligible at these energies and the error introduced by neglecting the binding energy is small.

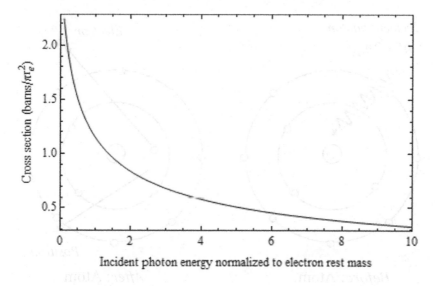

Figure 6-27. Total Compton scattering cross section versus incident photon energy

In competition with the incoherent scattering of photons by individual electrons is coherent (Rayleigh) scattering. When low-energy photons scatter at a small angle in a high-Z medium, the energy transferred to an electron is so small that even excitation of that atom is not possible. The energy is thus absorbed by the entire atom and even this small amount of energy will cause an atom to recoil. The energy loss of the incoming photon is considered to be negligible. Figure 6-23 shows that coherent scattering cross sections greatly exceed the incoherent scattering at low photon energies in a high-Z medium. In radiation shielding calculations, however, this type of scattering is usually neglected since the dominant method of energy attenuation is through photoelectric effect.

5.2.4 Pair Production (γ + Atom $\rightarrow$ e$^+$ + e$^-$ + Atom)

In this process, the incident photon is absorbed and an electron–positron pair is created (Fig. 6-28). The photon generates this electron–positron pair in the Coulomb field of the nucleus and this interaction has a photon threshold energy that is equal to the rest mass energies of two electrons, $2m_ec^2 = 1.022$ MeV. The same interaction can occur in the Coulomb field of an electron, with a threshold energy of $4m_ec^2 = 2.044$ MeV. The probability of pair production in the electric field of nucleus is however significantly higher and is the only interaction analyzed further.

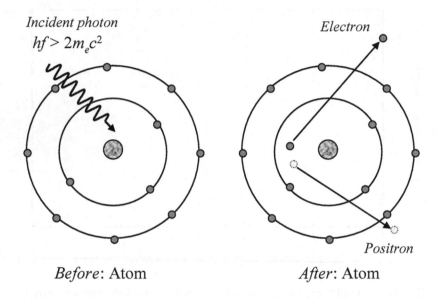

Figure 6-28. Pair production

In the pair production,
- *The total charge is conserved*: a photon with zero electric charge generates a pair which also has a total charge of zero (electron with negative charge and positron with positive charge)
- *According to the energy conservation law*:

  Incident photon energy = Energy used to generate the pair + Kinetic energy of the positron + Kinetic energy of the electron

  $$hf = 2m_ec^2 + T^+ + T^- = 1.022 + T^+ + T^- \quad \text{(MeV)} \tag{6-61}$$

- *According to the momentum conservation law*: pair production cannot take place in an empty space because some third entity must absorb the momentum ($p=h/\lambda =hf/c$) of the initial photon. The photon momentum is usually absorbed by an atomic nucleus. The following example shows why it is impossible for a photon to transfer all of its energy to a free electron. Consider a photon of energy hf and momentum hf/c. If the photon was to transfer all of its energy to an electron of mass m and velocity υ, then from the conservation of energy

$$hf = \frac{1}{2}m\upsilon^2$$

and conservation of momentum

$$h\frac{f}{c} = m\upsilon$$

Eliminating hf from these two equations gives

$$\upsilon = 2c$$

which is an impossible result since no particle can travel faster than light.

The electron and positron have energies equal to the difference between the initial photon energy and $2m_ec^2$. The energy spectra of the emitted electron and positron are continuous and are very similar to one another. The scattering angles of the positron and electron as well as the angular dependence on photon energy are complex and not easy to describe. This is due to the involvement of the nucleus in the momentum distribution after the interaction. For very high photon energies, the average scattering angle of the electron and the positron is proportional to 0.511 MeV/hf (MeV).

The total cross section for pair production per atom divided by Z^2 is graphically shown for various materials in Fig. 6-29. It can be observed that the cross section for all of the elements does not significantly change for incoming photon energies up to 10 MeV. For higher energies, the cross section for different materials starts to depart. The cross section for pair production in the electric field of an electron is also depicted in Fig. 6-23.

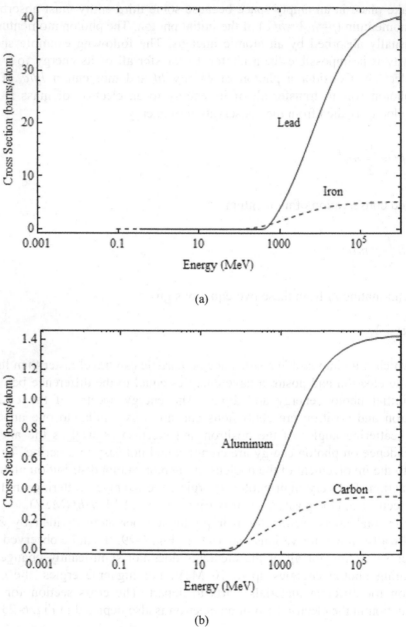

(a)

(b)

Figure 6-29. Cross section for pair production in different materials (cross section for pair production in the Coulomb field of the nucleus is divided by Z^2, cross section for the pair production in the Coulomb field of an electron is divided by Z): (**a**) in lead and iron; (**b**) in aluminum and carbon

The cross section depends on Z of the material and thus it can be considered important for low-Z media. The cross section also increases with incident photon energy. This dependence is nearly logarithmic

$$\sigma_{pp} \propto Z^2 \ln(hf) \qquad (6\text{-}62)$$

The positron that is formed very quickly disappears in an annihilation process that involves another electron. Annihilation ($e^+e^- \rightarrow \gamma$), which is the inverse of pair production ($\gamma \rightarrow e^+e^-$), occurs when a positron encounters an electron. The energy conservation equation for an annihilation event is

$$2m_ec^2 + K^+ + K^- = 2hf \qquad (6\text{-}63)$$

The first term represents the rest energy of the electron–positron pair, the second and third terms are the kinetic energies of the positron and electron before the collision and the term on the right-hand side represents the energy of the two photons created in the reaction, each having the same frequency f and energy hf. According to the energy conservation law, the value of hf must be at least $m_ec^2 = 0.511$ MeV.

NUMERICAL EXAMPLE

Photon Attenuation in Common Shielding Materials
Aluminum and lead are two materials commonly used in high-energy photon shielding. Using the data in Table 6-8, construct an attenuation plot of a 1 MeV photon beam passing through aluminum ($\rho = 2.7$ g/cm^3) and lead ($\rho = 11.34$ g/cm^3). Comment on the apparent effect of the material density on gamma attenuation (Fig. 6-30).

Solution in MATLAB
```
clear all
% Mass attenuation coefficients at 1MeV
mu_Al = 0.06146; %cm^2/g
mu_Pb = 0.0757;
rho_Al = 2.7; %g/cm^3
rho_Pb = 11.34;
mu = [mu_Al*rho_Al mu_Pb*rho_Pb];
x = linspace(0,30);
for j = 1:2
    for i = 1:100
        I(i,j) = exp(-mu(j)*x(i));
    end
end
```

```
figure
hold on
plot(x,I(:,1),'k')
plot(x,I(:,2),'k:')
xlabel('Distance (cm)')
ylabel('Fractional Intensity')
legend('Aluminum','Lead')
```

Results show that higher-*Z* materials are more effective as photon shields.

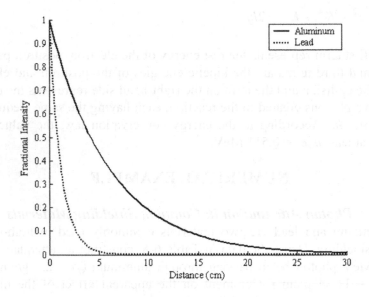

Figure 6-30. Attenuation of a 1 MeV photon beam in aluminum and lead

PROBLEMS

6.1. Expain whether α particles produce bremsstrahlung radiation?

6.2. Estimate the source energy of bremsstrahlung radiation from the lead container shielding 5 mCi source of ^{32}P (maximum β energy is 1.71 MeV). How much heat will be deposited in the wall of the container?

6.3. An energy of 35.5 eV is required to produce an ion pair. Estimate how many ion pairs are produced by a α particle with a 1.0 MeV kinetic energy and how much total charge is produced?

6.4. If the ionization potential of air is 33.9 eV, how many ion pairs are produced by a 5 MeV α particle? How many ion pairs would produce an β particle of the same energy? What would be the linear ranges of these two particles?

6.5. Determine the range of 2 MeV α particle in aluminum. Compare the value with that given in Table 5-3.

6.6. The half-value shielding layer is 5 mm thick. Calculate the shieled thickness to attenuate 99.912% of the incoming γ radiation.

6.7. Use the data from Table 5-8 to calculate the relative number of 1 MeV and 0.1 MeV γ rays that emerge from 15 cm thick water tank. Assume that the γ ray beam consists of equal number of both when entering the water tank.

6.8. Calculate the thickness of air, water and aluminum that will stop 20% of a beam of 1.5 MeV γ rays.

6.9. In the table http://physics.nist.gov/PhysRefData/contents.html find the values needed to determine and calculate the fraction of energy in a 30 keV X-ray beam deposited in 5 mm of soft tissue.

6.10. A γ ray (1.46 MeV) from ^{40}K is scattered through an angle of 30° and then again through an angle of 150°. Calculate the energy of γ ray after second scattering.

6.11. If the light of wavelength 400 nm is incident on a metal with a work function 5.5 V, calculate the external voltage that must be applied to the metal to have the electrons released from its surface?

6.12. A completely ionized carbon nucleus is accelerated through a potential difference of 7,000 V. What is the final kinetic energy of the carbon?

5.13. If the work function of a material is 10 eV, what is the lowest frequency photon that can cause electrons to be ejected?

6.14. For a 200 keV Compton photon scattered at 45° calculate its energy and the magnitude and direction of the momentum of the recoil electron.

6.15. Calculate and plot the linear and mass ranges of α particle, proton and electron as a function of energy in water, aluminum, lead and graphite.

6.16. Determine the linear energy loss resulting from the passage of a 0.1 MeV β particle through graphite (density = 2.25 g/cm³). Calculate the mass stopping power and the relative (to air) mass stopping power.

6.17. From http://physics.nist.gov/PhysRefData/contents.html determine the minimum energy that a proton must have to penetrate 30 cm of tissue (density 1 g/cm³), the approximate thickness of the human body. Using the same table calculate how much energy an α particle needs to penetrate 1 cm of the tissue layer.

6.18. In a Compton scattering experiment it is found that the fractional change in the wavelength is 1.0% when the scattering angle is 60°. What was the wavelength of the incident photons, and what would be the wavelength of the photons scattered through an angle of 90°?

6.19. Plot the Compton scattering energy of scattered beam (hf'/hf) for the initial photon energies of 0.05, 0.1, 0.2, 0.5, 1.0, 2.0, 5.0 and 10.0 MeV as a function of photon scattering angle.

6.20. Calculate the necessary shielding (glass with density of 2.23 g/cm³ and plastic with density of 1.03 g/cm³) to completely stop the β particles from ³H. The maximum β particle energy is 0.019 MeV and the average energy is 0.0057 MeV.

6.21. Tabulate the cut-off frequency for elements given in Table 6-10. Calculate the kinetic energy of ejected electrons.

6.22. Write the computer code to compute the maximum range of a proton in aluminum, air, silicon and water for the range of energies from 0.001 eV to 1 GeV.

6.23. Use the Bethe–Block formula and write the computer code to calculate energy loss of an α particle and proton in varying the ratio of particle velocity to the speed of light from zero to one.

6.24. Repeat the previous problem with electrons. What can you conclude from the results?

6.25. Discuss the head-on collision of charged particles.

6.26. Describe the inelastic scattering of charged particles with electrons.

6.27. How does inelastic scattering of charged particles take place with the nucleus?

6.28. Explain the bremsstrahlung radiation and define the bremasstrahlung hazard.

6.29. Explain the condition for electron–positron annihilation process.

6.30. Write Eq. (6-9) for non-relativistic α particles.

6.31. Use the equation from Problem 6.30 and plot $S' \equiv const \times (-dE/dx)$ versus Z for H, He, Al, Cu and Au.

6.32. Calculate the maximum energy of β spectrum in ^{14}C decay?

6.33. Review Chapter 3 to write the Fermi relation for nucleus radius. Then review Chapter 4 to write the wavelength of the γ ray. The so-called reduced wavelength is the wavelength divided by 2π. Comment on the ratio of the nucleus radius to the reduced wavelength of the emitted γ ray.

6.34. Find the decay of ^{49}Cr. Calculate the range of emitted particles.

6.35. Find the decay of ^{51}Cr. Calculate the range of emitted particles. Compare the decay of ^{51}Cr to the decay of ^{49}Cr (Problem 6.34).

6.36. Derive the equation for the kinetic energy in eV with which the nucleus recoils when a γ ray is emitted [*Hint*: Read APPLICATIONS at the end of Chapter 3 to obtain $(hf)^2/2Mc^2$].

6.37. Knowing that the emission of a characteristic Auger electron is more probable for lighter elements than heavy elements (electrons are more tightly bound to the nucleus) explain what the Auger electron spectroscopy is and what would be its use in science and technology.

6.38. Estimate the tritium (T) activity (Bq) in 100 kg person if the human body is about 10% hydrogen by weight and in the human body the $T:H = 3.3 \times 10^{-18}$.

6.39. What is the necessary thickness of aluminum to shield against 1.3 MeV electrons and what is the necessary thickness to shield against 20 MeV protons?

6.40. The photoeffect obviously does not occur with a free electron. By sketching the photon and photoelectron energy distributions as a function of momentum, comment on the reason why the photoelectric effect cannot physically take place with the unbounded electron.

6.41. First show at what energy a photon can lose at most one-half of its energy in Compton scattering with an electron. Then calculate the Al thickness to attenuate this energy to 90%.

6.42. Show how to calculate the thickness of air that will stop 15% of a beam of γ rays of known energy.

6.43. The α particles from the decay are detected in an ion chamber. If each α particle loses on average energy equal to E to ionize the medium, what would be the total charge that flows in the chamber, C, if elementary charge is q and the initial energy of α particle is T?

Chapter 7

NEUTRON PHYSICS
Interactions, Fission and Cross Sections

...I feel that I ought to let you know of a very sensational new development in nuclear physics. In a paper in the Naturwissenschaften Hahn reports that he finds when bombarding uranium with neutrons the uranium breaking up into two halves giving elements of about half the atomic weight of uranium. This is entirely unexpected and exciting news for the average physicist. The Department of Physics at Princeton, where I spent the last few days, was like a stirred-up ant heap. Apart from the purely scientific interest there may be another aspect of this discovery, which so far does not seem to have caught the attention of those to whom I spoke. First of all it is obvious that the energy released in this new reaction must be very much higher than in all previously known cases. It may be 200 million (electron-) volts instead of the usual 3–10 million volts. This in itself might make it possible to produce power by means of nuclear energy, but I do not think that this possibility is very exciting, for if the energy output is only two or three times the energy input, the cost of investment would probably be too high to make the process worthwhile. Unfortunately, most of the energy is released in the form of heat and not in the form of radioactivity.

I see, however, in connection with this new discovery potential possibilities in another direction. These might lead to a large-scale production of energy and radioactive elements, unfortunately also perhaps to atomic bombs. This new discovery revives all the hopes and fears in this respect which I had in 1934 and 1935, and which I have as good as abandoned in the course of the last two years...
Leo Szilard (1898– 1964) in his letter to Luis Strauss on January 25, 1939.

T. Jevremovic, *Nuclear Principles in Engineering*,
DOI 10.1007/978-0-387-85608-7_7, © Springer Science+Business Media, LLC 2009

1.   INTRODUCTION

Neutrons together with protons are the constituents of atomic nuclei. The neutron was discovered after more than two decades of speculation that electrically neutral particles exist in atoms (see Chapter 3). Because the neutron is electrically neutral, it easily interacts with nuclei and does not interact directly with electrons. Since the nucleus of an atom is about one ten-thousandth the size of the electron cloud, the chance of neutrons interacting with a nucleus is very small, allowing them to travel long distances through matter. As a free particle, the neutron is an important and yet unique *tool* used for various applications: in medicine to initiate powerful nuclear interactions whose products can directly destroy cancer cells (neutron capture therapy for example), for research on physical and biological materials, for imaging through easy allocation of light atoms especially hydrogen, to investigate properties of magnetic materials (neutrons possess a magnetic moment and thus act as small magnets), to track atomic movement (thermal neutron energies almost directly coincide with the energies of atoms in motion) and to maintain the fission chain reaction in nuclear reactors. Free neutrons are unstable (see Chapter 3) and decay in short time through β^- decay process into a proton, electron and neutrino. However, free neutrons will most likely interact with the surrounding matter and disappear through nuclear interactions long before they decay.

2.   NUCLEAR INTERACTIONS

A nuclear reaction involves interactions between nuclear particles (nucleons, nuclei); the outcome of which are other nuclear particles or γ rays. Assuming, for simplicity only two initial and two produced particles, a nuclear reaction is usually written as follows (Fig. 7-1):

$A + B \rightarrow C + D$

Usually particles A and C are light and B and D heavy.
Every nuclear interaction must obey the following laws:
- *Conservation of nucleons*: the total number of nucleons before and after a nuclear reaction is not changed
- *Conservation of charge*: the sum of the charges of all particles involved in the reaction before and after must be preserved
- *Conservation of linear and angular momentum*: the total momentum of interacting particles before and after the reaction is not changed

- *Conservation of energy*: energy, including the rest mass energies of particles, is not changed by a nuclear reaction.

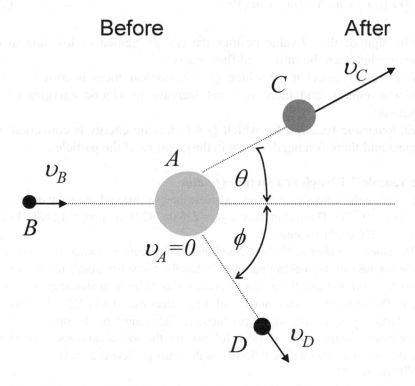

Figure 7-1. Schematics of a typical nuclear reaction seen in the laboratory system

The law of conservation of energy can be also used to predict whether a certain interaction is energetically possible. For the assumed interaction of particle A with particle B that produces two particles C and D, the sum of energies before and after the interaction takes into account the kinetic energies (E) and rest mass energies (mc^2) of each individual particle:

$$E_A + E_B + m_A c^2 + m_B c^2 = E_C + E_D + m_C c^2 + m_D c^2 \qquad (7\text{-}1)$$

Equation (7-1) may be rewritten as

$$(E_A + E_B) - (E_C + E_D) = [(m_A + m_B) - (m_C + m_D)]c^2 \qquad (7\text{-}2)$$

showing that the change in kinetic energies of the particles involved in a reaction is equal to the change in their rest mass energies. The change in rest mass energies of the particles involved in the reaction is known as the

Q-value of the reaction:

$$Q = [(m_A + m_B) - (m_C + m_D)]c^2 \qquad (7\text{-}3)$$

The sign of the *Q*-value defines the energy gained or lost due to the difference between the initial and final masses:

- *Exothermic reaction* for which $Q > 0$: nuclear mass is converted into kinetic energy and there is a net increase in kinetic energies of the particles.
- *Endothermic reaction* for which $Q < 0$: kinetic energy is converted into mass and there is a net decrease in the energies of the particles.

Example 7.1 Nuclear reaction *Q*-value

Complete the following reaction, calculate the *Q*-value and comment on its sign ^{235}U (*n*, ?) ^{236}U. The rest masses: $m_{235} = 235.0439231$ amu, $m_n = 1.0086649$ amu and $m_{236} = 236.0455619$ amu.

The atomic number of ^{235}U is 92 and that of the neutron is zero. The sum of the atomic numbers of the incident particles is thus 92. Since the atomic number of ^{236}U is also 92, it follows that the produced particle should have an atomic number equal to zero. The total atomic mass number of the incident particles is $235 + 1 = 236$. The total atomic mass number of the produced particles must be the same. Since the atomic mass number of ^{236}U is 236 it follows that the additional particle has atomic mass number zero. It therefore follows that the other particle is a γ ray:

^{235}U (n, γ) ^{236}U

For this reaction the *Q*-value is

$$Q = \left[(m_{235} + m_n) - (m_{236} + m_\gamma) \right] c^2 = 0.0070261 \, \text{amu}$$
$$Q = 0.0070261 \times 931.5 = 6.54 \, \text{MeV} > 0$$

thus showing that the reaction is exothermic.

Example 7.2 Nuclear reaction energy threshold

For the nuclear reaction ^{12}C + ^{14}N → ^{10}B + ^{16}A plot the dependence of the boron nucleus energy versus energy of the incident carbon nuclei for various scattering angles θ (Fig. 7-1).

Based on the Eq. (7-3) the reaction *Q*-value is −4.4506 MeV. From Fig. 7-1 it follows that the conservation of momentum can be expressed as

$$\sqrt{2m_A E_A} + 0 = \sqrt{2m_C E_C} \cos\theta + \sqrt{2m_D E_D} \cos\phi$$
$$\sqrt{2m_C E_C} \sin\theta = \sqrt{2m_D E_D} \sin\phi$$

giving Eq. (7-1) of the form

$$Q = E_C\left(1 + \frac{m_C}{m_D}\right) - E_A\left(1 - \frac{m_A}{m_D}\right) - \frac{2}{m_D}\sqrt{m_A m_D E_A E_D}\,\cos\theta$$

that can be solved for

$$\sqrt{E_D} = \frac{1}{m_A + m_D} \times$$

$$\left[\sqrt{m_A m_D E_A}\,\cos\theta \pm \sqrt{m_A m_D E_A \cos^2\theta + (m_A + m_D)\left[E_A(m_D - m_A) + Qm_D\right]}\right]$$

and can be plotted as a function of the energy of the incoming particle and the scattering angle.

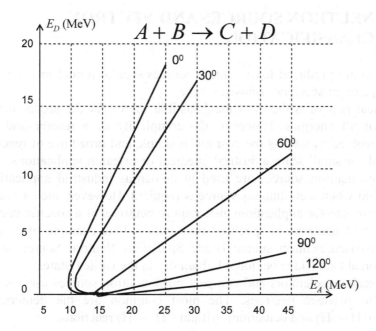

The schematic plot of the energy change as given helps us indicate the following:

- Since the Q-value for the reaction is negative there is a threshold for the incident particle as a function of the scattering angle below which the nucleus C (in this case boron) is not observed. The threshold energy is therefore

$$E_{tr} = \frac{-Qm_D(m_A + m_D)}{m_A m_D \cos^2 \theta + (m_A + m_D)(m_D - m_A)}$$

with the minimum value for $\theta = 0$ giving the absolute threshold value of the reaction (the smallest value of the incident particle energy for the reaction to take place). For this case the threshold energy is 8.26 MeV.

- The curves for all values of scattering angles cut the horizontal axis at the same point for which the energy of the particle C (boron in this case) is zero. That point (of 17.82 MeV for this reaction) is obtained as follows

$$E_A = \frac{-Qm_D}{m_D - m_A}$$

3.   NEUTRON SOURCES AND NEUTRON CLASSIFICATION

Neutrons are produced from neutron sources such as a nuclear reactor, a radioisotope or an accelerator-based source.

A nuclear reactor is the most inexhaustible source for the production of neutrons of all energies. However, the complexity of a reactor and the systems involved as well as the cost make simple and broad use of reactors impractical for small-scale industrial, medical or research applications. The radioisotope neutron sources are used in numerous industrial applications and are ideal when a continuous source is required. However, such a source is not appropriate for applications that require neutrons of a specific energy or emission of neutrons in specified time pulses. One example of a large accelerator-based neutron source is the Spallation Neutron Source under construction at Oak Ridge National Laboratory in the United States.

Small-scale accelerators and compact pulse neutron sources use nuclear reactions to produce neutrons. The most common are the deuterium–deuterium ($^2H - {}^2H$) and deuterium–tritium ($^2H - {}^3H$) reactions:

$^3H\,({}^2H, n)\,{}^4He;\ Q = 17.59$ MeV   and $^2H\,({}^2H, n)\,{}^3He;\ Q = 3.27$ MeV

These reactions produce 14.1 MeV and 2.5 MeV neutrons, respectively. Pulse neutron sources (also called pulse neutron generators) have found a number of applications in science, industry, medicine and technology. To name a few

– *Real-time analysis of bulk materials*: Materials such as cement and coal moving on conveyor belts are examples of bulk materials that are extensively examined by applying fast and thermal neutron beams for activation analyses. The purpose of such analysis is to measure the content and the amount of the elements present in the material. For example, the information obtained from neutron activation analysis of cement enables the optimal combination of raw material constituents as well as verification of chemical consistency. Another example is the application of neutron activation tests in on-line measurements of sulfur and the content of other elements in coal which are important for predicting its combustion efficiency and environmental impact.

– *Detection of explosive, chemical and nuclear materials*: Such materials may be accurately detected for fast security checks of airline cargo or other unknown packages.

– *Medical applications*: An accurate and simple measurement of the body's fat is achieved using neutron pulse generators. The measurement is based on neutron interactions with carbon and oxygen. By examining the quantity and distribution of carbon and oxygen, it is possible to evaluate the health of individuals with respect to obesity, aging and cardiovascular disease.

A very special interaction that results in a high production rate of neutrons of various energies is the interaction of an α particle with a beryllium atom:

$$^{9}\text{Be}\,(\alpha, n)\,^{12}\text{C} \qquad Q = 5.75 \text{ MeV}$$

Since the Coulomb repulsion force between the beryllium nucleus and the incoming α particle is not high, this reaction is very suitable for neutron production. The α particles are emitted through the radioactive decay of isotopes such as ^{226}Ra, ^{222}Rn, ^{210}Po, ^{239}Pu and ^{241}Am. Beryllium is the only naturally occurring isotope of beryllium and thus a neutron source utilizing this element is easy to realize, namely, powders of both beryllium and the α emitter are mixed together in ratios from 20:1 to 300:1 and the mixture is encapsulated. Such sources constantly emit neutrons and the energy spectrum is usually complex because decay products have different α energies and thus produce neutrons with different energies. Figure 7-2 depicts the neutron energy spectrum emitted from an americium–beryllium (AmBe) neutron source. Neutrons can also be produced in the reaction of γ rays with targets most commonly made of beryllium or deuterium (for example heavy water). Such reactions are referred to as photoneutron sources. The binding energy of neutrons in these light elements is low and a large amount of energy is therefore not required for the reaction to occur:

^{9}Be (γ, n) ^{8}Be; $Q = 1.63$ MeV   and ^{2}H (γ, n) ^{1}H; $Q = 2.23$ MeV

Neutrons produced by photodisintegration of nuclei are monoenergetic and such sources are reproducible (in terms of neutron energy). The most common sources of γ rays used for these interactions are the γ rays emitted in radioactive decays of ^{24}Na (E_γ=2.8 MeV, $T_{1/2}$=15 h) or ^{124}Sb (E_γ=1.67 MeV, $T_{1/2}$=60.9 days).

Neutrons are classified according to their energies because their interactions with matter are energy dependent. The most common classification is shown in Table 7-1.

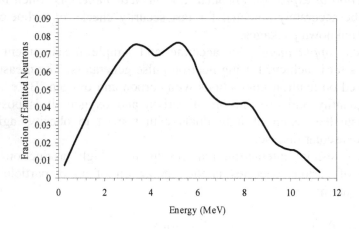

Figure 7-2. Typical neutron spectrum from an americium-beryllium source

Table 7-1. Classification of neutrons

Neutron energy	Name
0–0.025 eV	Cold
0.025 eV	Thermal
0.025 eV–0.4 eV	Epithermal
0.4 eV–0.6 eV	Cadmium
0.6 eV–1 eV	Epicadmium
1 eV–10 eV	Slow
10 eV–300 eV	Resonance
300 eV–1 MeV	Intermediate
1 MeV–20 MeV	Fast
> 20 MeV	Relativistic

Example 7.3 Nuclear reaction that revealed the existence of neutron

As described in Chapter 3, Rutherford was the first to correctly predict the existence of a neutral particle as a constituent of the nucleus as early as 1920. That idea has inspired many scientists around the world to start the search for other constituents of nuclei. Two German scientists, Bothe and Becker, studied the interaction that is today commonly used to produce neutrons: $^9Be(\alpha,n)^{12}C$. In their experiment they discovered that nearly 5 cm of lead reduced the radiation emerging from the reaction and attributed this phenomenon incorrectly to γ rays. Now, consider the same interaction and assume that a neutron produced in that interaction has energy of 5.3 MeV. Calculate the energy of the recoil proton if such a neutron encountered a head-on collision with a paraffin block (assume the collision is with a proton only).

Before the interaction, the neutron of mass m_n had a velocity υ_{n1} while the velocity of the proton of mass m_p was zero. After the interaction, the neutron moves with velocity υ_{n2} and the proton recoils with velocity υ_p. According to the law of conservation of energy

$$\frac{1}{2}m_n\upsilon_{n1}^2 = \frac{1}{2}m_n\upsilon_{n2}^2 + \frac{1}{2}m_p\upsilon_p^2 \tag{7-4}$$

The conservation of momentum for the head-on collision (see Chapter 3) gives

$$m_n\upsilon_{n1} = m_n\upsilon_{n2} + m_p\upsilon_p \tag{7-5}$$

Equations (7-4) and (7-5) can be simplified and combined assuming the mass of a proton is nearly equal to that of a neutron to give

$$\upsilon_p = \upsilon_{n1} \quad \upsilon_{n2} = 0 \tag{7-6}$$

This result shows that in a head-on collision a neutron is stopped by a proton, transferring all of its energy to the target. In our example, therefore, the energy of the target proton after the reaction is equal to the energy of the incident neutron, or 5.3 MeV.

4. NEUTRON ATTENUATION

4.1 Concept of the Cross Section

Microscopic cross section. The quantitative description of nuclear interactions requires known neutron cross section data. A rate at which a particular neutron interaction with a given target material will occur depends on the neutron energy and speed, as well as the nature of the target nuclei. The cross section of a target material for any given reaction thus represents the probability of a particular interaction and is a property of the nucleus and incident neutron energy. In order to introduce the concept of a neutron cross section, consider a parallel monoenergetic neutron beam falling on thin target of thickness x and area A, as shown in Fig.7-3. The intensity of the incident neutron beam is described with the number of neutrons per unit volume, n, and their velocity, υ, as

$$I_0 = n\upsilon \; [(\text{neutrons/cm}^3) \cdot (\text{cm/s}) = \text{neutrons/cm}^2/\text{s}] \qquad (7\text{-}7)$$

The total number of nuclei in the target of atomic density N is

Total number of nuclei in target $= NAx$ \qquad (7-8)

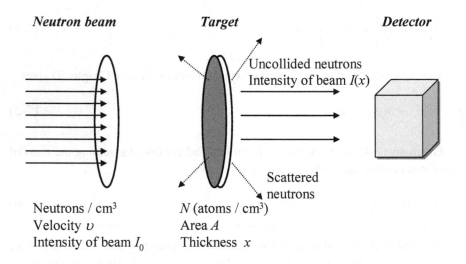

Figure 7-3. Concept of neutron microscopic cross section

The number of neutrons that collide with the target nuclei is proportional to the neutron beam intensity and the total number of nuclei in the target:

Number of neutron collisions per second in the whole target $= \sigma I_0 NAx$ \quad (7-9)

where σI_0 represents the number of neutron collisions with the single target's nuclei per unit time and σ is referred to as the effective cross sectional area, frequently called the *microscopic cross section*. It follows

$\sigma =$ number of neutron collisions per unit time with one nucleus per unit intensity of the incident neutron beam

The neutron microscopic cross section thus represents a visible area and for some interactions is closely equal to an actual area, πR^2 (see Example 7.4). The accepted unit of microscopic cross sections is the barn (b), which is equal to 10^{-24} cm^2. All neutron cross sections are functions of neutron energy and the nature of the target nucleus.

Macroscopic cross section. The probability of a neutron undergoing an interaction in the target as sketched in Fig. 7-3 is equal to the ratio of the reaction area to the total area:

$$\Sigma x = [\text{Reaction area}]/[\text{Total area}] \tag{7-10}$$

The reaction area of the target (of volume Ax) is defined as the number of nuclei in the target material, NAx, multiplied by the area of each nucleus, σ

$$\Sigma x = \frac{N\sigma Ax}{A} = N\sigma x \tag{7-11}$$

Thus, the relation between the microscopic (σ) and *macroscopic cross section* (Σ) is

$$\Sigma = N\sigma \ [\text{cm}^{-1}] \tag{7-12}$$

The number of nuclei in a target material made of a single element (also called the number density), N, is obtained from (see Chapter 2):

$$N = \frac{N_a \times \rho}{A} \tag{7-13}$$

where A is the atomic mass number and N_a is Avogadro's number.

Example 7.4 Microscopic and macroscopic cross sections for a single isotope

Calculate the microscopic cross section based on geometrical area and estimate the macroscopic cross section for ^{54}Fe, which has a density of 7.86 g/cm^3. Use the following empirical relation to estimate the radius of the nucleus, R: $R = (1.4 \, A) \times 10^{-16}$ m.

The microscopic cross section is estimated based on the nuclear radius calculated from the Fermi model of the nucleus (see Chapter 3):

$$R = 1.4A \times 10^{-16} \text{ m} = 1.4 \times 54 \times 10^{-16} \text{ m} = 75.6 \times 10^{-16} \text{ m}$$

$$\sigma = \pi R^2 = \pi \left(75.6 \times 10^{-16} \text{ m}\right)^2 = 1.79 \times 10^{-28} \text{ m}^2 = 1.79 \text{ b}$$

Figure 7-4 shows the neutron microscopic cross sections for ^{54}Fe and ^{55}Mn. It can be seen that the estimate is close to the measured value. The same empirical formula can be used for ^{55}Mn to estimate the microscopic cross section.

The number density of ^{54}Fe is

$$N = \frac{N_a \times \rho}{A} = \frac{(6.023 \times 10^{23})(7.86)}{54} = 8.77 \times 10^{22}\,\text{nuclei/cm}^3$$

The macroscopic cross section is thus $\Sigma = N\sigma = 0.157\text{cm}^{-1}$.

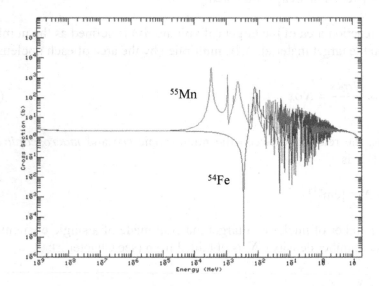

Figure 7-4. Microscopic cross section for neutron elastic scattering on ^{55}Mn and ^{54}Fe [from http://atom.kaeri.re.kr/]

Example 7.5 Microscopic and macroscopic cross sections for a mixture of elements

Calculate the microscopic and macroscopic absorption cross sections for natural uranium if $\sigma_{235} = 681$ b and $\sigma_{238} = 2.7$ b. The density of uranium is 19 g/cm^3.

The abundances (ε) of ^{238}U and ^{235}U in natural uranium (neglecting small amounts of ^{234}U) are 99.28% and 0.72%, respectively. The number densities are

$$N_{238} = \varepsilon_{238}\frac{N_a \times \rho}{A_{238}} = 0.9928\frac{(6.023 \times 10^{23})(19)}{238} = 4.77 \times 10^{22}\,\text{nuclei/cm}^3$$

$$N_{235} = \varepsilon_{235} \frac{N_a \times \rho}{A_{235}} = 0.0072 \frac{(6.023 \times 10^{23})(19)}{235} = 3.50 \times 10^{20} \text{nuclei/cm}^3$$

The macroscopic and microscopic cross sections of natural uranium are

$$\Sigma = N_{235}\sigma_{235} + N_{238}\sigma_{238} = 0.367 \text{cm}^{-1} \quad \text{and} \quad \sigma = \frac{\Sigma}{N_{235} + N_{238}} = 7.64\text{b}$$

Differential scattering cross section. The particular microscopic cross sections related to various neutron interactions with various materials are described in Section 4.5. It is of use, however, to introduce here the general definition of the scattering reaction and define the differential scattering cross section. In the scattering reaction as described in Section 4.5.1 the neutron will experience a great change in the direction of motion and its energy in reacting with light nuclei. While the microscopic cross section gives the information about the probability that such an interaction will occur with the given target nuclei at the given neutron energy, the information about the change in neutron energy, direction and scattering angle is not provided. The spatial distribution of scattered neutrons and their energy change (such as neutron slowing down in thermal reactors) is of great importance in nuclear reactor analysis and design. The *differential cross section* is therefore introduced to describe all these parameters. Figure 7-5 (a) defines the differential solid angle describing the neutron direction of motion. The scattering cross section that describes the probability that a scattering of a neutron with a nucleus will change its energy from E to E' is denoted as $\sigma_s(E \rightarrow E')$, Fig. 7-5 (b). It represents the distribution involving range of energies. Since by definition $\sigma_s(E)$ represents the scattering microscopic cross section or the probability that a neutron of energy E will be scattered from a given nucleus, it is apparent that it also represents the integral of the differential scattering cross section in the energy range from zero to infinity:

$$\sigma_s(E) = \int_0^\infty \sigma_s(E \rightarrow E') dE'$$

The scattering cross section that describes the probability that a scattering of a neutron with a nucleus will change its direction from $\vec{\Omega}$ to $\vec{\Omega'}$ is denoted as $\sigma_s(\vec{\Omega} \rightarrow \vec{\Omega'})$, Fig. 7-5 (b). The integral over the space will give the microscopic scattering cross section at the given energy:

$$\sigma_s\left(\vec{\Omega}\right)= \int_{4\pi}\sigma_s\left(\vec{\Omega}\rightarrow\vec{\Omega}'\right)\!d\vec{\Omega}'$$

It is important to notice that the scattering of a neutron on any nucleus in nuclear reactor applications will not basically depend on the incoming neutron direction. Therefore, the differential scattering cross section will not depend on the incoming neutron direction of motion but it will depend on the scattering direction (or the change of the direction). This dependence is described as a function of scattering angle (θ) as shown in Fig. 7-5 (b) or the cosine of the scattering angle ($\mu = \cos\theta$) as derived in Chapter 8 (Section 3.1). The so-called double differential scattering cross section is obtained by combining the energy and direction dependence of the scattering cross section as follows:

$$\sigma_s\left(E\rightarrow E',\vec{\Omega}\rightarrow\vec{\Omega}'\right)$$

that is related to the energy and direction-dependent differential scattering cross sections as follows:

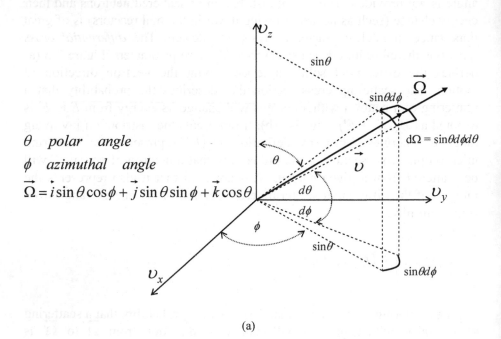

θ polar   angle

ϕ azimuthal   angle

$\vec{\Omega}=\vec{i}\sin\theta\cos\phi+\vec{j}\sin\theta\sin\phi+\vec{k}\cos\theta$

(a)

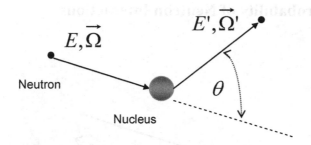

(b)

Figure 7-5. (a) Differential solid angle; (b) change in energy and direction of motion in neutron scattering

$$\sigma_s(E \to E') = \int_{4\pi} \sigma_s\left(E \to E', \vec{\Omega} \to \vec{\Omega'}\right) d\vec{\Omega'}$$

$$\sigma_s\left(E, \vec{\Omega} \to \vec{\Omega'}\right) = \int_0^\infty \sigma_s\left(E \to E', \vec{\Omega} \to \vec{\Omega'}\right) dE'$$

$$\sigma_s(E) = \int_{4\pi} \int_0^\infty \sigma_s\left(E \to E', \vec{\Omega} \to \vec{\Omega'}\right) dE' \, d\vec{\Omega'}$$

Although the calculation of such cross section in reactor physics is difficult and usually not done straightforwardly, there is a possibility to evaluate them for the case of neutron elastic scattering on stationary nuclei. For that purposes usually the differential scattering cross section is expressed as a product of the scattering cross section and the probability of neutron to be scattered in a given range of energies:

$$\sigma_s(E \to E') = \sigma_s(E) F(E \to E')$$

The probability $F(E \to E')$ can be evaluated analytically for a low-energy neutron interaction through potential scattering with light nuclei. The derivation is provided in Chapter 8, Section 3.1.

4.2   Probability of Neutron Interactions

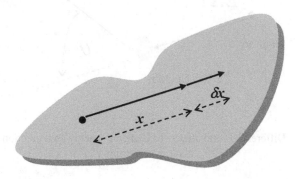

Figure 7-6. Neutron travel in a homogeneous medium

Neutrons travel with constant direction and speed until they interact with the medium. Considering only a homogeneous medium, the probability of a neutron interacting is a function of the distance at which a neutron will interact, x. This probability can be expressed as a MacLaurin series of distance x as

$$P_{reaction}(x) = a + bx + cx^2 + dx^3 + ... \tag{7-14}$$

where a, b, c, d, ... are the coefficients of expansion. Since the interaction of a neutron cannot occur at zero distance, the first term is equal to zero. For a sufficiently small distance δx (as sketched in Fig. 7-6), the series given by Eq. (7-14) reduces to

$$P_{reaction}(\delta x) = b \delta x \tag{7-15}$$

The probability that a neutron *will not* interact along the distance δx is

$$P_{non\text{-}reaction}(\delta x) = 1 - b \delta x \tag{7-16}$$

Since every interaction is independent of the previous interaction, the probability that a neutron will not interact along the distance $x + \delta x$ can be written as a product of two probabilities

$$P_{non\text{-}reaction}(x + \delta x) = P_{non\text{-}reaction}(x) P_{non\text{-}reaction}(\delta x)$$
$$= P_{non\text{-}reaction}(x)[1 - b \delta x] \tag{7-17}$$

Rearranging terms it follows

$$\frac{P_{non-reaction}(x+\delta x)-P_{non-reaction}(x)}{\delta x}=-bP_{non-reaction}(x) \qquad (7\text{-}18)$$

Taking the limit as $\delta x \to 0$ and replacing the constant b with Σ (macroscopic cross section)

$$\frac{dP_{non-reaction}(x)}{dx}=-\Sigma P_{non-reaction}(x) \qquad (7\text{-}19)$$

Integrating Eq. (7-19) gives the probability that a neutron does not interact and the probability that a neutron does interact along the distance x

$$P_{non-reaction}(x)=e^{-\Sigma x} \qquad P_{reaction}(x)=1-e^{-\Sigma x} \qquad (7\text{-}20)$$

The macroscopic cross section in the above equation is replaced with the linear attenuation coefficient (μ_l) in the case of γ ray attenuation (Chapter 6).

Example 7.6 Probability of neutron interactions
Calculate the probability that a neutron will travel 5 cm in a block of ^{54}Fe (see Example 7.4 for other data) without an interaction. What is the probability that the neutron will interact with the medium between 5 cm and 5.5 cm?
From Example 7.4, the macroscopic cross section for ^{54}Fe is $\Sigma = 0.157$ cm^{-1}. The probability of traveling 5 cm without an interaction is

$$P_{non-reaction}(5 \text{ cm})=e^{-\Sigma x}=e^{-0.157\times 5}=0.456$$

In order to calculate the probability of having an interaction between 5 cm and 5.5 cm, we first calculate the probability of traveling an additional 0.5 cm without interaction, or

$$P_{non-reaction}(5.5 \text{ cm})=e^{-\Sigma x}=e^{-0.157\times 5.5}=0.422$$

Thus the probability of a neutron interacting between 5 cm and 5.5 cm is

$$P_{non-reaction}(5 \text{ cm})-P_{non-reaction}(5.5 \text{ cm})=0.456-0.422=0.034$$

Alternatively, the product of two probabilities may be used: the probability that a neutron will not interact along the first 5 cm of travel and the probability that it will interact in the next 0.5 cm

$$P_{non\text{-}reaction}\left(5\text{ cm}\right) \times P_{reaction}\left(0.5\text{ cm}\right) = e^{-0.157 \times 5} \times \left[1 - e^{-0.157 \times 0.5}\right] =$$
$$= 0.456 \times \left[1 - 0.924\right] = 0.034$$

4.3   Neutron Mean Free Path

The neutron mean free path represents the average distance it travels in a medium without interacting. It is obtained from the probability that neutron will interact in the distance interval between x and $x+dx$. The probability is equal to the product of these two probabilities (Section 4.2 and Example 7.6)

- The probability that a neutron will not interact along the distance x

$$P_{non\text{-}reaction}\left(x\right) = e^{-\Sigma x} \tag{7-21}$$

- The probability that a neutron will interact along the distance dx

$$P_{reaction}\left(dx\right) = \Sigma dx \tag{7-22}$$

Therefore it follows that $P(x)dx$ is the probability that the neutron will pass through the medium at distance x and have an interaction along distance dx, i.e., between x and $x + dx$

$$P(x)dx = e^{-\Sigma x}\Sigma dx$$

If the medium is infinite the total probability of interaction will be equal to one meaning that the neutron will interact somewhere with absolute certainty

$$\int_0^\infty P(x)dx = \int_0^\infty e^{-\Sigma x}\Sigma dx = -e^{-\Sigma x}\Big|_0^\infty = 1$$

The mean free path has a continuous value and can be obtained by integrating the product of probabilities assuming the length of neutron travel can extend from zero to infinity (the probability $P(x)$ is used as a weighting factor)

$$\lambda \equiv \bar{x} = \frac{\displaystyle\int_0^\infty x e^{-\Sigma x} \Sigma dx}{\displaystyle\int_0^\infty e^{-\Sigma x} \Sigma dx} = \frac{1}{\Sigma} \qquad (7\text{-}23)$$

Example 7.7 Neutron mean free path

Calculate the mean free path and the time needed for a neutron with energy 100 eV to have its first interaction in a block of ^{54}Fe (see Example 7.4 for other data). The neutron mass is provided in Appendix 2.

The neutron mean free path is equal to the reciprocal of the macroscopic cross section of the medium, and the neutron velocity is obtained from its energy:

$$\lambda = \frac{1}{\Sigma} = \frac{1}{0.157 \text{cm}^{-1}} = 6.37 \text{cm}$$

$$\upsilon = \sqrt{\frac{2T}{m_n}} = \sqrt{\frac{2 \times 10^2 \times 1.6 \times 10^{-19}}{1.67492716 \times 10^{-27}}} = 1.38 \times 10^5 \text{m/s}$$

The time to the first interaction is therefore

$$t = \frac{\lambda}{\upsilon} = \frac{0.0637 \text{m}}{1.38 \times 10^5 \text{m/s}} = 0.46 \mu\text{s}$$

4.4 Reaction Rate and Concept of Neutron Flux and Neutron Current

In all situations involving the evaluation of neutron behavior the goal is to analyze neutron population as a whole and almost never the history of a single neutron. For the majority of applications (like neutron population behavior in nuclear reactors, transport of neutrons through shielding materials, or in biological media), it is important to determine the *neutron reaction rates*. A neutron interacts with the nuclei of a medium through scattering from one nucleus to another until it is absorbed or it escapes the boundary of a system. The mean free path that a neutron travels before it interacts can be defined as the mean free path for scattering, $\lambda_s = 1/\Sigma_s$, and for absorption, $\lambda_a = 1/\Sigma_a$. The total mean free path is thus equal to $\lambda_{tot} = 1/\Sigma_{tot}$. The *neutron density*, $n(\vec{r},t)$, representing an average number of neutrons in a given volume of a reactor core, is a required variable in determining the neutron reaction rates. The neutron density is as of now defined as spatially and time-dependent variable. Let's first assume that all neutrons in a reactor core have the same velocity υ, then the frequency with which neutron will experience a given interaction type is defined as $\Sigma_i \upsilon$, called the *interaction frequency*. The reaction rate per unit volume of the target material and unit time for an ith type of interaction is therefore

$$R_i\left(\vec{r},t\right)= \upsilon\Sigma_i n\left(\vec{r},t\right)= \frac{\Phi\left(\vec{r},t\right)}{\lambda_i} = \Phi\left(\vec{r},t\right)\Sigma_i = \Phi\left(\vec{r},t\right)N\sigma_i \qquad (7\text{-}24)$$

where $\Phi\left(\vec{r},t\right)$ represents the total distance that neutron travels in unit time and unit volume of a given target material. This variable is also called the *scalar neutron flux* and has a unit of a number of neutrons per unit time and unit area, neutrons/cm^2/s. Since the neutrons in a reactor core have different velocities and different energies, the neutron density, neutron flux, and neutron reaction rates depend on neutron energy (i.e., velocity)

$$R_i\left(\vec{r},E,t\right)= \upsilon\Sigma_i(E)n\left(\vec{r},E,t\right)= \frac{\Phi\left(\vec{r},E,t\right)}{\lambda_i} =$$
$$\Phi\left(\vec{r},E,t\right)\Sigma_i(E)= \Phi\left(\vec{r},E,t\right)N\sigma_i(E) \qquad (7\text{-}25)$$

Notice that the cross sections are also becoming energy dependent (Section 4.1).

It is now appropriate to comment on all the variables that define the neutron population existence in a reactor core. Namely a neutron in a reactor core is fully characterized with its position, $\vec{r}$, energy E (or its velocity that is related to its energy through $E = m\upsilon^2/2$), time t at which it is observed, and its direction of motion, $\vec{\Omega}$. The variables are sketched in Fig. 7-7. Thus, the most detailed quantities describing the overall behavior of neutron population in a reactor core are time-dependent angular neutron density and time-dependent angular neutron scalar flux as follows

Time-dependent angular neutron density $n\left(\vec{r},E,\vec{\Omega},t\right)$ and
time-dependent angular neutron flux
$$\Phi\left(\vec{r},E,\vec{\Omega},t\right)= \upsilon n\left(\vec{r},E,\vec{\Omega},t\right) \qquad (7\text{-}26)$$

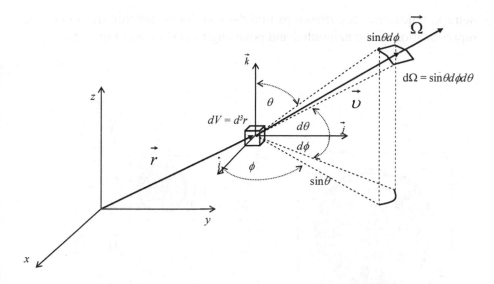

Figure 7-7. Spatial variables defining neutron position in nuclear reactor at a given time

The angular neutron density represents the average number of neutrons in a given volume dV about r, with energy dE about E and direction of motion Ω in solid angle $d\Omega$ at a given time t (Fig. 7-7). The time-dependent angular neutron flux gives the most detailed average description of the status of neutron population in a given time and given volume involving seven variables: three spatial coordinates, energy, two angle directions and time; the unit is neutrons/cm^2 s steradian per energy unit. The angular neutron flux represents the number of neutrons passing through an area of unit surface (1 cm^2) about r perpendicular to Ω, with energy dE about E and direction of motion Ω in solid angle $d\Omega$ at a given time t. In reactor analysis the angular flux is not usually a variable studied in all details but it is a starting variable in determining the reaction rates.

The physical interpretation of the angular neutron flux is now discussed using the example of a unit cell typical for thermal power reactors (refer to Chapter 8 for details on heterogeneous reactors and the meaning of boundary conditions). The visualizations are created to show the difference of the angular flux tendencies in the fuel region and in the moderator region as shown in Fig. 7-8. The spatial angular flux distribution is shown for two spatial locations, one in a fuel region and the other in a moderator region for the reflective (Fig. 7-8 (b)) and vacuum boundary conditions (in which case all neutrons leaving the unit cell boundary are considered lost). The angular flux distribution at the selected locations in the unit cell from Fig. 7-8 is discussed along different azimuthal and polar angles along which the neutron trajectories are *drawn* to find the solution to neutron transport. The representation of the azimuthal and polar angles is shown in Fig. 7-8 c).

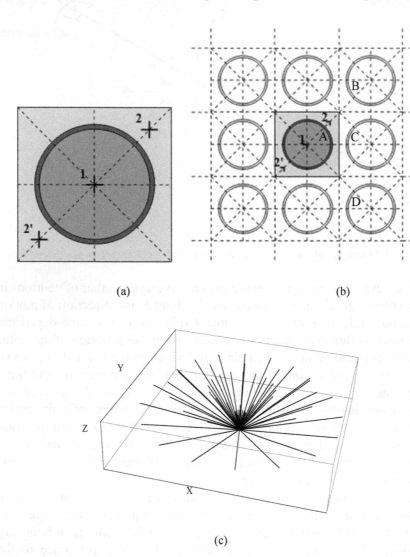

(a)   (b)

(c)

Figure 7-8. (a) Reactor unit cell (location 1 is in the fuel region, and locations 2 and 2' are in the moderator region), (b) interpretation of reflective boundary conditions [Refer to Chapter 8], (c) representation of azimuthal and polar directions of neutron trajectories

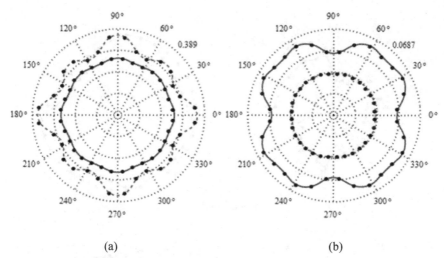

(a) (b)

Figure 7-9. Angular flux distribution along 32 azimuthal and 2 polar directions (angles of 60°
and 16°) in the unit cell from Fig. 7-8 (a) (location 1; reflective boundary conditions): (**a**) fast
neutrons, (**b**) thermal neutrons [Images created using AGENT, Jevremovic, 2004]

The angular flux distribution for the fast and thermal neutrons in the fuel
region is shown in Fig. 7-9. Since the fast neutrons are generated in the fuel,
the magnitude of the angular flux along the directions of closest distance
among the fuel pins (as indicated in Fig. 7-8 (b) as A–C direction) is
larger compared to other directions (such as A–B or A–D as indicated in
Fig. 7-8 (b)). The thermal neutrons are absorbed in moderator and thus
higher depressions in angular flux are observed along the directions of the
closest distance among the fuel pins. The difference between the polar
directions is explained later. Figure 7-10 shows the angular flux distributions
at two symmetric locations in the moderator region. Fast neutrons are born
in the fuel region and they move toward the moderator region; thermal
neutrons are created in the moderator region (due to the slowing down of
fast neutrons emerging from the fuel region) and they move toward the fuel
region (to cause more fission and therefore more fast neutrons). Thus the
larger magnitude of the fast neutron angular flux is along the A→2 direction
(or A→2' direction) since the A fuel pin is the closest to the selected
locations 2 and 2'. Due to a larger volume of a moderator along B→2 than
A→2 a larger thermal neutron angular flux is observed along the 225°
azimuthal direction than along 45°.

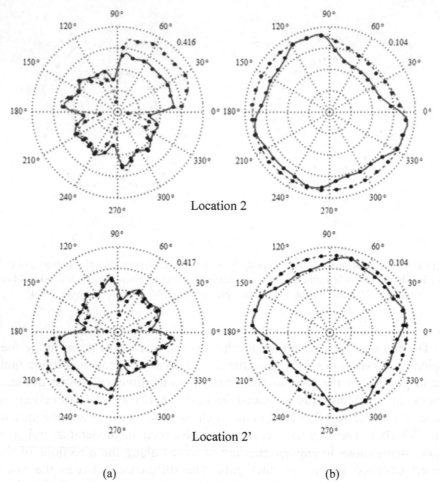

Location 2

Location 2'

(a)   (b)

Figure 7-10. Angular flux distribution along 32 azimuthal and two 2 directions (angles of 60°
and 16°) in the unit cell from Fig. 7-8 (a) (locations 2 and 2' with the reflective boundary
conditions show the symmetry in flux distribution): **(a)** fast neutrons, **(b)** thermal neutron,
[Images created using AGENT, Jevremovic, 2004]

Changing the reflective into vacuum boundary conditions, the magnitude
and the spatial distribution of the angular flux change as shown in Fig. 7-11.
At the center of the unit cell (location 1), the fast neutron angular flux is
almost isotropic as a result of isotropic neutron scattering in the fuel (heavy
nuclei) and isotropic emission of neutrons from the fission (Section 5). The
thermal neutron flux, however, depends on the amount of a moderator
present toward the unit cell boundary, thus the highest fluxes are observed
along 45°, 135°, 225° and 315°, while the lowest fluxes are seen along 0°,
90°, 180° and 270° directions. At the location 2, the direction of fast neutron

transport is from the cell toward the moderator, while the direction and magnitude of the thermal neutron angular flux is directly related to the amount of a moderator present in the direction toward selected location 2. From Figs. 7-9, 7-10 and 7-11 we could also see that the angular flux values differ for different polar angle.

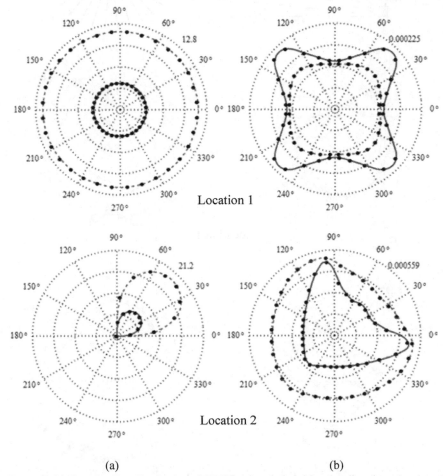

(a)   (b)

Figure 7-11. Angular flux distribution along 32 azimuthal and 2 polar directions (angles of 60° and 16°) in the unit cell from Fig. 7-7 (a) (locations 1 and 2 with the vacuum boundary conditions): (**a**) fast neutrons, (**b**) thermal neutrons, [Images created using AGENT, Jevremovic, 2004]

Figure 7-12 compares the angular flux at the locations 1 and 2 in the unit cell for eight different polar angles.

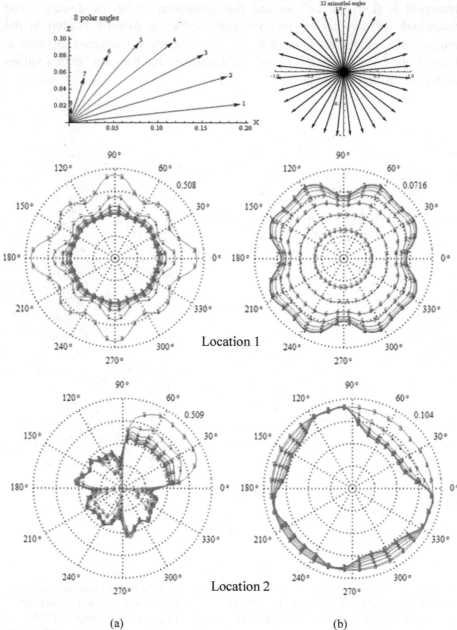

Figure 7-12. Angular flux distribution along 32 azimuthal and 8 polar directions in the unit cell from Fig. 7-8 (a) (locations 1 and 2 with the reflective boundary conditions); polar and azimuthal angle distributions are also shown; (**a**) fast neutrons, (**b**) thermal neutrons [Images created using AGENT, Jevremovic, 2004]

Because the small polar angle produces the longer neutron trajectory, the angular fast neutron flux in the fuel region (location 1) will increase with reduced polar angles. However, because the fuel is the absorber of thermal neutrons, the longer track in the fuel will produce the stronger absorptions of thermal neutrons and thus the thermal neutron angular flux at smaller polar angles will be lower, which is inversed compared to the fast neutrons. At location 2, the fast neutron angular flux is larger for small polar angles, because the larger fast neutron flux is coming from the fuel region. The thermal neutrons are generated in the moderator through elastic scattering, so the longer track in the moderator will generate more thermal neutrons causing thermal neutron angular flux to be larger at small polar angles at location 2. Since in these examples the scattering is assumed to be isotropic, more scattering along the longer path at smaller polar angles generates more isotropic distribution of the angular flux.

The angular interaction rates can be defined based on Eq. (7-25) and Eq. (7-27).

The scalar neutron flux and corresponding reaction rates (integrated over the neutron directions of motions) are the variables of interest to analyze and are usually used in reactor design and safety studies. The time-dependent angular neutron density and time-dependent angular neutron flux are correlated to energy and spatially dependent variables through appropriate integration:

$$n\left(\vec{r},E,t\right)= \int_{4\pi} n\left(\vec{r},E,\vec{\Omega},t\right)d\vec{\Omega}$$

$$n\left(\vec{r},t\right)= \int_0^\infty n(\vec{r},E,t)dE = \int_0^\infty dE \int_{4\pi} n\left(\vec{r},E,\vec{\Omega},t\right)d\vec{\Omega}$$

$$\Phi\left(\vec{r},E,t\right)= \int_{4\pi} \Phi\left(\vec{r},E,\vec{\Omega},t\right)d\vec{\Omega} \qquad (7\text{-}27)$$

$$\Phi\left(\vec{r},t\right)= \int_0^\infty \Phi(\vec{r},E,t)dE = \int_0^\infty dE \int_{4\pi} \Phi\left(\vec{r},E,\vec{\Omega},t\right)d\vec{\Omega}$$

If the time-dependent angular neutron density and thus flux are independent of neutron direction of motion (which means isotropic) then it follows

$$n\left(\vec{r},E,\vec{\Omega},t\right)=\frac{1}{4\pi}n\left(\vec{r},E,t\right)$$

$$\Phi\left(\vec{r},E,\vec{\Omega},t\right)=\frac{1}{4\pi}\Phi\left(\vec{r},E,t\right) \qquad (7\text{-}28)$$

The spatial distribution of energy-dependent neutron scalar flux is shown in Fig. 7-13 for one simplified nuclear reactor (refer to Chapter 8 for more discussion on neutron scalar flux).

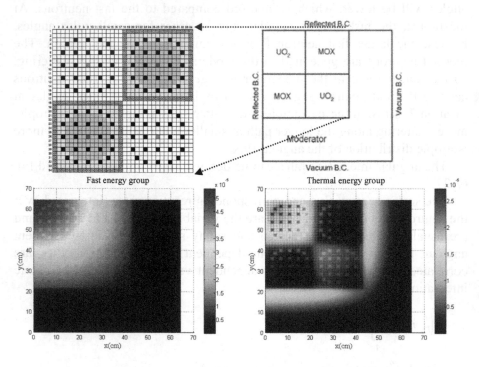

Figure 7-13. Spatial distribution of energy-dependent scalar neutron flux distribution in a quarter reactor core consisting of four fuel assemblies surrounded with water [Refer to Chapter 8] [Images created using AGENT, Jevremovic, 2004, 2008]

The spatially integrated neutron density and scalar flux are given with the following integrals:

$$n(t) = \int_V n(\vec{r},t)dV' = \int_x \int_y \int_x n(x,y,z,t)dx'\,dy'\,dz'$$

$$(7\text{-}29)$$

$$\Phi(t) = \int_V \Phi(\vec{r},t)dV' = \int_x \int_y \int_x \Phi(x,y,z,t)dx'\,dy'\,dz'$$

If the average velocity of a neutron population and average microscopic cross sections are used

$$\bar{\upsilon} = \frac{\int\limits_0^\infty n(t,\upsilon)\upsilon d\upsilon}{\int\limits_0^\infty n(t,\upsilon)d\upsilon}; \overline{\sigma}_i = \frac{\int\limits_0^\infty \sigma_i(\upsilon)n(t,\upsilon)\upsilon d\upsilon}{\int\limits_0^\infty n(t,\upsilon)\upsilon d\upsilon} \quad (7\text{-}30)$$

the scalar neutron flux and the reaction rates become

$$\Phi(t) = n(t)\bar{\upsilon}; \quad R_i(t) = \Phi(t)N\overline{\sigma}_i \quad (7\text{-}31)$$

In a stationary (steady-state) condition corresponding to normal reactor operation all these variables become independent of time, and derivatives in respect to time are zero.

The angular neutron flux is used to define the angular neutron current. The number of neutrons passing through an incremental surface area dA (which normal to the surface is shown in Fig. 7-14) with energies between E and $E + dE$ that are going in a direction $\vec{\Omega}$ during the time interval from t to $t + dt$ is

$$\begin{aligned}
\vec{J}\left(\vec{r}, E, \vec{\Omega}, t\right)dAdEdt &= \upsilon \vec{n} \cdot \vec{\Omega} n\left(\vec{r}, E, \vec{\Omega}, t\right)dAdEdt \\
&= \vec{n} \cdot \vec{\Omega}\Phi\left(\vec{r}, E, \vec{\Omega}, t\right)dAdEdt
\end{aligned} \quad (7\text{-}32)$$

The angular neutron current and the net number of neutrons with energies between E and $E + dE$ that are passing through area dA in the direction of positive normal regardless of direction $\vec{\Omega}$ during the time dt are defined as follows:

$$\begin{aligned}
\vec{J}\left(\vec{r}, E, \vec{\Omega}, t\right) &= \vec{\Omega}\Phi\left(\vec{r}, E, \vec{\Omega}, t\right) \\
\int \vec{n} \cdot \vec{\Omega}\Phi\left(\vec{r}, E, \vec{\Omega}, t\right)&dAdEdtd\vec{\Omega}
\end{aligned} \quad (7\text{-}33)$$

which gives the relation for neutron current density integrated over the directions of motions as

$$\vec{J}\left(\vec{r}, E, t\right) = \int\limits_{4\pi} \vec{J}\left(\vec{r}, E, \vec{\Omega}, t\right)d\vec{\Omega} = \int\limits_{4\pi} \vec{\Omega}\Phi\left(\vec{r}, E, \vec{\Omega}, t\right)d\vec{\Omega} \quad (7\text{-}34)$$

The energy-integrated current is defined by the appropriate integration of the neutron current density or neutron angular current as follows:

$$\vec{J}(\vec{r},t)= \int\limits_{0}^{\infty} \int\limits_{4\pi} \vec{J}(\vec{r},E,\vec{\Omega},t)d\vec{\Omega}dE = \int\limits_{0}^{\infty} \vec{J}(\vec{r},E,t)dE \tag{7-35}$$

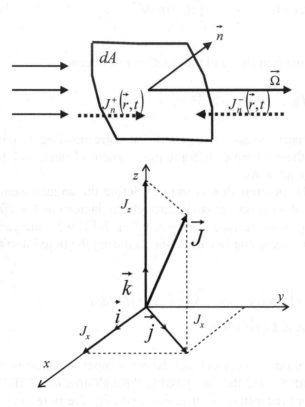

Figure 7-14. Neutron incident on differential area *dA* and definition of neutron current vector [Refer to Chapter 8]

The units of $\vec{J}(\vec{r},t)$ and $\Phi(\vec{r},t)$ are same, number of neutrons/cm^2 s. However, the difference is that the current is a vector describing the net rate at which neutrons (or any particles) pass through a surface oriented in a given direction, while neutron flux describes the total rate at which neutrons pass through a unit area regardless of the orientation of neutron direction of motion. Figure 7-15 shows a neutron current distribution across the plane in a simplified reactor core. There are situations when the neutron current is defined through partial neutron currents, through the currents corresponding to the neutrons passing a given surface in a positive and negative direction:

$$J_n\left(\vec{r},E,t\right) \equiv \vec{n}\cdot\vec{J}\left(\vec{r},E,t\right) = \int\limits_{4\pi} \vec{n}\cdot\vec{\Omega}\Phi\left(\vec{r},E,t\right)d\vec{\Omega}$$

$$J_n\left(\vec{r},E,t\right) = J_n^+\left(\vec{r},E,t\right) - J_n^-\left(\vec{r},E,t\right)$$

$$J_n^+\left(\vec{r},E,t\right) = \int\limits_{2\pi+} \vec{n}\cdot\vec{\Omega}\Phi\left(\vec{r},E,t\right)d\vec{\Omega} \quad J_n^-\left(\vec{r},E,t\right) = \int\limits_{2\pi-} \vec{n}\cdot\vec{\Omega}\Phi\left(\vec{r},E,t\right)d\vec{\Omega}$$

where $2\pi\pm$ indicates the angular integration that is performed along the surface normal (positive sign) or in the opposite direction (negative sign). The partial currents are shown in Fig. 7-14.

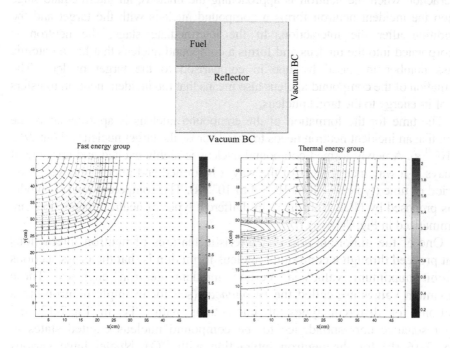

Figure 7-15. Neutron current distribution in a simplified arbitrary reactor core (the arrows represent the neutron current, the lines represent the scalar neutron flux). Notice the return of thermal neutrons from the reflector into the fuel region. [Refer to Chapter 8]

Example 7.8 Reaction rate and flux

In a medium consisting of 10^{20} atoms of fissile material, a total neutron flux, Φ, is sustained at 5×10^{14} neutrons/cm²/s. If the reaction rate is 1.5×10^{13} reactions/cm³/s, calculate the macroscopic and microscopic cross sections of the medium.

The cross sections can be obtained from the following relations:

$$\Phi = nv$$

$$R_i = nvN\sigma_i = \Phi\Sigma_i \quad \rightarrow \quad \Sigma_i = \frac{R_i}{\Phi} = 0.03 \text{ cm}^{-1} \quad \rightarrow \quad \sigma_i = \frac{\Sigma_i}{N} = 300 \text{ b}$$

4.5   Neutron Interactions

Neutron interactions can be described in three steps: the condition before the interaction when the neutron is approaching the nucleus, an intermediate stage when the incident neutron forms a compound nucleus with the target and the condition after the interaction. In the intermediate stage, the neutron is incorporated into the nucleus and forms a compound nucleus that has an atomic mass number increased by one in comparison to the target nucleus. The formation of the compound nucleus also means that the incident neutron transfers all of its energy to the target nucleus.

The time for the formation of the compound nucleus is approximately the time that an incident neutron needs to travel across the target nucleus (about $2R/c$ $\sim 10^{-21}$ s). A newly formed compound nucleus is highly excited and unstable. It decays in a way independent of the way it is formed and after a relatively long period of time (typically from 10^{-19} s to 10^{-15} s). The compound nucleus model was proposed by Niels Bohr in 1936. Different types of neutron interactions are summarized in Fig. 7-16 (a).

One of the important and characteristic features of neutron interactions that proceed through a compound nucleus formation is that the cross sections exhibit maximum values at certain incident neutron energies. These maximum values are called the *resonances* (see for example neutron cross sections for different types of interactions with ^{56}Fe in Fig.7-17 and the resonance correspondence to the compound nucleus excited states in Fig. 7-16 (b) for the neutron interaction with ^{16}O). Nuclei have various excited states that correspond to different configurations of the nucleons within the nucleus (see Chapter 3). An incident neutron and a target nucleus are more likely to combine and form a compound nucleus if the energy of the incident neutron is such that the compound nucleus is produced in one of its excited states. These resonances appear in the cross section because it is necessary to form the compound nucleus before the interaction can proceed.

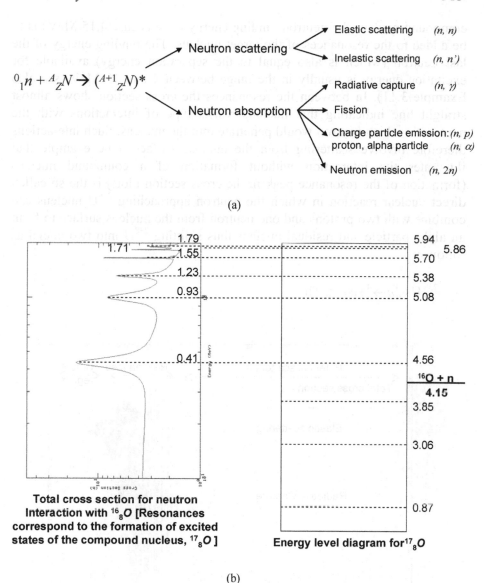

(a)

Total cross section for neutron Interaction with $^{16}_{8}O$ [Resonances correspond to the formation of excited states of the compound nucleus, $^{17}_{8}O$]

Energy level diagram for $^{17}_{8}O$

(b)

Figure 7-16. (**a**) Different types of neutron interactions; (**b**) total microscopic cross section for the neutron interaction with ^{16}O [From http://atom.kaeri.re.kr/] and the energy level diagram of the compound nucleus (energies are given in MeV)

The excitation energy of the compound nucleus is equal to the kinetic energy of the incident neutron plus the separation (binding) energy of the neutron in the compound nucleus. The total cross section for neutron interaction with ^{16}O shown in Fig. 7-16 (b) exhibits the resonance peaks that correspond to the formation of the excited states in ^{17}O (the compound nucleus). In order to obtain the energies for the excited states of the compound nucleus the neutron binding energy (in this case 4.15 MeV) is to be added to the resonances of the target nucleus. The binding energy of the last neutron (which is also equal to the separation energy) available for excitation energy is usually in the range between 4 and 15 MeV (see also Example 3.11). In between the resonances the cross section shows almost straight line indicating that nucleus resists type of interactions with the neutron in which neutron would penetrate into the nucleus. Such interactions therefore involve scattering from the nucleus surface. One example that illustrates the interaction without formation of a compound nucleus (formation of the resonance peak in the cross section plots) is the so-called direct nuclear reaction in which the neutron approaching ^{238}U nucleus can combine with two protons and one neutron from the nucleus surface to form an alpha particle and residual nucleus thus splitting ^{238}U into two nuclei as follows:

$$_{92}^{238}U + _{0}^{1}n \rightarrow _{2}^{4}He + _{90}^{235}Th$$

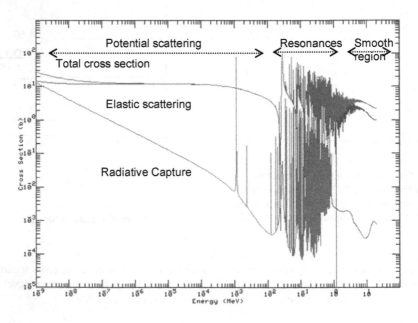

Figure 7-17. Microscopic cross section for various interactions of a neutron with ^{56}Fe [From http://atom.kaeri.re.kr/]

As explained in Section 3 the number of excited states increases with the distance between them reduced for the higher atomic mass number of a nucleus. Thus, neutron cross section for interactions with heavy nuclei will show larger number of resonances. The resonance will start at lower energies for higher mass number of a nucleus (as shown in Fig. 7-31). In addition with the increased atomic mass number the width of the resonances (therefore excited states also) becomes narrower. This indicates that the states are more stable because there are more nucleons for incoming neutron in the compound nucleus to share the energy. Therefore, the overall probability for nucleons to escape the nucleus is reduced.

In the following sections the reactions shown in Fig. 7-16 (a) are discussed with the exception of the fission reaction, which is described in Section 5.

4.5.1 Elastic Scattering (*n, n*)

Neutron has no charge and there is no need to overcome the Coulomb potential barrier, thus it can easily interact with matter. That is why the neutrons of very low energies (close to zero) will scatter elastically from the nucleus. There are two possible ways for a neutron to scatter elastically from a nucleus.

- *Resonance or compound elastic scattering*: the neutron is absorbed by the target nucleus to form a compound nucleus followed by re-emission of a neutron, schematically depicted in Fig. 7-18 (a).
- *Potential elastic scattering*: the neutron is scattered away from the nucleus surface by the short-range nuclear force, schematically depicted in Fig. 7-18 (b).

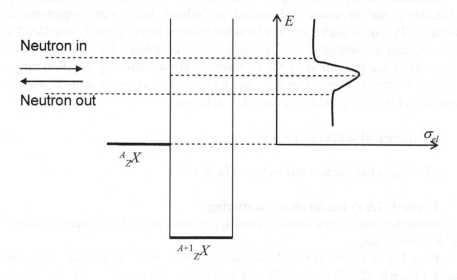

(a)

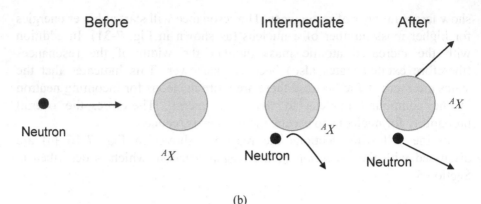

(b)
Figure 7-18. (**a**) Energy level diagram for the resonance elastic neutron scattering [Adopted from Duderstadt and Hamilton, 1976]; (**b**) schematics of the potential elastic neutron scattering

Potential scattering is the most common form of neutron elastic scattering. The more unusual of the two interactions is the resonance elastic scattering which is highly dependent on initial neutron kinetic energy. The cross section for this interaction exhibits a resonance region as shown in Fig. 7-17. Near the resonance energies there is a quantum mechanical interference between the potential and resonance scattering. The interference is constructive above and destructive below the resonance energy. More discussion on elastic scattering cross section is provided in Section 4.5.6.

Potential elastic scattering is more common and can be understood by visualizing the neutrons and nuclei as billiard balls with impenetrable surfaces. Potential scattering in which the neutron never actually touches the nucleus and a compound nucleus is not formed takes place with incident neutrons of energies up to ~1 MeV. Neutrons are scattered by the short range nuclear forces as they approach the nucleus. The cross section is almost constant (Fig. 7-17) and is expressed by relation

$$\sigma_{el} \text{ (potential scattering)} = 4\pi R^2 \tag{7-36}$$

where R is the nuclear radius (see Chapter 3).

Example 7.9 Potential elastic scattering
Using the experimental elastic scattering data from Fig. 7-17, estimate the radius of the ^{56}Fe nucleus.

From Fig. 7-17, the potential microscopic elastic cross section has a constant value of nearly 12 b from about 0.03 eV to 0.6 MeV. Thus, $4\pi R^2 = 12 \times 10^{-24}$ cm^2. Solving for R, we obtain $R = 9.77 \times 10^{-13}$ cm.

An elastic scattering reaction between a neutron and a target nucleus does not involve energy transfer *into* a nucleus. Momentum and kinetic energy are, however, conserved and there is usually some transfer of kinetic energy from the neutron to the target nucleus. The target nucleus thus gains the amount of kinetic energy that the neutron loses and moves away at an increased speed. If the neutron collides with a massive nucleus it rebounds with almost the same speed and loses a negligible amount of energy. However, light nuclei will gain a significant amount of energy from such a collision and will therefore be more effective in slowing down neutrons.

The largest energy transfer occurs in a *head-on collision* in which the neutron does not change its initial direction. Neutrons lose most of their incident energy when they interact elastically with light elements such as hydrogen. This is because the hydrogen nucleus has a mass (of one proton) nearly equal to that of the neutron. Materials with a large content of hydrogen, such as water or paraffin, are therefore very important in the slowing down of neutrons (see Chapter 8). For example, in the case of hydrogen, the energy of a head-on scattered neutron will be zero, which means that the neutron transferred all of its energy to the hydrogen nucleus (see Example 7.3).

4.5.2   Inelastic Scattering (*n, n'*)

In order for a neutron to undergo inelastic scattering with a nucleus its incident energy must be adequate to place the target nucleus in one of its virtual, i.e., excited states. As a result, the inelastic cross section exhibits threshold energy (and is zero up to that value). In general, the energy levels of the excited states of a nucleus decrease with increasing mass number. Elements of high and moderate mass number usually have minimum excitation energy in the range of 0.1–1 MeV. Elements of lower mass number have increased nuclear excitation energies. This is why neutron inelastic scattering is more probable for heavier nuclei and thus the inelastic cross section is non-zero over a large energy region for heavier nuclei.

At energies well above the threshold value, the inelastic cross section is nearly equal to the elastic cross section. Three examples of inelastic cross sections in heavy, moderate and light elements are shown in Figs. 7-19, 7-20, and 7-21. For example, it can be seen that the threshold energy for oxygen is around 6 MeV while for ^{238}U it is only 44 keV. Neutrons cannot undergo inelastic scattering in hydrogen or deuterium (see Section 4.5.5). Magic numbered nuclei behave like light nuclei with respect to inelastic scattering for the same reason.

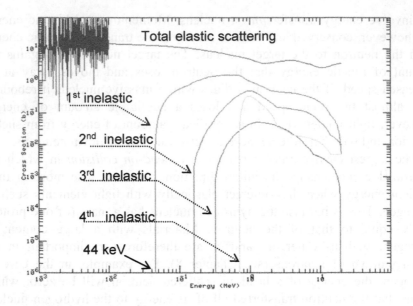

Figure 7-19. Inelastic microscopic cross section for ^{238}U [From http://atom.kaeri.re.kr/]

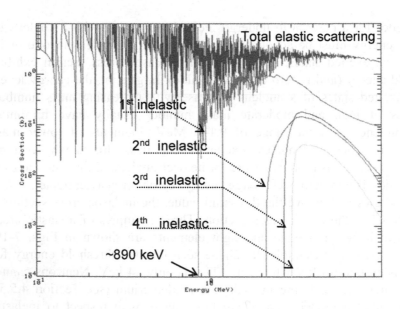

Figure 7-20. Inelastic microscopic cross section for ^{56}Fe [From http://atom.kaeri.re.kr/]

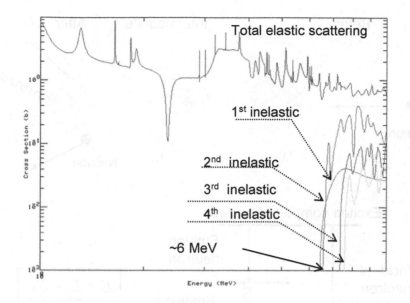

Figure 7-21. Inelastic microscopic cross section for ^{16}O [From http://atom.kaeri.re.kr/]

Inelastic scattering proceeds in two steps as depicted in Fig. 7-22. The interaction involves formation of a compound nucleus as an intermediate stage of the interaction process. The compound nucleus is formed in an excited state due to the energy brought by the incident neutron. In the next step, a neutron of lower kinetic energy is released from the nucleus leaving the nucleus in a lower excited state. The nucleus then reclaims stability, usually by emitting the excess energy in the form of γ rays.

The energy of the emitted γ rays is equal to the excess energy of the excited state of the target nucleus. The total incident neutron energy, E_0, is distributed between the emitted γ ray, E_γ, and the expelled neutron, E

$$E = E_0 - E_\gamma \tag{7-37}$$

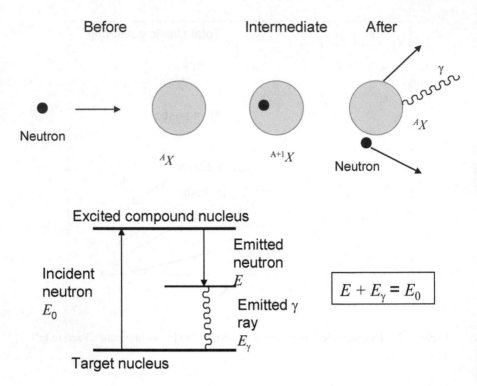

Figure 7-22. Schematics of neutron inelastic scattering

4.5.3   Radiative Capture (n, γ)

Neutron capture (absorption of a neutron) is often called radiative capture because γ rays are produced in the majority of these reactions. In this reaction neutrons form an isotope with mass number increased by one from the original nucleus, Fig. 7-23(a). The newly formed nucleus can be radioactive and will decay. The neutron capture reaction does not require any specific neutron energy and the reaction can occur with the neutron of any energy. These reactions are almost always exothermic (positive Q-value) because the binding energy of the newly formed nucleus is larger than the sum of the binding energies of the neutron and the original nucleus.

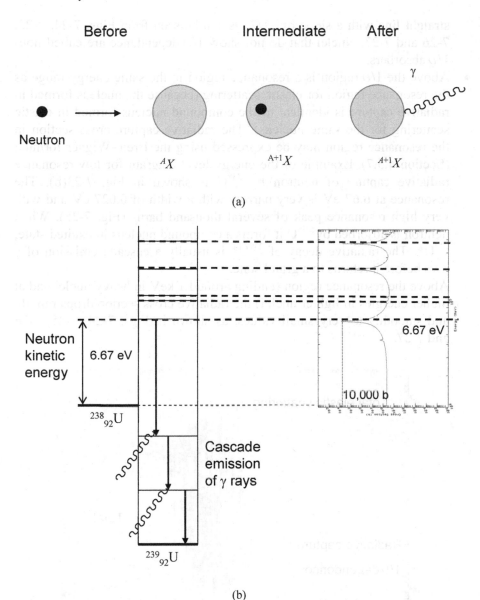

(b)

Figure 7-23. (a) Schematics of radiative neutron capture; (b) energy level diagram for radiative capture in ^{238}U resonance at 6.67 eV [Adopted from Duderstadt and Hamilton, 1976]

The radiative capture cross section is usually divided into three regions:

- In the low-energy region, for most nuclei, the radiative capture cross section varies as the inverse square root of incident neutron energy. Since the neutron speed is proportional to the square root of energy, the radiative cross section is said to vary as $1/\upsilon$. Since the cross sections are usually plotted on a log–log scale the $1/\upsilon$ dependence appears as a straight line with a slope of $-1/2$, as can be seen from Figs. 7-24, 7-25, 7-26 and 7-27. Nuclei that do not show $1/\upsilon$ dependence are called non-$1/\upsilon$ absorbers.

- Above the $1/\upsilon$ region is a resonance region in the same energy range as the resonance region for elastic scattering (because the nucleus formed in radiative capture is identical to the compound nucleus formed in elastic scattering for the same nucleus). The radiative capture cross section in the resonance region may be expressed using the Breit–Wigner formula (Section 4.5.7). Example of the energy level diagram for low-resonance radiative capture of neutron by ^{238}U is shown in Fig. 7-23(b). The resonance at 6.67 eV is very narrow with a width of 0.027 eV and with very high resonance peak of several thousand barns (Fig. 7-25). When neutron is absorbed by ^{238}U it forms a compound nucleus in excited state, $^{239}U^*$. The radiative decay of $^{239}U^*$ is usually a cascade emission of γ rays before it returns to a grund state.

- Above the resonance region (ending around 1 keV in heavy nuclei and at higher energies in lighter nuclei) the radiative cross section drops rapidly and smoothly to very small values, as shown in Figs. 7-24, 7-25, 7-26 and 7-27.

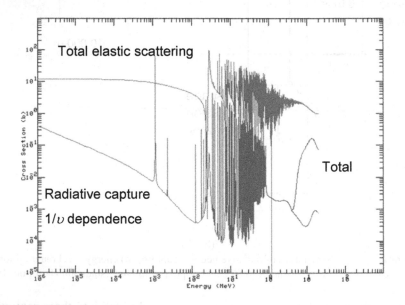

Figure 7-24. Radiative capture microscopic cross section for ^{56}Fe in comparison with its total and elastic scattering cross section [From http://atom.kaeri.re.kr/]

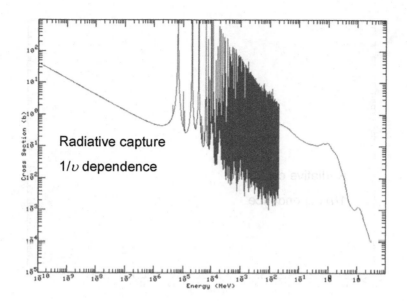

Figure 7-25. Radiative capture microscopic cross section for ^{238}U [From
http://atom.kaeri.re.kr/]

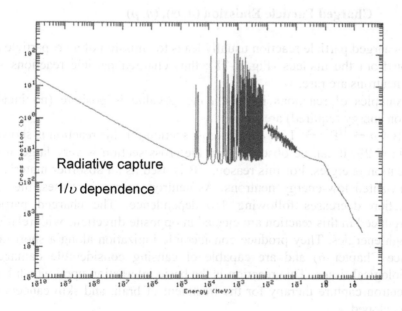

Figure 7-26. Radiative capture microscopic cross section for ^{93}Nb [From
http://atom.kaeri.re.kr/]

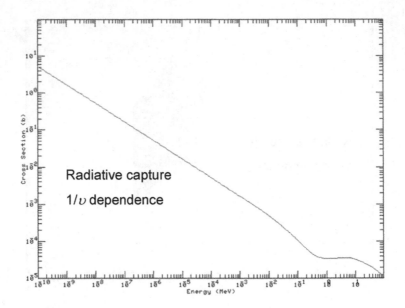

Figure 7-27. Radiative capture microscopic cross section for ^{1}H [From
http://atom.kaeri.re.kr/]

4.5.4   Charged Particle Emission (n, α), (n, p)

A charged particle reaction usually leads to emission of an α particle or a
proton from the nucleus (Fig. 7-28); thus charged particle reactions with
slow neutrons are rare.

Examples of reactions in which the Q-value is positive (no incident
neutron energy required) are

- $^{10}B + n \rightarrow {}^{11}B^* \rightarrow {}^7Li + \alpha$: The cross section for this reaction is shown in
  Fig. 7-29. It can be observed that the cross section is very large at low
  neutron energies. For this reason, ^{10}B is used as an absorber material for
  unwanted low-energy neutrons. As neutron energy increases, the cross
  section decreases following $1/\upsilon$ dependence. The charged particles
  produced in this reaction are ejected in opposite directions with relatively
  high energies. They produce considerable ionization along a short range
  (see Chapter 6) and are capable of causing considerable damage to
  biological tissue. This reaction is the basic interaction upon which boron
  neutron capture therapy for the treatment of brain and skin cancers was
  developed.

- $^6Li + n \rightarrow {}^7Li^* \rightarrow {}^3H + \alpha$: This reaction is similar to the previous one and
  also shows strong $1/\upsilon$ dependence. The remaining nucleus is tritium, a
  β^- emitter and an isotope of special interest in fusion science. This

reaction is used for the production of tritium.

- $^{16}O + n \rightarrow {}^{16}N + p$: is an endothermic reaction of interest in reactor design since it represents the source of radioactivity when water is used as a moderator.

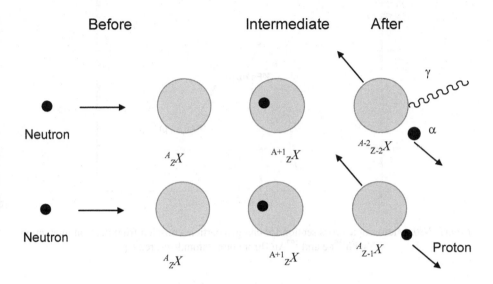

Figure 7-28. Schematics of charged particle emission

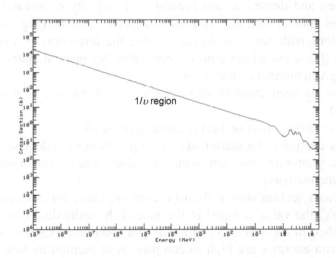

Figure 7-29. Microscopic cross section for $^{10}B + n$ interaction [From http://atom.kaeri.re.kr/]

The majority of the interactions involving charged particle emission, however, are threshold reactions requiring the neutron to possess a minimum amount of energy. The cross sections tend to be small, especially for heavy nuclei (Fig. 7-30).

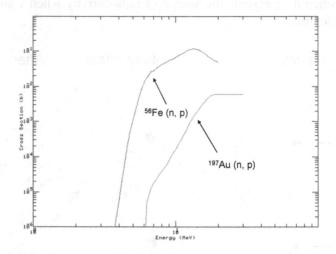

Figure 7-30. Microscopic cross section for charged particle emission from neutron interaction with ^{56}Fe and ^{197}Au [From http://atom.kaeri.re.kr/]

4.5.5   Hydrogen and Deuterium

Hydrogen and deuterium are present in a majority of nuclear reactors. These nuclei interact with the neutrons in a specific manner (Fig. 7-31):

- interactions with neutrons do not involve the formation of a compound nucleus (these nuclei have no excited states because all states are filled for the given number of nucleons)
- there are no resonances (because there is no formation of a compound nucleus)
- elastic scattering cross section is constant up to 10 keV
- radiative capture cross section at all energies shows $1/\upsilon$ dependence, and
- inelastic scattering does not occur (because there is no formation of a compound nucleus).

Water cross section shows almost a constant value for neutron energies above ~1 eV; the value is equal to the sum of the individual cross sections for two hydrogen and one oxygen atoms (for total cross section of ~45 b). When neutron energies are high in comparison to thermal motion of nuclei neutron "sees" them as free and stationary centers of interactions. That is why the total cross section is a simple summation of the individual cross sections of the nuclei involved. The cross section increases from ~1 eV to zero neutron energies because neutron starts to "see" the chemical bonding of the atoms and not anymore as the individual centers of scattering. Therefore, in the low neutron energy region the interaction is taking place with the water molecule of atomic mass 18. This means that the cross section is proportional to the reduced mass of the scattering molecule:

$$\sigma_s \prec \left(\frac{mM}{m+M} \right)^2 = \left(\frac{A}{A+1} \right)^2$$

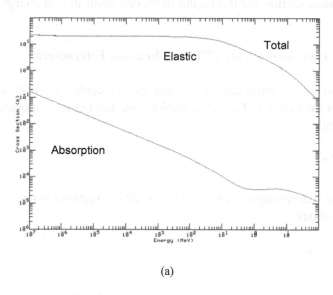

(a)

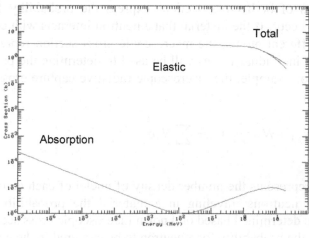

(b)

Figure 7-31. Neutron microscopic cross sections for (a) ^{1}H and (b) ^{2}H [From http://atom.kaeri.re.kr/]

The free-atom cross section and the bound-atom cross section are thus related as

$$\sigma_{free} = \sigma_{bound}\left(\frac{A}{A+1}\right)^2$$

For example, the cross section for neutron interaction with the free hydrogen atom (nucleus) is ~20 b in the low-energy region, Fig. 7-31(a). Thus, the cross section for the bound hydrogen atom in that energy region is 80 b.

4.5.6   Cross Sections for Different Neutron Interactions

The concept of microscopic, σ, and macroscopic, Σ, cross sections is described in Section 4.1. The cross sections for any neutron interaction, i, are related as follows:

$$\Sigma_i = N\sigma_i \tag{7-38}$$

The total macroscopic cross section for all interactions is a summation of individual values:

$$\Sigma = \sum_i N\sigma_i \tag{7-39}$$

The values of both cross sections express the probability for neutron interaction to occur. If the material that a neutron interacts with consists of a mixture of different atoms, the macroscopic cross section representing the summation of individual values will be used to determine the probability of interaction. For example, the macroscopic radiative capture cross section of the material is

$$\Sigma_\gamma = N_1\sigma_{\gamma,1} + N_2\sigma_{\gamma,2} + ... = \sum_j N_j\sigma_{\gamma,j} \tag{7-40}$$

where N_j represents the number density of nuclei of each constituent in a material. For neutrons traveling in a material the probability of certain interactions is determined based on known macroscopic cross section values. For example, the probability for a neutron to be captured in the next collision with the atoms of material j is given by

$$\frac{N_j \sigma_{\gamma,j}}{\Sigma} \tag{7-41}$$

In the analysis of cross sections and interactions it is common to group similar interactions. For example, the absorption cross section relates to all interactions that terminate the neutron history: capture interaction, fission and charged particle interactions

$$\sigma_a = \sigma_\gamma + \sigma_f + \sigma_p + \sigma_\alpha + \ldots \tag{7-42}$$

Example 7.10 Probability of neutron interactions in a homogeneous medium

Estimate the probability of a neutron interacting with ^{235}U to be captured if the microscopic cross sections are $\sigma_\gamma = 98.6$ b, $\sigma_f = 582.2$ b and $\sigma_s = 13.8$ b.

Since the medium is homogeneous and thus composed of only one type of atom, the probability can be computed using the microscopic cross sections. The probability that the neutron will be captured is therefore

$$\frac{\sigma_\gamma}{\sigma_\gamma + \sigma_f + \sigma_s} = \frac{98.6}{98.6 + 582.2 + 13.8} = 0.142$$

Neutron reaction cross sections vary with neutron energy, neutron interaction type and isotope type. Those interactions that do not exhibit threshold values, such as capture and fission in ^{235}U or capture in ^{238}U, have large cross sections at low neutron energy. A threshold interaction observes zero cross section values up to certain energy, such as fission in ^{238}U for which fission becomes significant only if the neutron energy is above 1 MeV. The energy of interest in reactor physics ranges from the high energy that fission neutrons are born with to thermal neutron energies in thermal nuclear reactors. Across this wide span of energies, the cross sections for different neutron interactions show different dependence. The mean value of the fission neutron spectrum (the energies with which fission neutrons are born – see Section 5) is around 2 MeV, while neutrons that are slowed to the thermal region have energies of 0.025 eV. The high-energy neutrons are moving at a high speed relative to the nuclei in a medium; therefore the dominant interactions are scattering in which neutrons slow down. In materials that have a large scattering cross section (like hydrogenous medium) neutrons lose most of their energy after only a few interaction events and come into thermal equilibrium with the nuclei of the medium. Since the nuclei themselves are in thermal motion there is an exchange of momentum in scattering interactions. Such neutrons have a

Maxwellian spectrum (see Section 4.6) dependent on the temperature of the medium. Therefore, the neutron population in a reactor has a complicated spectrum that is a mixture of fast, intermediate and slow neutrons. The particular spectrum characteristics are determined by the materials present in the medium. For example, in a medium with a high scattering to absorption cross section ratio, the spectrum of neutrons will fall predominantly in the thermal energy region. However, in the opposite case of a medium consisting of materials with high absorption to scattering ratios, the neutron spectrum will not differ much from the source spectrum.

At low energies the total microscopic cross section for the non-threshold interactions behaves as

$$\sigma_{tot} = 4\pi R^2 + \frac{C}{\sqrt{E}} \tag{7-43}$$

where C is a constant, E is the neutron energy and R represents the radius of a nucleus.

The first term in Eq. (7-43) represents the elastic cross section, while the second term gives the cross section for radiative capture or other exothermic reactions possible at that energy. If the first term dominates over the second term, then the total cross section is constant at low energies. An example is shown in Fig. 7-32 for ^{56}Fe for which the total cross section is constant at low energies. If the second term dominates over the first term, the total cross section behaves as $1/\upsilon$. An example is shown in Fig. 7-32 for ^{239}Pu for which the cross section varies with the inverse square root of neutron energy at low energies. Cross section data libraries usually give the capture and fission cross sections for thermal energy neutrons traveling at the speed of $\upsilon_p = 2,200$ m/s. This velocity corresponds to neutron energy of 0.025 eV at the ambient temperature of 293 K. For the nuclei for which $1/\upsilon$ dependence of the absorption cross section is valid, the absorption cross section at any other energy E of up to few eV can be estimated from

$$\sigma_a(E) = \sigma_a(E_p)\sqrt{\frac{E_p}{E}} \qquad \sigma_a(\upsilon) = \sigma_a(\upsilon_p)\frac{\upsilon_p}{\upsilon} \tag{7-44}$$

In the resonance region, elastic scattering, radiative capture and inelastic scattering, and thus total cross section, all exhibit resonances in same energy region. The cross section at the peak values can be as high as a few thousand barns. The resonances correspond to the discrete energy levels of the compound nucleus formed after neutron interaction (see Chapter 3).

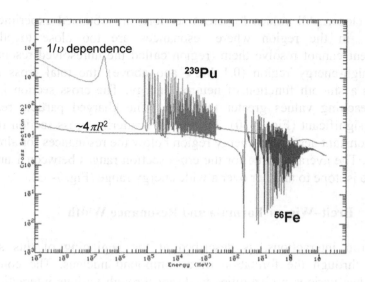

Figure 7-32. Total microscopic cross section dependence on neutron energy for ^{56}Fe and ^{239}Pu [From http://atom.kaeri.re.kr/]

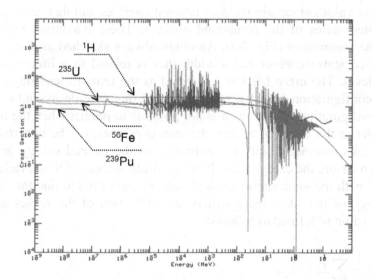

Figure 7-33. Comparison of elastic microscopic scattering cross sections for ^{1}H, ^{56}Fe, ^{235}U and ^{239}Pu [From http://atom.kaeri.re.kr/]

Neutrons with energy comparable to the energy levels of a compound nucleus have a high probability of interaction. The lowest energy at which resonances begin to appear is around 0.5 eV and the maximum is about 0.1 MeV. As can be seen from Fig. 7-33, as energy is increased the resonances become closer. All values for cross sections are obtained experimentally; however, in the region where resonances are too close together an experiment cannot resolve them (region called the unresolved resonances). In the high-energy region (0.1 MeV and above), the total cross section becomes a smooth function of neutron energy. The cross section is small rarely reaching values greater than 5 b. The charged particle reactions become significant (Fig. 7-30). The elastic scattering cross section remains almost constant across the energy region below the resonances for almost all isotopes. The average value for the cross section ranges between 1 and 10 b from one isotope to another over a wide energy range (Fig. 7-33).

4.5.7   Breit–Wigner Formula and Resonance Width

Neutron interactions, as described at the beginning of this section, proceed through the formation of a compound nucleus. The compound nucleus has various probabilities to decay through various interaction-type channels, such as elastic or inelastic scattering or absorption (fission, emission of charged particles, or two neutrons). The important characteristic of a compound nucleus formation is that the neutron cross sections exhibit maximum values at certain incident neutron energies and that correspond to the excited states of the compound nucleus. These maximum values are called the *resonances* (Fig. 7-7). Although always sketched as a single line the excited state however has a width that is related to a lifetime of that energy level. The mean lifetime is defined as the time nucleus stays in that energy configuration before it rearranges and therefore decays into a more stable configuration. However, it is not possible to predict when the nucleus will undergo the decay to release the excess of energy. The uncertainty in time is thus associated with the existence of the excited energy level. In order to measure the energy state there is a finite amount of time available. If N nuclei with the same energy excited state are measured to find the value of the energy of that state there will be a distribution of the values and the average value is defined as follows:

$$\overline{E} = \frac{1}{N} \sum_{i=1}^{N} E_i \qquad\qquad (7\text{-}45)$$

Then the spread in the measured energy levels, Γ, can be defined as

$$\Gamma = \sqrt{\frac{1}{N} \sum_{i=1}^{N} \left(E_i^2 - \overline{E}^2 \right)} \qquad\qquad (7\text{-}46)$$

This is called the total resonance width and is sketched in Fig. 7-34. The width defines the half the maximum of the resonance peak in the cross section value at the resonance energy E_R. If λ is the decay constant of the excited energy level defined as inverse of mean life time, τ (see Problem 5.37, Chapter 5), then according to the Heisenberg uncertainty principle the total level width, Γ, is related to the decay constant as

$$\Gamma = \frac{\hbar}{\tau} = \hbar\lambda \tag{7-47}$$

Thus according to this, for example, the mean life time of the 6.67 eV energy level in ^{238}U (which corresponds to the compound nucleus of ^{239}U*) is $\sim 10^{-14}$ s [Rydin, 1977].

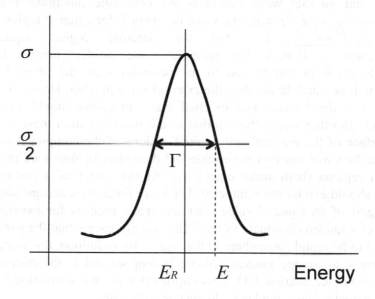

Figure 7-34. Resonance width

The total level width is proportional to the total probability of the compound nucleus decay and is therefore equal to the sum of the partial widths, i.e., the probabilities for all decay channels (scattering, capture, fission, α emission, etc.)

$$\Gamma = \sum_i \Gamma_i = \Gamma_n + \Gamma_\gamma + \Gamma_f + \Gamma_\alpha + \dots \tag{7-48}$$

In order to understand why the resonances are formed in the compound nucleus-type interactions of a neutron with the nucleus, the neutron is considered to be a quantum particle in a square potential well (see Chapter 4, Sections 3.5–3.7). Therefore the interaction of a neutron with nucleus is described using the Schrödinger wave equation inside and outside of the compound nucleus with properly defined potential energy of the well. Neutrons are electrically neutral and thus there is no Coulomb barrier to be overcome. That is why the potential well of the compound nucleus can be represented as a square potential well, Fig. 7-35(a). Outside the nucleus the potential energy is $U = 0$. Thus the total external energy is equal to the kinetic energy of the incoming neutron. Inside of the nucleus there is a potential well created by attractive nuclear forces. The potential energy of attraction is equal to the binding energy of the incident neutron, $U = -BE$. For this case the solution to the wave equation given by Eq. (4-23) is to be found in both regions, outside and inside the nucleus, i.e., potential well. The external and internal wave functions are both sine functions; they are proportional to $\sin kr / r$ with the wave numbers k, for external region equal to $k_{ext} = \sqrt{2mE} / \hbar$ and for the internal region equal to $k_{int} = \sqrt{2m(E - (-BE))} / \hbar$. For example if the incident neutron kinetic energy is small in comparison to the potential well, the internal wave number will be much larger than the external wave number. However, at the point $r = R$ the internal and external wave functions should show the continuity. In other words the solution to the wave equation must match at the interface of the external and internal regions of the compound nucleus such that the wave function magnitude and derivative (a slope) are the same for both regions (both wave equations). At the center of a nucleus the solution should exhibit the symmetry. The wave function is defined such that the integral of its squared value over the space such as for example the volume of a nucleus is equal to unity. That simply means that the particle is supposed to be found somewhere in that space. By definition, the probability of a particle to occupy nuclear volume is proportional to the microscopic cross section (see Section 4.1). This implies that the wave function is large inside the nucleus and small outside the nuclear volume.

Figure 7-35(b) illustrates three distinctive cases of the wave function (representing incident neutron) of the compound nucleus. The wave function shape is symmetric around the nucleus centerline. The external wave amplitude is determined by the value of neutron kinetic energy, E. The external wave function frequency is much smaller than the internal one, and thus the external wave function amplitude is larger than the amplitude of internal wave function. The two wave functions amplitudes meet at the particular incident neutron kinetic energy. At that point the derivative for both external and internal wave functions is zero. The energy for which this condition is satisfied is exactly the energy of the resonance of a compound nucleus and therefore corresponds to the maximum cross section! This condition varies with the radius of a nucleus and the neutron binding energy.

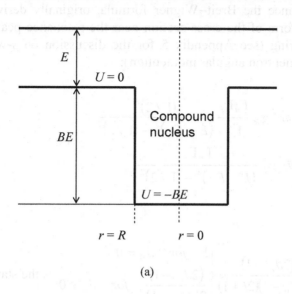

(a)

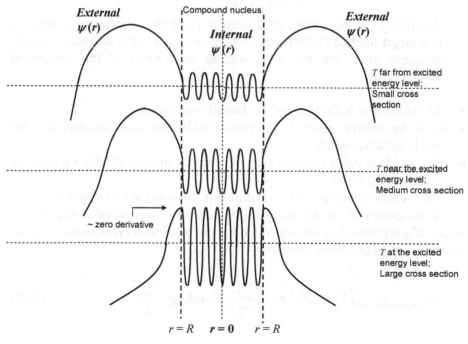

(b)

Figure 7-35. (**a**) Square potential well for a compound nucleus formation: *BE* is the binding energy of a neutron, *E* is the kinetic energy of a neutron, *U* is the potential well; (**b**) compound nucleus wave function according to Schrödinger equation [Adapted from Rydin, 1977]

As sketched in Fig. 7-16(b) the excited energy levels in a compound nucleus show the resonance peaks in the cross section values. For a single isolated resonance the Breit–Wigner formula, originally derived in 1936, describes the form of the cross section near the resonance peak for s-wave neutron scattering (see Appendix 5 for the discussion on s-wave neutron scattering and neutron angular momentum):

$$\sigma_j(E) = 4\pi\lambda^2 g_R \frac{\Gamma_n \Gamma_j}{\Gamma^2} \frac{(\Gamma/2)^2}{(E - E_R)^2 + (\Gamma/2)^2}$$

$$\text{or} \qquad (7\text{-}49)$$

$$\sigma_j(E) = \pi\lambda^2 g_R \frac{\Gamma_n \Gamma_j}{(E - E_R)^2 + (\Gamma/2)^2}$$

where

- $g_R = \dfrac{(2J_C + 1)}{2(2J_0 + 1)(2l + 1)} \begin{cases} 1 & for \quad J_0 = 0 \\ \dfrac{(2J_C + 1)}{2(2J_0 + 1)} & for \quad J_0 \neq 0 \end{cases}$ is the statistical factor

  describing the probability of a neutron (with the angular momentum l) and target nucleus (with the angular momentum J_0) to add their angular momenta such that the total angular momentum of the compound nucleus is J_C

- $\lambda^2 = h/2\pi m\upsilon$ is the reduced De Broglie wavelength
- Γ_n is the neutron width, Γ_j the partial width for decay channel j, Γ the total resonance width
- E_R – single isolated resonance appears at this energy called the resonance energy (Fig. 7-34)
- E is the kinetic energy of the incoming neutron in the *COM* system

At the energy of the incoming neutron equal to the resonance energy, the Breit–Wigner formula gives the value for the maximum cross section for the formation of a compound nucleus:

$$\sigma_{CompoundNucleus}(E_R) = \pi\lambda^2 g_R \frac{\Gamma_n^R \Gamma}{(\Gamma/2)^2} = 4\pi\lambda^2 g_R \frac{\Gamma_n^R}{\Gamma} \qquad (7\text{-}50)$$

If the resonance is very narrow then the cross section at the neutron energy $E = E_R \pm \Gamma/2$ (Fig. 7-34) is approximated with

$$\sigma_{CompoundNucleus}\left(E_R \pm \Gamma/2\right) = \pi\lambdabar^2 g_R \frac{\Gamma_n\Gamma_j}{\left(E_R \pm \Gamma/2 - E_R\right)^2 + \left(\Gamma/2\right)^2}$$

$$= \pi\lambdabar^2 g_R \frac{\Gamma_n^R\Gamma}{2\left(\Gamma/2\right)^2} = 2\pi\lambdabar^2 g_R \frac{\Gamma_n^R}{\Gamma} = \frac{\sigma_{CompoundNucleus}\left(E_R\right)}{2} \qquad (7\text{-}51)$$

For the *s*-wave neutron radiative capture, the cross section around the single isolated resonance is described with

$$\sigma_\gamma(E) = \pi\lambdabar^2 g_R \frac{\Gamma_n\Gamma_\gamma}{\left(E - E_R\right)^2 + \left(\Gamma/2\right)^2} \qquad (7\text{-}52)$$

The full width at half maximum of the resonance peak is Γ. As Figs. 7-28, 7-29 and 7-30 show, well below the resonance the cross section follows $1/\upsilon$ dependence; the wavelength of slow neutrons is much greater than the size of the nucleus and thus all such neutron interactions are *s*-wave (angular momentum is zero and the collision is a head-on type), Appendix 5. For the *s*-wave elastic scattering of a neutron from the nucleus, the Breit–Wigner formula for the cross section of an isolated resonance is given with

$$\sigma_s(E) = 4\pi R^2 + \pi\lambdabar^2 g_R \frac{\left(\Gamma_n\right)^2}{\left(E - E_R\right)^2 + \left(\Gamma/2\right)^2} +$$

$$4\pi\lambdabar^2 g_R \frac{\Gamma_n\left(E - E_R\right)R}{\left(E - E_R\right)^2 + \left(\Gamma/2\right)^2} \qquad (7\text{-}53)$$

The cross section for a neutron elastic scattering consists of three terms: the potential elastic scattering term (practically constant for a low-energy region), the compound nucleus formation term responsible for the resonance peak value and the interference term that explains the asymmetry in the cross section around the resonance. Figure 7-36 shows the cross section asymmetrical trend around the single resonance. The interference phenomena is caused by the partial neutron beam resonance scattering (partial beam is passing through the nucleus) and partial beam scattering off the nucleus surface (potential scattering); in other words it is due to the competition between the potential and resonance elastic scattering. As shown in Fig. 7-36 the interference effect is destructive at energies below the resonance energy and constructive at energies above the resonance energy.

This is a characteristic of a resonance reaction sometimes observed in actual cross section measurements. The potential scattering provides constant background cross section, while the tendency toward compound nucleus formation non-symmetrically curves the cross section.

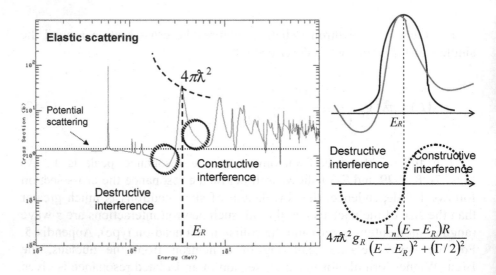

Figure 7-36. Interference phenomena in resonance microscopic cross section of a *s*-wave neutron elastic scattering

4.6   Maxwell–Boltzmann Distribution

In a medium in which neutrons are not absorbed and from which neutrons cannot escape, the only possible interaction is scattering with the nuclei of the atoms. The scattering interactions reduce the neutron energy. However, an endless slowing-down process is not possible because of the thermal motion of the atoms. Due to that fact they cannot be assumed to be stationary, which is usual approximation in analyzing neutron interactions. When neutron energy becomes comparable to the energy of thermal motion of the atoms, the neutrons come to a thermal equilibrium. It means that the probability that a neutron will gain or lose energy in a collision with the nuclei is equal. When neutron loses its energy it is slowed down and this is called the *downscattering* unlike the situations when neutrons gain the energy in collisions referred to as the *upscattered*. The average kinetic energy of thermal motion of the atoms (according to the kinetic theory of gases) is given with

$$\overline{E} = \frac{3}{2}kT \tag{7-54}$$

where k is the Boltzmann constant (1.380662×10^{-23} J/K) and T is the temperature of the medium (in Kelvin). In a thermal equilibrium state that requires no losses or absorption of neutrons, the neutrons can gain or lose kinetic energy ($mv^2/2$) by exchanging it with the nuclei of atoms in the medium. In such an ideal medium without absorption and leakage, the neutron energy distribution will be the same as that of the atoms in thermal motion. The thermal neutrons, even at a specific temperature, do not all have the same energy or velocity. Such spectrum is called a *Maxwellian–Boltzmann distribution*, or referred as a *Maxwellian distribution*. Although such conditions are not satisfied in a real reactor system, it is useful to assume that neutrons become thermalized to the extent that they follow the Maxwellian distribution:

$$\frac{n(E)}{n} = \frac{2\pi}{(\pi kT)^{3/2}} e^{-E/kT} E^{1/2} \iff \frac{n(v)}{n} = \frac{4\pi v^2}{(2\pi kT/m)^{3/2}} e^{-mv^2/2kT} \tag{7-55}$$

where
n – thermal neutron population per unit volume
m – neutron rest mass
T – temperature in K
$n(E)$ and $n(v)$ – Maxwellian energy (or velocity) distribution of neutrons per unit volume and unit energy (or velocity) interval
The left side of Eq. (7-55) represents the fraction of neutrons having energies (or velocities) within a unit energy interval (or velocity interval) and the right side represents the Maxwellian distribution, Fig. 7-37.
The most probable neutron velocity, v_p, is found by setting the derivative of $n(v)$ with respect to velocity equal to zero:

$$\frac{dn(v)}{dv} = \frac{8\pi v n}{(2\pi kT/m)^{3/2}} e^{-mv^2/2kT} - \frac{4\pi v^2 n}{(2\pi kT/m)^{3/2}} \frac{2mv}{2kT} e^{-mv^2/2kT} = 0 \tag{7-56}$$

$$v_p = \sqrt{\frac{2kT}{m}} \tag{7-57}$$

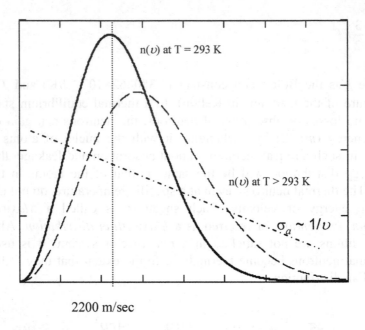

2200 m/sec

Figure 7-37. Maxwellian velocity distribution of neutrons

The kinetic energy of thermal neutrons at the most probable velocity is

$$E(\upsilon_p) = \frac{m\upsilon_p^2}{2} = \frac{m}{2}\frac{2kT}{m} = kT \tag{7-58}$$

Note that the most probable (as well as average) kinetic energy is independent of particle mass. For thermal neutrons at 20°C (or 293 K) the most probable velocity and the corresponding kinetic energy are

$$\upsilon_p(T) = \sqrt{\frac{2kT}{m}} = \sqrt{\frac{2 \times 1.38 \times 10^{-23}\,\text{J/K} \times 293\text{K}}{1.66 \times 10^{-27}\,\text{kg}}} = 2200 \text{ m/s}$$

$$E(\upsilon_p) = kT = 1.38 \times 10^{-23}\,\text{J/K} \times 293\text{K} = 4.043 \times 10^{-21}\,\text{J} \times \frac{1}{1.6 \times 10^{-19}\,\text{J/eV}}$$

$$= 0.025\text{eV}$$

The most probable energy can be obtained by setting the derivative of $n(E)$ with respect to energy equal to zero. The solution leads to the value of

$E_p = kT/2$. The velocity of thermal neutrons that corresponds to this energy is $\upsilon(E_p) = \sqrt{kT/m}$. The values for microscopic cross sections provided on most charts and tables are measured for the neutron velocity of 2,200 m/s, which corresponds to an ambient temperature of 68°F (see Section 4.5.6) and energy of 0.025 eV. The average neutron velocity is obtained from

$$\bar{\upsilon} = \frac{\int\limits_0^\infty n(\upsilon)\upsilon d\upsilon}{\int\limits_0^\infty n(\upsilon)d\upsilon} = \frac{\int\limits_0^\infty \frac{4\pi n\upsilon^3 e^{-m\upsilon^2/2kT}}{(2\pi kT/m)^{3/2}}d\upsilon}{\int\limits_0^\infty \frac{4\pi n\upsilon^2 e^{-m\upsilon^2/2kT}}{(2\pi kT/m)^{3/2}}d\upsilon} = \sqrt{\frac{8kT}{\pi m}} \tag{7-59}$$

The ratio of the average velocity to the most probable velocity of neutrons in the Maxwellian spectrum is

$$\frac{\bar{\upsilon}}{\upsilon_p} = \frac{\sqrt{8kT/\pi m}}{\sqrt{2kT/m}} = \frac{2}{\sqrt{\pi}} = 1.128$$

The cross section at these velocities changes accordingly; the neutron flux for the Maxwellian distribution of neutrons is given by

$$\phi(E) = \upsilon n(E) = \frac{2\pi n}{(\pi kT)^{3/2}}e^{-E/kT}E\sqrt{\frac{2}{m}} \tag{7-60}$$

The average absorption cross section for this population of neutrons assuming $1/\upsilon$ dependence can be estimated:
- As described in Section 4.4.6 for the $1/\upsilon$ absorption cross section dependence, the following correlation between the cross sections holds

$$\sigma_a(E) = \sigma_a(E(\upsilon_p))\sqrt{\frac{E(\upsilon_p)}{E}} \quad \text{where } E(\upsilon_p) = kT \tag{7-61}$$

- The average absorption cross section is then

$$\overline{\sigma_a(E_p)} = \frac{\int_0^\infty \sigma_a(E)\phi(E)dE}{\int_0^\infty \phi(E)dE}$$

$$= \frac{\sigma_a(E_p)\int_0^\infty \sqrt{\frac{E_p}{E}}\frac{2\pi n}{(\pi kT)^{3/2}}\sqrt{\frac{2}{m}}Ee^{-E/kT}dE}{\int_o^\infty \frac{2\pi n}{(\pi kT)^{3/2}}\sqrt{\frac{2}{m}}Ee^{-E/kT}dE} \qquad (7\text{-}62)$$

$$= \frac{\sigma_a(E_p)\int_0^\infty \sqrt{\frac{kT}{E}}Ee^{-E/kT}dE}{\int_0^\infty Ee^{-E/kT}dE} = \frac{\sqrt{\pi}}{2}\sigma_a(E_p)$$

The values for microscopic absorption cross sections at a higher temperature are lower than the tabulated value (which is generally for the most probable neutron velocity at ambient temperature) and any cross sections which involve absorption (fission, capture) must be corrected for the existing temperature. The average absorption cross section at the average neutron velocity and temperature, T, higher than the ambient is given by

$$\overline{\sigma_a\big(E(\upsilon_p),T\big)} = \frac{\sqrt{\pi}}{2}\sigma_a\big(E(\upsilon_p),293\text{ K}\big)\sqrt{\frac{293}{T}} \qquad (7\text{-}63)$$

Example 7.11 Average and temperature-corrected $1/\upsilon$ absorption cross section

The absorption cross section for ^{235}U at the most probable neutron velocity is 680.8 b. Assuming the cross section follows the $1/\upsilon$ rule determine the average cross section at the temperatures of 293 K and 600 K.

The average absorption cross section at the most probable neutron energy is

$$\overline{\sigma_a\big(E(\upsilon_p)\big)} = \frac{\sqrt{\pi}}{2}\sigma_a\big(E(\upsilon_p)\big) = 603.3\text{ b}$$

If the temperature of neutron population is increased to 600 K, the average absorption cross section will change as

$$\overline{\sigma_a\left(E\left(\upsilon_p\right),T\right)} = \frac{\sqrt{\pi}}{2}\sigma_a\left(E\left(\upsilon_p\right),293\text{ K}\right)\sqrt{\frac{293}{T}} = \frac{\sqrt{\pi}}{2}\times603.3\times\sqrt{\frac{293}{600}} = 373.6\text{ b}$$

However, the absorption cross sections of some materials important in reactor neutronic design do not exhibit exact $1/\upsilon$ dependence. Examples are ^{235}U, ^{238}U and ^{239}Pu. In these cases, an empirical factor, $g(T)$, based on actual cross section measurements is introduced to correct for the departure from $1/\upsilon$ behavior. The actual thermal cross section corrected for the average absorption temperature is then

$$\overline{\sigma_a\left(E\left(\upsilon_p\right),T\right)} = \frac{\sqrt{\pi}}{2}g(T)\sigma_a\left(E\left(\upsilon_p\right),293\text{ K}\right)\sqrt{\frac{293}{T}} \qquad (7\text{-}64)$$

Example 7.12 Average and temperature-corrected non-1/υ absorption cross section

The radiative capture cross section for ^{235}U at the most probable neutron velocity is 98.81 b. From the table of nuclides (http://atom.kaeri.re.kr/), the $g(T)$ factor is found to be 0.9898. Calculate the average radiative capture cross section at energy 0.0253 eV. Calculate the value of the cross section at 600 K.

The average radiative capture cross section for a Maxwellian distribution of the neutron population is

$$\overline{\sigma_a\left(E_0\right)} = \frac{\sqrt{\pi}}{2}g(T)\sigma_a\left(E_0\right) = \frac{\sqrt{\pi}}{2}\times0.9898\times98.81 = 86.67\text{ b}$$

If the temperature of the neutron population is increased to 600 K, the average radiative capture cross section becomes

$$\overline{\sigma_a\left(E_0,T\right)} = \frac{\sqrt{\pi}}{2}g(T)\sigma_a\left(E_0,293\text{ K}\right)\sqrt{\frac{293}{T}} = 86.67\times\sqrt{\frac{293}{600}} = 60.57\text{ b}$$

In thermal reactors, it is not possible to obtain a neutron spectrum that will follow exactly the Maxwellian distribution. The reasons for this are
1. Neutrons produced by the fission process are high-energy neutrons that are (in thermal reactors) slowed down by primarily elastic collisions with moderator (light) nuclei. The proportion of neutrons of higher energy is greater than that required by the Maxwellian distribution (Fig. 7-38). This is because neutrons that are absorbed or that leak out of the reactor do not have a chance to slow down. This shift of the neutron energy spectrum from Maxwellian toward the high neutron energy region is called absorption hardening.
2. In the low-energy region, a real neutron spectrum approaches the

Maxwellian distribution. The departure depends on the absorption and leakage (escape from the geometrical boundaries) rate in the system, as neutrons may either be absorbed or lost before they come to equilibrium with the moderator atoms. In this energy region, the absorption cross section is inversely proportional to the neutron speed. In spite of these facts, the neutron spectrum in the thermal region is usually approximated by the Maxwellian distribution at a temperature somewhat higher than the moderator temperature. This temperature is called an effective neutron temperature.

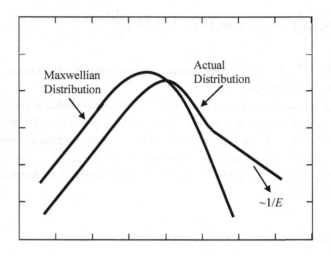

Figure 7-38. Energy spectrum of thermal neutrons (departure from Maxwellian distribution)

4.7   Doppler Broadening

Cross sections are commonly associated with the energy of the incoming neutron. However, the cross sections actually depend on the *relative energy* of the interacting neutron and nucleus. The relative energy is identical to the neutron energy only if the nucleus is at rest. When the nucleus is at rest only neutrons with the energy that is equal to the resonance energy will be absorbed. However, the nuclei in a solid are "vibrating" about fixed points, and this energy of vibration increases with temperature.

At some given temperature, the vibration energies tend to follow a Maxwellian distribution over a wide range of the energy spectrum. Therefore, even for monoenergetic neutrons, the energies relative to the target nuclei vary over a wide range of values (below and above the incident neutron energy). This phenomenon is called the *Doppler effect* because of the similarity with the change in wavelength observed with a moving source of light or sound with constant frequency. Since the vibration energies increase with temperature, the range of the neutron–nucleus relative energies also increases. This means that there is a wider range of neutron velocities that coincide with the absorption peak. As a result of the Doppler effect, the width of a resonance peak increases with temperature (Fig. 7-39), an effect known as *Doppler broadening*. The increase in the resonance width means the increase in energy range at which neutrons can be captured by the nucleus. This is accompanied by a reduction in resonance height, while the area under the resonance remains constant.

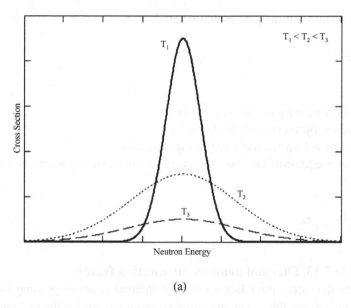

(a)

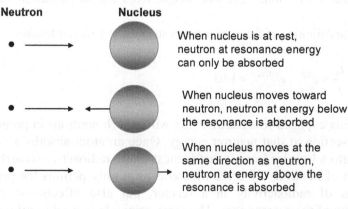

(b)

Figure 7-39. Doppler broadening: (**a**) the cross section change with the temperature; (**b**) relation between neutron and nucleus in respect to resonance absorption

The capture cross section at the resonance energy is reduced with increased temperature while the overall probability of a capture is increased. The total rate of neutron absorptions in the resonance region (a product of neutron flux and cross section) increases with temperature. The overall effect of Doppler broadening is the increase of neutron resonance absorptions in fuel. This aspect is important in reactor safety in analyzing the temperature reactivity coefficients.

4.8   Neutron Beam Attenuation and Neutron Activation

Neutron beam attenuation is determined from

$$I = I_0 e^{-\Sigma x} \qquad\qquad (7\text{-}65)$$

where
I_0 = initial intensity of the neutron beam
Σ = macroscopic cross section (cm^{-1})
x = thickness (cm) of the attenuating material
The above equation can be expressed in terms of the attenuation factor, AF, given as

$$AF = \frac{I_0}{I} = e^{\Sigma x} \qquad\qquad (7\text{-}66)$$

Example 7.13 Thermal neutron attenuation factor
Calculate the attenuation factor (AF) for thermal neutrons passing through a layer of water 2.5 cm thick. The macroscopic cross section for thermal neutrons is 0.02 cm^{-1}.
From the definition of the neutron beam attenuation factor, it follows

$$AF = \frac{I_0}{I} = e^{\Sigma x} = e^{0.02 \times 2.5} = 1.05$$

Materials exposed to a neutron flux will absorb neutrons in proportion to the cross section at that neutron energy. Once an atom absorbs a neutron it changes into a heavier isotope that is most likely radioactive (unstable). The absorption of neutrons by certain materials not only permits the production of sources of radioactivity in a reactor, but also affects the structural components of the reactor core. The same principle is used to infer the level of neutron flux at points of interest in a reactor core using neutron-absorbing foils [Foster, Arthur R. & Wright, Robert L., 1968]. The activity of the foils following irradiation is proportional to the neutron flux in which the foil was placed. For example, if an isotope formed in neutron flux, Φ, is unstable, it will start to decay as soon as it is produced. Assuming there are N nuclei of a newly formed isotope and N_0 nuclei of the original target isotope, the rate of change of new nuclei can be obtained from

$$\frac{dN}{dt} = \Phi N_0 \sigma_a - \lambda N = \Phi \Sigma_a - \lambda N \tag{7-67}$$

where λ is the decay constant of the newly formed, unstable isotope. The above equation can be rearranged to obtain the first-order differential equation as

$$\frac{dN}{dt} + \lambda N = \Phi \Sigma_a \tag{7-68}$$

the solution of which is of the form

$$N = \frac{1}{e^{\int \lambda dt}} \int e^{\int \lambda dt} \Phi \Sigma_a dt - \frac{C}{e^{\int \lambda dt}} \tag{7-69}$$

where C is a constant of integration. Equation (7-69) can be rearranged to obtain

$$N = \frac{1}{e^{\lambda t}} \int e^{\lambda t} \Phi \Sigma_a dt - \frac{C}{e^{\lambda t}} = \frac{\Phi \Sigma_a}{\lambda} - \frac{C}{e^{\lambda t}} \tag{7-70}$$

The constant of integration is obtained from the initial condition

$$N(t = 0) = 0 \quad \Rightarrow \quad 0 = \frac{\Phi \Sigma_a}{\lambda} - \frac{C}{1} \quad \Rightarrow \quad C = \frac{\Phi \Sigma_a}{\lambda} \tag{7-71}$$

Thus

$$N = \frac{\Phi \Sigma_a}{\lambda} \left(1 - e^{-\lambda t}\right) \tag{7-72}$$

The buildup of a radioactive isotope during irradiation in a neutron flux is depicted in Fig. 7-40. The material will decay with its characteristic half-life once removed from the neutron flux.

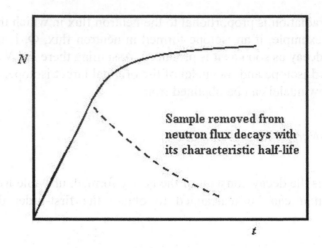

Figure 7-40. Buildup of radioactive isotope during irradiation in neutron flux [Adopted from Foster, Arthur R. & Wright, Robert L., 1968]

Example 7.14 Neutron activation

A cylinder made of ^{59}Co has a volume of 1 cm^3. It was placed in a reactor core with a flux of 10^8 n/cm^2/s for 1 year. Calculate the activity of the sample on removal from the reactor and the activity of the same sample 1 year following later. The temperature of the sample during irradiation was 200° C. After the absorption of a neutron, ^{59}Co forms ^{60}Co in its isomeric state. The unstable ^{60}Co decays in two ways, through either internal conversion or beta minus decay as shown in Fig. 6-36. The density of a sample is 8.71 g/cm^3. The atomic weight of ^{59}Co is 58.94 gr/gr-atom.

Let's first determine the decay constants for the two isomers (data are given in Fig. 7-41):

$$\lambda_1 = \frac{\ln 2}{T_1} = \frac{\ln 2}{10.6} \times 60 = 3.97 \text{ h}^{-1} \qquad \lambda_2 = \frac{\ln 2}{T_2} = \frac{\ln 2}{1925.1 \times 24} = 1.50 \times 10^{-5} \text{ h}^{-1}$$

The number of target nuclei is

$$N_0 = \frac{V \rho N_a}{A} = \frac{(1 \text{ cm}^3)(8.71 \text{g}/\text{cm}^3)(6.023 \times 10^{23} \text{ at/gr-atom})}{58.94 \text{ gr/gr-atom}} =$$

$$= 0.890 \times 10^{23} \, ^{59}\text{Co atoms}$$

$^{59}Co + n \longrightarrow {}^{60}Co^*$

$^{60}Co^* \longrightarrow$ Internal conversion to ^{60}Co, $T_1 = 10.5$ minutes, $\sigma_1 = 18$ barns

β- decay to ^{60}Ni, $T_2 = 1925.1$ days, $\sigma_2 = 19$ barns

Figure 7-41. Neutron absorption by ^{59}Co

Since the temperature of the sample was higher than 293 K, the average cross sections must be corrected

$$\overline{\sigma_{a1}(E_p)} = \frac{\sqrt{\pi}}{2}\sigma_{a1}(E_p) = \frac{\sqrt{\pi}}{2}\times 18 = 15.95 \text{ b}$$

$$\overline{\sigma_{a1}(E_p,T)} = \frac{\sqrt{\pi}}{2}\sigma_{a1}(E_p, 293\text{K})\sqrt{\frac{293}{T}} = 15.95\times\sqrt{\frac{293}{473}} = 12.55 \text{ b}$$

$$\overline{\sigma_{a2}(E_p)} = \frac{\sqrt{\pi}}{2}\sigma_{a2}(E_p) = \frac{\sqrt{\pi}}{2}\times 19 = 16.84 \text{ b}$$

$$\overline{\sigma_{a2}(E_p,T)} = \frac{\sqrt{\pi}}{2}\sigma_{a2}(E_p, 293\text{K})\sqrt{\frac{293}{T}} = 16.84\times\sqrt{\frac{293}{473}} = 13.25 \text{ b}$$

The 1-year irradiation period will saturate the short-lived isomer and the second term in Eq. (7-72) can be neglected to give

$$N_1 = \frac{\Phi N_0 \overline{\sigma_{a1}(E_p,T)}}{\lambda_1} = \frac{(10^8 \text{ n/cm}^2\text{s})(0.89\times 10^{23}\text{ atoms})(12.55\times 10^{-24}\text{ cm}^2)}{3.97\dfrac{1}{3600\text{ s}}}$$

$$N_1 = 1.01\times 10^{11}\ {}^{60}\text{Co atoms}$$

The concentration of the longer lived isomer is

$$N_2 = \frac{\Phi N_0 \overline{\sigma_{a2}(E_p,T)}}{\lambda_2}\left(1 - e^{-\lambda_2 t}\right)$$

$$= \frac{(10^8 \, n/cm^2 \, s)(0.89 \times 10^{23} \, atoms)(13.25 \times 10^{-24} \, cm^2)}{1.50 \times 10^{-5} \, \dfrac{1}{3600 \, s}} (1 - e^{-1.50 \times 10^{-5} \times 1 \times 365 \times 24})$$

$N_2 = 3.48 \times 10^{15} \; ^{60}\text{Co atoms}$

The activity on removal is

$$A_0 = N_1 \lambda_1 + N_2 \lambda_2 = 1.11 \times 10^8 \, Bq + 1.45 \times 10^7 \, Bq = 1.26 \times 10^8 \, Bq$$

$$A_0 = 1.26 \times 10^8 \, Bq \times \frac{1}{3.7 \times 10^{10} \, Bq/Ci} = 3.4 \, mCi$$

One year following removal from the reactor core, the activity of the sample will be the activity of the long-lived isotope since the short-lived will have decayed away

$$A = A_0 \left(e^{-\lambda_2 t} \right) = (0.39 \, mCi)(e^{-1.50 \times 10^{-5} \times 365 \times 24}) = 0.34 \, mCi$$

5.   FISSION

5.1   Mechanism of the Fission Process

Fission represents a class of nuclear interactions in which the original target nucleus splits into smaller nuclei. Fission also represents the class of neutron interactions that produces neutrons and energy and as such is a basic principle of nuclear power generation. Fission can be a spontaneous process. For example, ^{240}Pu and ^{252}Cf decay by spontaneous fission (described in Section 5.7); however, such nuclei are rare and the decay rate is very low.

In the neutron-induced fission process, a neutron interacts with the target nucleus creating a compound nucleus that is unstable and splits into smaller nuclei releasing two or more neutrons and energy. The compound nucleus thus temporarily contains all of the charge and mass involved in the reaction and exists in an excited state. The excitation energy added to the compound nucleus is equal to the sum of the binding energy of the incident neutron and its kinetic energy. A schematic of the fission process is illustrated in Fig. 7-42(a) for neutron interaction with ^{235}U. The fission process cannot be explained based on the behavior of one nucleon within the nucleus. The collective behavior of all nucleons is required to be analyzed and is the best explained with the liquid drop model (described in Chapter 2, Section 3.1). According to this model the compound nucleus is approximated with the incompressible fluid drop that has roughly constant density of nucleons with the binding energy per nucleon nearly the same except for the nucleons at the surface (due to surface tension and different electrostatic repulsion). In spherical and symmetric nucleus (i.e., non-deformed) the surface tension tends to stabilize the nucleus by outweighing the disruptive effect of the protons repulsion. When liquid drop is subjected to the mechanical disturbance it is set into oscillation. The collective system of nucleons in a compound nucleus is set to a perturbation mode due to the addition of a neutron (notice that the energy of excitation is not concentrated on a single nucleon). The perturbation causes the compound nucleus to vibrate which is deforming the sphericity and thus the nucleus symmetry. This leads to a non-uniform distribution of the electric charge and diluted effect of the short-range strong nuclear force due to increased surface area, Fig. 7-42(b). During the perturbation two lobes of different sizes are formed. The lobes tend to approach the highest stability around the magic numbers as indicated in Fig. 7-42 b). The neck area also contains nucleons that make the balance of the total number created after the absorption of a neutron. The actual point of break is not known and is statistically distributed over the neck area. The break is caused by the Coulomb repulsion between the protons. The Coulomb energy for two just-touching spherical nuclei as depicted in Fig. 7-42(b) is given with

$$E_{Coulomb} = \frac{kZ_1Z_2e^2}{(R_1 + R_2)}$$

representing the minimum energy that is required for a compound nucleus ($Z = Z_1 + Z_2$, $A = A_1 + A_2$) to break into two smaller nuclei. The available energy for fission to take place is equal to the energy of the reaction, Q, plus the energy supplied to the compound nucleus with the incoming neutron. The Q-value is equal to

$$Q = [M(Z, A) - (M(Z_1, A_1) + M(Z_2, A_2))]c^2$$

and the energy available for fission is defined as follows:

$$E_{available} = Q + E_{supplied} \geq E_{Coulomb}$$

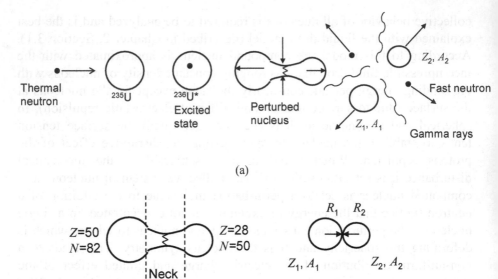

Figure 7-42. (**a**) Schematics of a fission process for ^{235}U; (**b**) perturbed nucleus at the moment of fission according to the liquid drop model

From what has been explained by now, it is clear that fission requires energy to be supplied to stretch the nucleus from its spherical and symmetrical shape into elongated shape and cause the break. If the energy is not sufficient, the perturbed nucleus will return to the spherical form by releasing the extra energy through emission of a γ ray or neutron. That is why fission requires the so-called critical energy (E_{crit}) to be available for a compound nucleus to split into smaller nuclei. The process is illustrated in Fig. 7-43. At the extreme right, fission fragments are apart and the potential energy of the system is virtually zero. As fission fragments become closer, there is an increase in potential energy due to the electrostatic repulsion force acting between the protons. The potential energy reaches its maximum value when the fission fragments are in contact with one another. At this point, the attractive nuclear forces become dominant and the potential energy decreases to a certain value that corresponds to the ground state of the compound nucleus. In order for fission to occur, the nucleus must transition from the left to the right side in Fig. 7-43. The energy difference between the maximum value and the energy that corresponds to the ground state of the compound nucleus represents the critical energy (also called the activation energy) for fission. According to the liquid drop model, the critical energy for fission decreases as Z^2/A increases. This is explained by the fact that repulsion between nucleons (which favors fission) increases with Z^2, while the attraction force is nearly proportional to A: for $Z^2/A < 35$, the critical energy is so large that neutrons (or other particles) of high energy are required to cause fission; for $Z^2/A > 35$, the critical energy is on the order of the binding energy of the incident neutron and thus fission can be caused by a low-energy neutron.

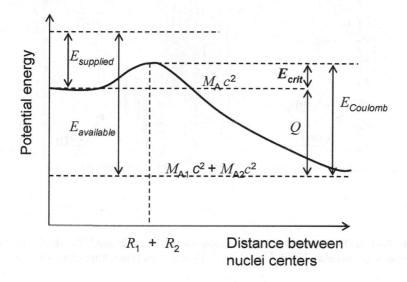

Figure 7-43. Critical energy for fission [Adapted from Rydin, 1977]

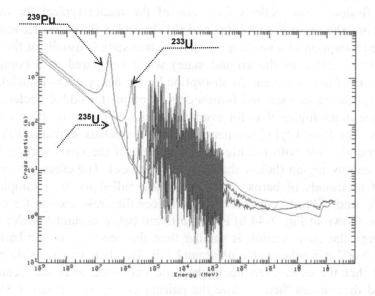

(a)

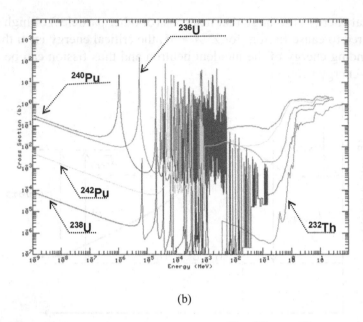

(b)

Figure 7-44. (**a**) Fission cross section for *fissile* nuclei ^{233}U, ^{235}U and ^{239}Pu; (**b**) fission cross section for *fertile* nuclei ^{236}U, ^{238}U, ^{232}Th, ^{239}Pu and ^{240}Pu [From http://atom.kaeri.re.kr/]

The smaller nuclei formed after the compound nucleus decays are called fission products or fission fragments. They are usually radioactive and decay by β^- decay (Section 5.5).

The fission cross sections for some of the nuclei typical for nuclear reactors are shown in Fig. 7-44(a) and (b). In odd-*A* fissile nuclei resulting from the absorption of a neutron the first excited state is usually at the lower energy level (close to the ground state) when compared to a compound nucleus formation from neutron absorption by the heavy even-*A* nuclei. That means that the absorption and fission cross sections for odd-*A* nuclei at low energy are much higher than for even-*A* nuclei which can be observed from Fig 7-44. The lowest lying resonance for odd-*A* nucleus shown in Fig. 7-44 (a) is around 1 eV with the highest magnitude of the cross section in the thermal energy region (below the lowest resonance). The cross section is in order of thousands of barns. Such nuclei are called *fissile*. Examples are ^{235}U, ^{233}U and ^{239}Pu. Except for the resonances the cross section for even-*A* nuclei as shown in Fig. 7-44 b) is insignificant below around 1 MeV; below that energy the cross section is smaller than the one-tenth of the barn, and above 1 MeV it reaches value of around 1 b. This behavior is explained by the fact that the even–even nuclei become even–odd after the neutron is absorbed that causes them to lose the pairing energy (see Chapter 3). This requires for the incoming neutron to possess a kinetic energy sufficient to create compound nucleus capable of decaying through the fission channel.

Such nuclei are considered to be unfashionable beyond the threshold energy that is approximately around 1 MeV and they are called *fertile* nuclei. Examples are ^{236}U, ^{238}U, ^{232}Th, ^{240}Pu and ^{242}Pu.

5.2   Fission Rate and Reactor Power

As described in Section 4.4, the rate of any interaction involving monoenergetic neutrons is equal to $\Sigma\phi$. For fission reactions it follows

$$\text{Fission rate} = \Sigma_f \Phi \ [\text{fissions/m}^3/\text{s}] \tag{7-73}$$

where
$\Sigma_f = N\sigma_f$
$\Phi = nv$
N – number of fissile nuclei [nuclei/m^3]
σ_f – fission cross section [m^2/nucleus]
n – neutron density [neutrons/m^3]
v – neutron speed [m/s]

In a reactor, neutrons are not monoenergetic; they cover a wide range of energies. Neutron flux and cross sections, and thus reaction rates, are energy dependent. At a given neutron energy, the neutron flux at a given time varies with the spatial position in a reactor. Also, the spatial distribution of fissile material is not entirely uniform initially and is not uniform after a reactor has been operating for a certain time. In order to determine a fission rate at a given time, Eq. (7-73) has to be integrated over all neutron energies and spatial positions in a reactor which is in practice done using computer codes.

However, an approximate method can be used to roughly estimate the reactor thermal power. In thermal reactors, the majority of fissions occur in the thermal energy region where flux and macroscopic cross sections are both very large. The fission rate can be estimated assuming the average values (space and energy) for flux and cross section. Therefore, in a reactor of volume V [m^3]

$$\text{Total number of fissions} = V \overline{\Sigma_f \Phi} \tag{7-74}$$

Assuming that the reactor has been operating for enough time that nearly all of the radioactive decay energy is being deposited as heat, and that a fission rate of 3.1×10^{10} fissions/s is required to produce 1 watt of thermal power, thermal reactor power can be approximated with

$$P_{th} = \frac{V\overline{\Sigma_f \Phi}}{3.1\times10^{10}}[W]$$
(7-75)

Example 7.15 Reactor power

A water-moderated reactor contains 100,000 kg of uranium dioxide enriched to an average of 2.5% by weight in ^{235}U. The atomic ratio $H/^{235}U$ is 200. Calculate the approximate (spatial) average thermal neutron flux for a thermal power of 3,000 MWth with an average moderator temperature of 310° C. *(Adopted from: Glasstone and Sesonske, 1994)*

For ^{235}U at $T = 300$ K, the total fission cross section at 0.0253 eV is 584.4 b and $g(T) = 0.9786$:

$$\overline{\sigma_f(E_p)} = g(T)\frac{\sqrt{\pi}}{2}\sigma_f(E_p) = 0.9786\times\frac{\sqrt{\pi}}{2}\times584.4 = 506.8 \text{ b}$$

If the temperature of the neutron population is increased to 310°C (583 K), the average radiative capture cross section becomes

$$\overline{\sigma_f(E_p,T)} = g(T)\frac{\sqrt{\pi}}{2}\sigma_f(E_p,300K)\sqrt{\frac{300}{T}}$$

$$= 506.8\times\sqrt{\frac{300}{583}} = 363.5 \text{ b}$$

The fraction of ^{235}U in $^{235}UO_2$ is 235/(235+(2×16) = 235/267. Therefore, the mass of ^{235}U is

$$m = 10^5\text{ kg}\frac{235}{267}\frac{2.5}{100} = 2.2\times10^3\text{ kg}$$

The total number of ^{235}U nuclei in a reactor is

$$NV = \frac{mN_a}{A} = \frac{(2.2\times10^3)(6.023\times10^{23})}{235\times10^{-3}} = 5.6\times10^{27}\text{ nuclei}$$

and the neutron flux is

$$\overline{\Phi} = \frac{3.1\times10^{10}P_{th}}{NV\overline{\sigma_f(E_p,T)}} = 4.6\times10^{17}\text{ n/cm}^2\text{ s}$$

5.3   Fission Neutrons

In the first 10^{-14} s following the fission process, 99% of the product neutrons are emitted; these are called the prompt neutrons. The prompt neutrons accompany the emission of fission fragments and prompt γ rays. Over a period of several minutes, the unstable fission fragments emit so-called delayed neutrons. The role of each group of neutrons, prompt and delayed, in reactor kinetics is explained in Chapter 7.

The average number of neutrons emitted per each neutron absorbed that causes a fission reaction is usually denoted as ν and for thermal reactor fuel is ~ 2.5 (Table 7-2). This number is not an integer because it represents the average value over a number of fission events (each single fission event emits an integer number of neutrons).

Table 7-2. Number of neutrons emitted per fission, ν, and per neutron absorbed, η

Neutron energy	^{233}U		^{235}U		^{238}U		^{239}Pu	
	ν	η	ν	η	ν	η	ν	η
0.025 eV	2.50	2.30	2.43	2.07	-	-	2.89	2.11
1 MeV	2.62	2.54	2.58	2.38	-	-	3.00	2.92
2 MeV	2.73	2.57	2.70	2.54	2.69	2.46	3.11	2.99

The number of neutrons emitted per each neutron absorbed (in fission and all other interactions) in the fissile materials is denoted as η

$$\eta = \nu \frac{\Sigma_f}{\Sigma_a} \qquad (7\text{-}76)$$

where, for a single fissile material, the macroscopic cross sections for fission and absorption can be replaced with microscopic values

$$\frac{\Sigma_f}{\Sigma_a} \rightarrow \frac{\sigma_f}{\sigma_a} \qquad (7\text{-}77)$$

This ratio represents the fraction of neutrons that are absorbed and subsequently cause fission in a given material and is usually written as

$$\frac{\sigma_f}{\sigma_a} = \frac{\sigma_f}{\sigma_f + \sigma_\gamma} = \frac{1}{1+\alpha} \quad \leftarrow \quad \alpha = \frac{\sigma_\gamma}{\sigma_f} \qquad (7\text{-}78)$$

where α represents the capture-to-fission ratio, an energy-dependent parameter of great importance in reactor core design, as explained later. The number of neutrons emitted per each neutron absorbed can now be expressed in terms of α

$$\eta = \frac{\nu}{1+\alpha} \tag{7-79}$$

However, the reactor core consists of more than one single fissile material, and thus Eq. (7-76) must be written in a more generalized form

$$\eta_{fuel} = \frac{\sum_i \left(\nu\Sigma_f\right)_i}{\Sigma_a} \tag{7-80}$$

where the numerator represents the sum over all fissile nuclides and the denominator represents the total absorption cross section for all materials present in fuel. For example, for a thermal reactor in which the fuel is in the form of uranium oxide and the uranium is a mixture of ^{235}U and ^{238}U, the above equation reduces to

$$\eta_{fuel} = \nu^{235} \frac{\Sigma_f^{235}}{\Sigma_a^{235} + \Sigma_a^{238}}$$

knowing that the only fissile material is ^{235}U and that the absorption cross section for oxygen is small enough to be neglected.

The number of neutrons emitted per each neutron absorbed in the fissile materials, η, is preferred to be as large as possible. The usual reactor-type classification based on the η value is [Rydin, 1977]

- $\eta \geq 1$ for a chain reaction to be feasible
- $2 > \eta > 1$ for converting fertile fuel to fissile
- $\eta \geq 2$ for fuel breeding

Prompt neutrons are emitted with different energies. As a result, the population of prompt neutrons exhibits a distribution or the so-called energy spectrum. The prompt neutron energy spectrum is depicted in Fig. 7-45 for thermal fission of ^{235}U. The spectrum shows a peak (most probable value) at energy of approximately 1 MeV and an average value of 2 MeV. The energy spectrum is important because, in addition to the fissile material present in a reactor core, there is usually an amount of fertile materials (such as ^{238}U or ^{232}Th) for which the fission cross sections have a threshold value below which the fission cross section is zero. The prompt neutron energy spectrum can be described by the following equations:

Cranberg – Frye: $\chi(E) = 0.453e^{-1.036E} \sinh(2.29E)^{0.5}$, Fig.7-44

Maxwellian: $\chi(E) = 0.77e^{-0.776E} \sqrt{E}$ (7-81)

Watt: $\chi(E) = 0.453e^{-E} \sinh(2E)^{0.5}$

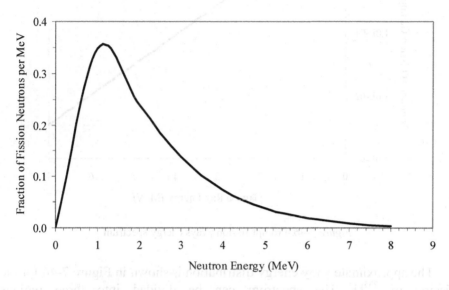

Figure 7-45. Prompt fission neutron energy spectrum for thermal fission of ^{235}U

The fission neutron spectrum is normalized such that

$$\int_0^\infty \chi(E)dE = 1$$

5.4 Fission γ Rays

The γ radiation emitted per each fission event is divided into two groups: prompt and delayed. The prompt γ rays are emitted within 0.1 μs of the fission event (arbitrarily defined time) with an average energy of 1 MeV. One portion of the prompt γ rays is emitted at about the same time as the prompt neutrons, and another portion is represented by the γ rays from the decay of fission fragments with short half-lives. The delayed γ rays come from the decay of fission fragments having half-lives longer than the arbitrarily defined time of 0.1 μs.

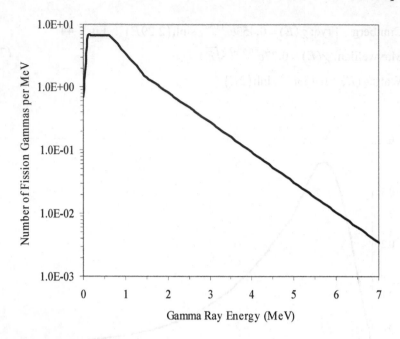

Figure 7-46. Prompt fission γ rays energy spectrum

The approximate γ ray energy distribution is shown in Figure 7-46 for the fission of ^{235}U. The spectrum can be divided into three regions approximately represented by the following relations:

$$\Gamma(E) = \begin{bmatrix} 6.6 & E = 0.1 \sim 0.6 \text{ MeV} \\ 20.2e^{-1.78E} & E = 0.6 \sim 1.5 \text{ MeV} \\ 7.2e^{-1.09E} & E = 1.5 \sim 10.5 \text{ MeV} \end{bmatrix} \qquad (7\text{-}82)$$

The total energy of the prompt γ rays is close to 7.3 MeV per fission event with an average value of around 0.9 MeV.

5.5   Fission Products

5.5.1   Fission Yield

The majority of fission events produce two fission products. The pair formed per single fission varies from event to event giving a broad distribution of isotopes. For example, a detailed study of the thermal neutron fission of ^{235}U has shown that about 80 different isotopes are created. Some of these fission products are shown in Table 7-3. The yield represents the proportion (percentage) of all nuclear fissions that form isotopes of a given mass. The ^{235}U fission yield is plotted in Fig. 7-47 versus atomic mass number. This plot is shown to illustrate that most of the fission fragments are radioactive and decay via β decay, which changes the atomic number but not the atomic mass number. The fission yield curve indicates that the maximum yield for any one isotope is less than 7%. It also indicates that fission products fall into two broad groups: a group of light nuclei with mass number between 80 and 110, and a heavy group with mass numbers between 125 and 155.

Table 7-3. ^{235}U thermal fission products

Element	A	Half-life	Fission yield (%)
Strontium	89	51 days	4.8
Strontium	90	28 years	5.8
Yttrium	91	58 days	5.4
Zirconium	95	65 days	6.3
Ruthenium	103	40 days	3.0
Ruthenium	106	1 years	0.4
Antimony	125	2 years	0.02
Tellurium	127	105 days	0.04
Tellurium	129	37 days	0.35
Cesium	137	30 years	6.2
Cerium	141	33 days	6.0
Cerium	144	280 days	6.0
Promethium	147	2.6 years	2.4
Samarium	151	80 years	0.44

The most probable isotopes to be produced have mass numbers between approximately 95 and 139, each having a yield of about 6.4%. The kinetic energy of fission fragments per fission event is also distributed according to the fission fragment distribution given in Fig. 7-47. For the most abundant isotope in the heavy group, the kinetic energy is around 67 MeV and around 98 MeV for the isotopes in light group. The ratio of these two energies (98/67) is 1.46, which is equal to the ratio of their masses (139/95). The nuclide chain with $A = 135$ leading to ^{135}Xe fission product has a yield of nearly 6 %. With this high yield and the cross section of few million barns in thermal energy region, the xenon poisoning is important effect in thermal reactors (described in Chapter 9).

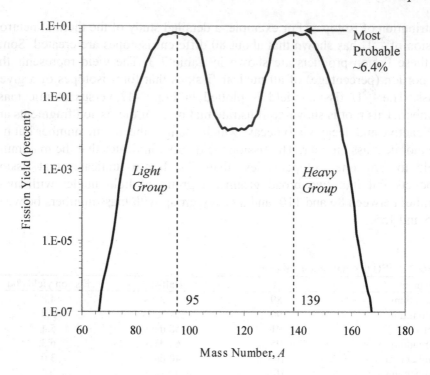

Figure 7-47. ^{235}U thermal fission yield versus atomic mass number

The fission is an asymmetric process as described in Section 5.1 because each lobe of the compound nucleus tends to be doubly magic. Many different modes of fission are possible (which Fig. 7-47 shows) because the break can take place at any distance between the two lobes. Therefore the fission yield of each nucleus depends on the probability distribution of the breakage in the neck region of the compound nucleus and varies with neutron energy (Fig. 7-43).

5.5.2   Formation and Removal of Fission Products in a Reactor

The amounts and activities of individual fission products are important in reactor design because:

- it is necessary to evaluate the potential hazards associated with an accidental release of fission products into the environment
- it is necessary to determine a proper cooling time for the spent fuel (before it becomes ready for reprocessing), which depends on the decay times of fission products
- it is necessary to estimate the rate at which heat is released as a result of radioactive decay of the fission products after the shut-down of a reactor
- it is necessary to calculate the poisoning effect of the fission products

(the parasitic capture of neutrons by fission products that accumulate during the reactor operation)

The rate at which the concentration of a nuclear species (N_i) in a reactor core changes with time is given by (Fig. 7-48)

$$dN_i/dt = \text{Formation rate} - \text{Destruction rate} - \text{Decay rate} \qquad (7\text{-}83)$$

The *formation* of a nuclide i (atomic mass number A, atomic number Z) is defined by fission, neutron capture in nuclide j (atomic mass number $A - 1$, atomic number Z) and radioactive decay (usually β decay) of nuclide k (atomic mass number A, atomic number $Z - 1$). It can be expressed as

$$\text{Formation rate} = \gamma_i N_f \sigma_f \Phi + N_j \sigma_j \Phi + \lambda_k N_k \qquad (7\text{-}84)$$

Where γ_i is the fission yield of that nuclide,

N_f, N_j and N_k are the nuclear number densities of the fissile nuclides,

σ_f is the fission cross section of the fissile material,

σ_j is the capture cross section of the nuclide j,

Φ is the neutron flux and

λ_k is the radioactive decay constant of nuclide k.

The *destruction* of a nuclide i (atomic mass number A, atomic number Z) by neutron capture is defined as

$$\text{Destruction rate} = N_i \sigma_i \Phi \qquad (7\text{-}85)$$

Destruction of a nuclide also occurs through its own β decay, which is expressed as

$$\text{Decay rate} = \lambda_i N_i \qquad (7\text{-}86)$$

Thus the rate at which the concentration of a nuclear species (N_i) in a reactor core changes with time becomes

$$dN_i/dt = \gamma_i N_f \sigma_f \Phi + N_j \sigma_j \Phi + \lambda_k N_k - N_i \sigma_i \Phi - \lambda_i N_i \qquad (7\text{-}87)$$

This equation can be solved for N_i, at any time, assuming that all other concentrations and constants are known. This equation develops into coupled differential equations for which the exact solution is obtained using computer codes.

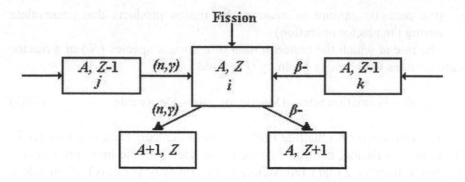

Figure 7-48. Formation and removal of fission products

After a certain time of reactor operation, a concentration of any fission fragment should reach an equilibrium (saturation) value. At that point, the rate of production is equal to the rate of removal of that nuclide. For many important fission products, like ^{90}Sr or ^{131}I, only the first and the last terms in Eq. (7-87) are significant

$$dN_i/dt \sim \gamma_i N_f \sigma_f \Phi - \lambda_i N_i \tag{7-88}$$

At the reactor start-up ($t = 0$), we may assume that the concentration of fission fragment i is zero. Also, a good assumption is that the neutron flux reaches a constant value shortly after start-up. At that point in time, a saturation concentration (density) of the nuclide i is

$$N_{i,sat} \approx \frac{\gamma_i N_f \sigma_f \Phi}{\lambda_i} \tag{7-89}$$

It follows that for a given flux and fission fragment cross section (σ_f), the saturation number density is increased by large fission yield (γ_i) and long half-life (small λ_i). If other conditions are equal, the saturation number density increases with the neutron flux.

The ratio of the number density at any time to the saturation value is then given by

$$\frac{N_i}{N_{i,sat}} \sim 1 - e^{-\lambda_i t} \tag{7-90}$$

When the reactor is shut-down, the neutron flux becomes negligible and the rate at which the concentration of a nuclear species (N_i) in a reactor core changes with time reduces to

$$dN_i/dt = \lambda_k N_k - \lambda_i N_i \qquad (7\text{-}91)$$

If the rate of decay of k into i is larger than the decay rate of i, then the nuclear density of the fission fragment i increases with time. If i decays faster than it is generated by the decay of k, then the fission fragment concentration decreases with time. In this case, however, after a certain period of time, the decrease in k decay will produce a situation where the concentration of i attains its maximum value (when $\lambda_k N_k = \lambda_i N_i$). After that, the concentration decreases again with time.

The activity after shut-down may be determined using a semi-empirical approach (for times 10 s to 100 days after shut-down), which gives a close estimate to detailed calculations using computational methods. The total rate of β emission is given by the following semi-empirical relation

$$\text{Rate of β emission/fission event} \sim 3.2\ t^{-1.2}\ [1/s] \qquad (7\text{-}92)$$

where t is given in seconds after the fission event. If every fission product is assumed to emit a β particle when it decays, the activity may be determined by

$$\text{Fission product activity/fission} \sim 3.2\ t^{-1/2}\ [\text{Bq}] \sim 8.6 \times 10^{-11}\ t^{-1/2}\ [\text{Ci}] \qquad (7\text{-}93)$$

5.6 Energy Released in Fission

The energy released by fission can be calculated based on the difference in mass between the masses of the neutron and the fissile nucleus before fission and the fission fragments and fission neutrons after fission. There is a variation in the total energy released per fission that depends on the fissionable isotope and the products of the fission event. On average, some 200 MeV is released per thermal fission. This energy is distributed as shown in Table 7-4.

Table 7-4. Fission energy distribution

Fission product	Energy (MeV)
Kinetic energy of fission fragments	165 ± 5
Instantaneous gamma rays	7 ± 1
Kinetic energy of neutrons	5 ± 0.5
Beta particles from product decay	7 ± 1
Gamma rays from product decay	6 ± 1
Neutrinos from product decay	10
Total	200 ± 6

Example 7.16 Energy released per thermal fission event

In a typical thermal fission of ^{235}U as shown below, calculate the instantaneous fission energy.

$$_{0}^{1}n + {}_{92}^{235}U \rightarrow {}_{92}^{236}U^{*} \rightarrow {}_{55}^{140}Cs + {}_{37}^{93}Rb + 3\left({}_{0}^{1}n\right)$$

$$m_{\text{reactants}} = m_{{}_{92}^{235}U} + m_{{}_{0}^{1}n} = \left(235.043924 + 1.008665\right) = 236.052589 \text{ amu}$$

$$m_{\text{products}} = m_{{}_{55}^{140}Cs} + m_{{}_{37}^{93}Rb} + 3\left(m_{{}_{0}^{1}n}\right) =$$

$$\left(139.90910 + 92.91699 + 3.02599\right) = 235.85208 \text{ amu}$$

The instantaneous fission energy is the energy released immediately after the fission process. It is equal to the energy equivalent of the mass lost in the fission process. It can be calculated as follows

$$\Delta m = m_{\text{reactants}} - m_{\text{products}} = 236.052589 - 235.85208 = 0.200509 \text{ amu}$$

$$E = 0.200509 \text{ amu} \left(\frac{931.5 \text{MeV}}{\text{amu}}\right) = 186.8 \text{ MeV}$$

5.7 Spontaneous Fission

Spontaneous fission is a characteristic mode of radioactive decay for very heavy nuclei. However, the spontaneous fission is less energetically probable than other modes of the decay such as the α decay. The mathematical condition for spontaneous fission to occur, $Z^2/A > 45$, is developed based on the liquid drop model of fission process (Section 5.1).

Radioisotopes for which spontaneous fission is a nonnegligible decay mode are used as neutron sources. One such example is ^{252}Cf that decays by spontaneous fission with 3.09% probability (the remaining decay mode is by α emission with 96.91% probability). The half-life is 2.645 years. The ^{252}Cf neutron emission is 2.314×10^6 sec^{-1} μg^{-1}, with specific activity of 0.536 mCi/μg. The neutron energy spectrum is similar to a fission reactor spectrum; the most probable energy is 0.7 MeV and the average energy is 2.1 MeV. The spontaneous fission rates for few heavy nuclei are listed in Table 7-5. The α decay for these isotopes is ten to hundred million times more probable mode of a radioactive decay.

Table 7-5. Spontaneous fission yields [Shults, Kenneth J. & Faw, Richard E. (2002)]

Nucleus	Half-life	Fission probability/decay [%]	Neutrons per fission	Neutrons /g s
^{235}U	7.04×10^8 years	2.0×10^{-7}	1.86	3.0×10^{-4}
^{238}U	4.47×10^9 years	5.4×10^{-5}	2.07	0.0136
^{239}Pu	2.41×10^4 years	4.4×10^{-10}	2.16	2.2×10^{-2}
^{240}Pu	6,569 years	5.0×10^{-6}	2.21	920
^{252}Cf	2.638 years	3.09	3.73	2.3×10^{12}

APPLICATIONS

Neutron Activation Analysis (NAA)

The NAA is developed based on the fact that there are over 50 elements present in nature that have radioactive isotopes with one neutron more than their stable isotopes. NAA represents a sensitive and accurate technique that is used to determine the chemical content of a material sample by identifying the presence of elements through neutron irradiation (Section 4.8). The important advantage of this technique is that it preserves the sample. Samples can be irradiated in a reactor or by using any other portable neutron source (Section 3, Section 5.7).

The neutron irradiation of a sample induces radioisotopes formation that following irradiation start decaying through the emission of particles or gamma rays which are characteristics of the elements. As described in Section 4.5 when neutron, electrically neutral particle, passes through the material it can be absorbed by the atom nucleus and create a new atom that is usually radioactive. The most common interaction is a radioactive capture which means the gamma rays will be emitted. It is important to notice that the electrical neutrality of a neutron assures that the center of the material sample is equally well "seen" as its surface.

NAA is nowadays used in environmental studies to differentiate pollutants and verify their sources, in measuring traces of impurities in semiconductors and pharmaceuticals, in analyzing the evidence in the investigation of criminal cases, in studying the historical artifacts and in nutritional studies.

An example of NAA use was in measuring the iridium in soil that initiated a theory that the dinosaurs vanished 65 million years ago due to the collision of an asteroid with the Earth. Such a collision is speculated to have caused the sunlight reduction with a noticeable impact on life on Earth because the dust has been spread all over the globe; the first effect was that fewer plants grew. The observable reduction in plant food has affected the number of animals. The reduction is assumed to be so vast that most species, including the dinosaurs, became extinct. The regular content of iridium in

Earth crust is 1 µg/kg and in asteroids is around 500 µg/kg. Thus any increase in iridium content in Earth soil points toward the possibility of asteroid impact on Earth because iridium is not soluble in water and does not migrate to underground water beds.

Reactor Flux Profiling with Foil Irradiation

The neutron flux levels are profiled (measured) using the techniques called the foil irradiation. Usually the gold, indium, cadmium or copper foils or wires are placed in the neutron field of the reactor to map the neutron flux levels. The wires would give a three-dimensional flux profile while foils are used at the specific points to verify the measurement. Indium has a high resonance with the cross section of 29,400 b for radiative capture of neutrons at energy of 1.44 eV, Fig. 7-49. Thus ^{115}In will be activated by neutrons below and at the resonance. ^{113}Cd has high resonance capture at the upper end of thermal energy region. Notice that the cross section rapidly drops below 10 b at neutron energies below the 1.44 eV ^{115}In resonance. When the foil (indium or gold for example) is covered with cadmium and irradiated at the same position, since cadmium is transparent to the epithermal neutrons and absorbent for the thermal neutrons, the activity of the covered foil will be mainly from the epithermal neutrons. Thus the difference between the covered and un-covered (bare) foils will indicate the activity due to thermal neutron flux. Usually the so-called *cadmium ratio* is defined as the ratio of activity of the bare to the cadmium-covered foil.

Neutron Radiography and Neutron Radioscopy

Neutron radiography like X-ray radiography (described in Chapter 5) is a static two-dimensional non-destructive imaging technique providing detailed information about material types and structures inside an object. The neutron beam to which the object is exposed attenuates through both scattering and absorption according to the attenuation law described in Section 4.9 and depends on material cross sections. Neutrons are attenuated by the light materials such as hydrogen, boron, beryllium, oxygen, nitrogen or lithium but penetrate many heavy materials. Thus using the neutron radiography it is possible to distinguish between the different isotopes especially the light elements. In the X-ray radiography the beam attenuation increases regularly with the atomic mass number (interaction with electron cloud), while the attenuation in neutron radiography is rather random function of atomic number (interaction with nucleus). Main application is in visualizing the two-phase fluid flow, migration of fluids into porous media or detection of explosive materials.

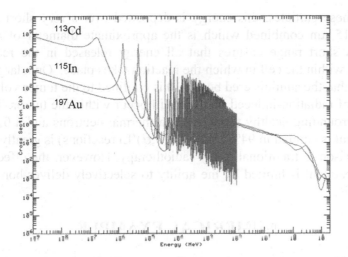

Figure 7-49. Radiative capture cross section for ^{113}Cd, ^{115}In and ^{197}Au [From http://atom.kaeri.re.kr/]

Neutron radioscopy represents a continuous visualization of the attenuation of a neutron beam using real-time detectors. Neutron radiography images are obtained on a photographic film and therefore are two-dimensional images; neutron radioscopy requires scintillators and video camera producing time-dependent 3D images of dynamic events in complex systems.

Boron Neutron Capture Therapy

The radiation binary targeted therapies are modern treatment options that are developed to assure a high degree of killing effect to the cancer tumor, while maximizing the spearing of the healthy tissue. Boron neutron capture therapy is one such method proposed in 1936. From its first trials in 1950s to these days the BNCT has been the topic of ongoing clinical investigation for the treatment of advanced glioma, head and neck cancers and melanoma. The theoretical concept of BNCT is to selectively deliver boron-10 (isotope of high absorption cross section, see Section 4.5.4) to tumor cells, which are then irradiated with a beam of thermal (or epithermal) neutrons. The resulting reaction releases an alpha particle and a lithium ion, both massive charged particles with high linear energy transfer (LET):

$$ {}^{10}B + {}^{1}n \rightarrow {}^{11}B^* \rightarrow {}^{7}Li + \alpha + 2.79 \text{ MeV} $$

Of the 2.79 MeV released in the reaction, 1.47 MeV and 0.84 MeV are imparted to the α and the lithium ion, respectively, in the form of kinetic energy [Mundy, D., Harb, W. and Jevremovic, T., 2006]. Due to the high

LET of these particles, this energy is deposited over a *very* short range of only 10–15 μm combined which is the approximate diameter of a human cell. This short range ensures that all energy released in the reaction is deposited within the cell in which the reaction takes place. On a larger scale, provided that the administered boron-10 is confined to the tumor volume, the majority of radiation-induced damage will occur within the tumor. The dose to the surrounding healthy tissue (due to thermal neutrons and a 0.48 MeV gamma that is emitted in 94% of the $^{10}B(n,\alpha)^7Li$ reactions) is greatly reduced in comparison to traditional X-ray radiotherapy. However, the effectiveness of this treatment is limited by the ability to selectively deliver boron to the tumor cells.

NUMERICAL EXAMPLE

Neutron Attenuation in Common Moderator Materials

Graphite and water are two common materials used as moderators in nuclear reactors. Using the given data, construct an attenuation plot of two beams of thermal neutrons passing through water and graphite. Comment on the apparent effectiveness of each material as a moderator.

	ρ (g/cm^3)	σ_t (b)	M (g/mole)
Water	1.0	5.33	18.015
Graphite	1.6	103.66	12.000

Solution in MATLAB:

```
clear all
Na = 6.022e23; % Avogadro's Number
% Total thermal microscopic cross section
sigma_C = 5.33; %b
sigma_Water = 103.66; %b
s = 10^-24*[sigma_C sigma_Water];
rho(1) = 1.6; % carbon g/cm^3
rho(2) = 1; % water g/cm^3
M = [12 18.015];
for i = 1:2
   Sigma(i) = s(i)*rho(i)*Na/M(i);
end
x = linspace(0,15);
for j = 1:2
```

```
for i = 1:100
    I(i,j) = exp(-Sigma(j)*x(i));
end
end
figure
hold on
plot(x,I(:,1),'k')
plot(x,I(:,2),'k:')
xlabel('Distance (cm)')
ylabel('Fractional Intensity')
legend('Graphite','Water')
```

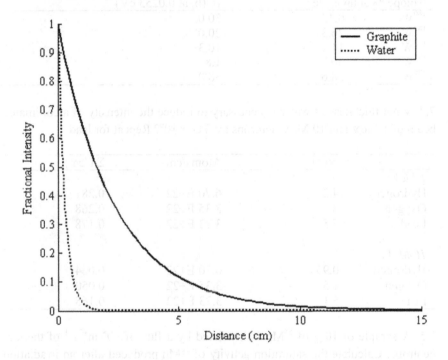

Figure 7-50. Neutron attenuation in graphite and water

Water appears to be a generally more effective moderating material (Fig. 7-50).

PROBLEMS

7.1. Uranium oxide (UO_2) has a theoretical density of 10.96 g/cm³. Calculate the number density (nuclei/cm³) of uranium and oxygen if a sample of UO_2 has a density equal to theoretical density. Calculate the number densities for the fuel in a reactor that has usually a density equal to 0.95 the theoretical density.

7.2. Calculate the macroscopic absorption and scattering cross section for 4.8 atom percent enriched UO_2 fuel both at 20° C and 300° C. Assume the density of UO_2 is 92% theoretical density (see Problem 7.1).

7.3. The microscopic cross sections at 0.0253 eV for tungsten are given in the table below. Calculate the capture cross section for the element tungsten. Which isotope contributes the most to the capture cross section? If only the isotopes 180, 184 and 186 produce a radioactive daughter by the reaction (n,γ) what is the activation cross section for tungsten?

Isotope % abundance		σ_c (b, at 0.0253 eV)
^{180}W	0.12	30.0
^{182}W	26.3	20.0
^{183}W	14.3	10.3
^{184}W	30.7	1.8
^{186}W	28.6	38.0

7.4. What thickness of water is necessary to reduce the intensity of a collimated beam of 1 MeV and 10 MeV neutrons by factor 10^6? Repeat for lead.

	σ (b)	Atoms/cm^3	Σ (/cm)
1 MeV:			
Hydrogen	4.2	6.70 E+22	0.281
Oxygen	8	3.35 E+22	0.268
Lead	5.5	3.23 E+22	0.178
10 MeV:			
Hydrogen	0.95	6.70 E+22	0.064
Oxygen	1.5	3.35 E+22	0.050
Lead	5.1	3.23 E+22	0.165

7.5. A sample of 10 g of ^{55}Mn is irradiated by a flux of 10^8 m^{-2}s^{-1} of thermal neutrons. Calculate the saturation activity of ^{56}Mn produced after an irradiation time of 7 h. The cross section for $^{55}Mn(n,\gamma)^{56}Mn$ is 13.41 b and the half-life of ^{56}Mn is 2.6 h.

7.6. If 10 g of gold sample is inserted into the reactor at neutron flux of 10^9 n/cm^2 s, how many atoms of ^{198}Au will be formed after 30 min? What is the activity of the sample after it is removed from the reactor assuming none of the gold atoms decays until removed from the reactor.

7.7. Discuss the following two nuclear reactions:

$$^{35}\text{Cl}(n,\alpha)^{32}\text{P} \qquad ^{32}\text{S}(n,p)^{32}\text{P}$$

7.8. Calculate the threshold energy for the reaction $^{13}\text{C}(n,\alpha)^{10}\text{Be}$. The atomic masses in amu are $M(^{13}\text{C}) = 13.0033548$; $M(^{10}\text{Be}) = 10.0135337$.

7.9. A parallel beam of 0.25 MeV neutrons impinges on target of aluminum that is 1 cm thick. Calculate what fraction of neutrons will undergo a neutron capture event on their first collision in the last 1 mm of a target (σ_{tot} (Al) = 3b and σ_γ(Al) = 1b).

7.10. If an isotropic source is placed in the center of a sphere what is the probability (in percent) that a neutron will be emitted in a cone with a solid angle of 0.30 steradians?

7.11. Prove Eq. (7-57). Calculate the most probable energy for neutrons with Maxwell–Boltzmann distribution and explain why it is not the energy corresponding to the most probable velocity?

7.12. Evaluate the nuclear reaction $^4\text{N} + {}^4\text{He} \rightarrow {}^{17}\text{O} + {}^1\text{H}$. Is it endothermic or exothermic? Calculate the energy (in MeV), Q, of the reaction. Masses in amu: H = 1.007825; neutron = 1.008665; He = 4.00260; ^{14}N = 14.00307; and ^{17}O = 16.99914.

7.13. Some stars at the end of their lives collapse combining their protons and electrons to form a so-called neutron star. Such a star could be approximated by a giant atomic nucleus. Assume its mass is equal to that of the Sun (2×10^{30} kg) and that it collapsed into neutrons (1.67×10^{-27} kg), what would be the radius of this star?

7.14. Boron is a common material used to shield against thermal neutrons. Calculate the thickness of boron required to attenuate an incident thermal neutron beam to 0.1% its intensity. Use the thermal cross section of 103 cm^{-1}.

7.15. Calculate the fission rate density to produce a thermal power density of 400 kW/litter (typical for fast breeder reactors), assuming that the main fissile isotope is ^{239}Pu.

7.16. Follow the numerical example as given and calculate and plot the neutron beam attenuation through beryllium.

7.17. Plot the fission cross section to show that ^{232}Th requires a very fast neutron to induce fission. Compare it to the conditions for fission on ^{238}U.

7.18. ^{236}U fissions into ^{102}Mo and ^{131}Sn isotopes. Knowing that the number of nucleons must be conserved write the reaction. How many free neutrons are produced in this and how many in the fission process that produces ^{88}Br and ^{140}La?

7.19. If one of fission fragments for the ^{240}Pu fission is ^{90}Sr what is the second element? If the mass number of the second element is 142, how many free neutrons are produced? Write the equation.

7.20. For the fission reaction ^{235}U + n → ^{142}Cs + ^{90}Rb + $4n$ estimate the energy released per reaction.

7.21. A borated-steel sheet (relative density 7.8) which is used as a control rod in a reactor is 2 mm thick and contains 2% boron by weight. The atomic masses of boron and iron are 10.8 and 55.9 and their nuclear absorption cross sections for thermal neutrons are 755×10^{-28} m^2 and 2.5×10^{-28} m^2, respectively. Assuming that the thermal neutrons strike the sheet at normal incidence, what fraction of them is absorbed?

7.22. The nuclide ^{256}Fm decays through spontaneous fission with a half-life of 158 min. If the energy released is about 220 MeV per fission, calculate the fission power produced by 1 μg of this isotope.

7.23. The thermal fission cross section for ^{235}U is 577 b while its thermal capture (non-fission) cross section is 101 b. The isotope ^{238}U does not fission for neutrons with thermal energies but does have a small capture cross section of 2.75 b. Naturally occurring uranium is 99.3% ^{238}U and 0.7% ^{235}U. Given that an average of 2.44 fast neutrons is produced per fission calculate how many of these fast neutrons are produced for each thermal neutron absorbed in natural uranium.

7.24. A 100 MW reactor consumes half its fuel in 3 years. How much ^{235}U does it contain?

7.25. A beam of thermal neutrons is incident upon a thick layer of cadmium (density 8,650 kg/m^3, cross section 24,506 b). Find the absorption length (i.e., the distance in which the beam is reduced by a factor $1/e$).

7.26. A free neutron decays into a proton, electron and antineutrino. Assuming the latter to be massless and the original neutron to be at rest, calculate the maximum momentum that could be carried off by the electron and compare this with the maximum momentum which the antineutrino could have.

7.27. A spectrum of β particles are emitted during the fission process. How far will a 9 MeV β travel in a water-moderated reactor? (Recall the radiation interactions with matter described in Chapter 6).

7.28. Cadmium and boron are strong neutron absorbers and are the most common materials used in control rods. Write neutron absorption reaction in boron and calculate the Q-value for this reaction.

7.29. Find the energy of a hydrogen atom moving at speed of 2.2×10^6 cm/s. What is the kinetic energy of the thermal neutron at room temperature moving at the speed of 2,200 m/s?

7.30. Show that the number of neutrons per absorption, η, for ^{235}U homogeneous thermal reactor is ~2.08.

7.31. Calculate the neutron density from a reactor thermal flux of 10^{12} n/cm^2 s. Compare it with the number of particles 1 cm^3 contains at standard temperature and standard pressure.

7.32. Estimate the reactor power for which the fuel is made of 5% enriched uranium metal. The total weight of the fuel is 100 kg. The average neutron flux is 10^{13} n/cm^2 s. Assume the density of the fuel is 18.7 g/cm^3. The microscopic fission cross section for ^{235}U is 549 b.

7.33. Determine the probability that a 2 MeV neutron will undergo its first collision in 0.476 cm diameter UO$_2$ fuel rod enriched to 4% in ^{235}U. Assume that the neutron is born in the center of the fuel rod and that it travels radially toward the fuel boundary. The fuel density is 94% theoretical density (10.96 g/cm^3).

7.34. For the interference term in elastic scattering cross section explain the effect of increased temperature. Illustrate the trend (if any) and explain it.

7.35. Define the differential microscopic cross sections by appropriate integrals. Can the same concept be applied to macroscopic cross sections? If so, write the appropriate mathematical expressions.

7.36. Define the time-dependent neutron angular flux. Write to show why $\Phi(\vec{r}, E, \vec{\Omega}, t) dVdEd\Omega dt$ represents the total of the path lengths traveled during the time dt by all neutrons in the incremental phase space volume $dVdEd\Omega$.

7.37. (a) Write the excitation energy in the compound nucleus after the absorption of a neutron. [Answ. $M(Z,N) + M_n - M(Z,N+1)$]

(b) Calculate the difference in excitation energy for these two compound nucleus formations:

$$^{235}_{92}U + n \rightarrow ^{236}_{92}U^*$$

$$^{238}_{92}U + n \rightarrow ^{239}_{92}U^*$$

7.38. Calculate the approximate kinetic energy of a nucleon inside the nucleus. State all the assumptions. [*Hint*: Because of the strong nuclear force acting between the nucleons, each nucleon occupies a volume with the radius approximately equal to a half range of the strong nuclear force, $a \sim 2 - 3 \times 10^{-13}$ cm; the radius is equal to the reduced de Broglie wavelength of the nucleon.]

7.39. What atom% of uranium is needed to assure the thermal factor η of 1.85?

7.40. A thermal reactor is fueled with ^{235}U mixed with ^{10}B. Calculate what percentage of the ^{10}B has been consumed at the time of 50% consumption of ^{235}U.

7.41. The elastic scattering of a neutron on hydrogen is predominantly forward scattering in the laboratory system (the system where we measure and observe the interaction events). This means that the angular differential cross section is strongly peaked forward. Sketch this angular differential scattering from the point of interaction.

7.42. Complete the nuclear reaction and calculate the Q-value: $^{6}_{3}Li + ^{2}_{1}H \rightarrow ^{4}_{2}He +?$ Comment.

7.43. For a ^{235}U atom that splits up into two nuclides with mass number 117 and 118, estimate the amount of energy that is released

7.44. Free neutron decays with half-life of 11.7min. Calculate the relative probability that a thermal neutron will decay before it is absorbed in an infinite medium of water. How does neutron decay?

7.45. The fission obtained from ^{239}Pu is empirically represented with

$$\chi(E) = 0.6739\sqrt{E}e^{-E/1.41}$$

Calculate the average energy of the fission neutrons.

Chapter 8

NEUTRON TRANSPORT

Time-Independent and Time-Dependent Neutron Transport
Theory – Concepts and Examples

When we have carried out the indicator experiments that proved that barium was present, I wrote some personal letters to Lise Meitner, telling her of our results. In my letter of 19 December I wrote: ... The thing is there is something so odd about the 'radium isotopes' that for the moment we don't want to to tell anyone but you. The half-lives of three isotopes are pretty accurately determined; they can be separated from all elements except barium; all reactions are correct. Except for one – unless there are some very weird accidental circumstances involved: the fractionation doesn't work. Our Ra isotopes behave like Ba... Strassmann and I agree that for time being nobody should know but you. Perhaps you can put forward some fantastic explanation... *Otto Hahn* (1879–1968)

1.   INTRODUCTION

Design of a reactor core requires detailed prediction of the *balance* between neutron production and neutron loss. The rates of neutron production, transport and absorption are key information not only for core design and analysis but also for thermal–hydraulic, heat–mass transfer, accident scenarios and radioactivity release estimates. After neutrons are born in fission reactions, they move through the reactor core and undergo collisions of various types (absorption and scattering). There are two main absorption processes which may occur, radiative capture and fission. In fission, the target isotope splits and releases additional neutrons. In radiative capture, the neutron is parasitically absorbed and does not contribute to sustaining the chain reaction. In scattering collisions (elastic or inelastic)

T. Jevremovic, *Nuclear Principles in Engineering*,
DOI 10.1007/978-0-387-85608-7_8, © Springer Science+Business Media, LLC 2009

neutrons change their energy, spatial position and direction of motion in a process known as *slowing down*. In general, the interaction of neutrons with nuclei in medium may be considered as neutrons being *transferred or transported* from one location to another, from one energy to another and from one direction of motion to another. A schematic diagram of the various paths for a neutron born in a thermal reactor is depicted in Fig. 8-1. The details of neutron interactions as well as the concept of the sustained (and controlled) chain reaction are described in the succeeding sections.

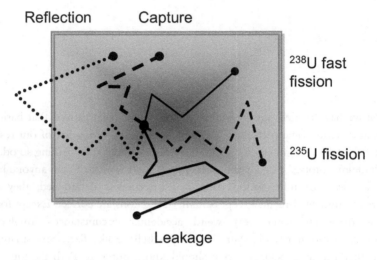

Figure 8-1. Schematic diagram of the history of a neutron born in thermal nuclear reactor

2.   CONCEPT OF TIME-INDEPENDENT NEUTRON TRANSPORT

2.1   The Nuclear Chain Reaction

Seven months after the discovery of the neutron and more than 6 years before the discovery of uranium fission (September 1932) Leo Szilard postulated that a controlled release of nuclear power may be possible if materials that would sustain the neutron chain production could be identified. In 1934 he filed a patent application for a weapon based on the release of nuclear power from such materials and also defined the concept of critical mass. A year later, he received the patent which made him the legal inventor of the nuclear bomb. The reason he patented the idea was to protect the use of such powerful weapon and prevent the destruction Second World

War brought to humanity. After the discovery of neutrons, many scientists across the world developed a number of experiments to analyze the effects of bombarding different materials with this new particle.

In late 1938, Otto Hahn and Lise Meitner were able to develop a theoretical interpretation of experiments involving neutron interactions with uranium. On December 21, 1938, Hahn submitted a paper to a German journal, *Naturwissenschaften*, in which he showed convincing evidence of the fission leading to production of radioisotopes from uranium irradiated with thermal neutrons. Soon after the concept of the fission was understood, a number of trials followed to find the method to produce a self-sustained reaction in which neutrons born in fission would induce fission in other uranium nuclei. As described in Chapter 7, on average 2.5 neutrons are emitted per thermal fission event. In order to sustain a fission reaction, at least one should be conserved to continue the fission process. The essential problem in achieving a sustained nuclear fission reaction is related to the neutron economy.

A history of a single neutron born from fission in enriched uranium is schematically depicted in Fig. 8-1. Interaction of ^{235}U with a neutron of any energy will split the nucleus into two smaller nuclei and release a few fast neutrons. However, ^{238}U can absorb neutrons in the non-fission reaction, i.e., radiative capture that removes neutron from the chain reaction. This is why the fuel in thermal reactors is enriched in ^{235}U by increasing its content from the natural value of 0.7%. The typical enrichment in nuclear power reactors is about 5%. Fast neutrons can produce fission of ^{238}U nucleus and the probability of this interaction depends on reactor core structure, fuel type and fuel composition. Along with being absorbed (radiative capture by fuel or other materials present in a core) neutrons can be removed by escaping the physical boundaries of the system. As long as more neutrons are produced than lost the chain reaction will be sustained and the fission process will generate additional neutrons and energy.

In nuclear weapon, the chain reaction is uncontrolled and a giant amount of energy is generated in a short period of time leading to an explosion. In nuclear reactors the control and sustainability of the chain reaction is achieved by introducing materials which absorb neutrons.

2.2 Fick's Law

The number of neutrons per unit volume is a function of neutron energy, neutron spatial position and its direction of motion, and is referred as *angular neutron density*. The definition of neutron density, flux and current is provided in all details in Chapter 7, Section 4.4.

Neutron balance is described by the *neutron transport equation* which expresses the distribution of the neutron population in space, energy and time. In a steady-state condition, the neutron density is assumed to be constant with respect to time.

The neutron transport equation is also called the *Boltzmann equation* because it is derived from the kinetic theory of gases developed by Boltzmann in the later part of the 19th century.

Neutrons of a given energy moving in a given direction collide with nuclei of atoms in a reactor core producing other neutrons that have a wide range of energies and directions of motion. It is thus necessary to describe neutron transport by integrating over all neutron energies and spatial directions. The neutron transport equation is therefore an integro-differential equation which can be solved exactly for only a few simple cases. For practical applications, various simplifications and computational methodologies are developed and solutions are produced using complex software packages.

One of the simplest approximations to transport theory is *diffusion theory*. The name is given because it involves relationships similar to Fick's law of gas diffusion. Diffusion theory is explained in detail in the following sections. Section 4.6 focuses on what is currently the most attractive and advanced deterministic approach in neutron transport modeling in complex geometries (method of characteristics) including representative examples for various reactor types. Diffusion is defined as the random walk (Brownian motion) of a group of particles from a region of higher concentration to a region of lower concentration. This means that the diffusing particles flow in the direction of decreasing concentration and such a flow rate is proportional to the negative concentration gradient (Appendix 6).

Fick's law defines diffusion of particles from the region of higher concentration to the region of lower concentration as (see Fig. 8.2)

$$J_x = -\chi \frac{dC}{dx} \tag{8-1}$$

where J_x [cm^{-2}s^{-1}] represents the net current in direction x, C is the particle concentration, x is position and χ is the diffusivity constant [cm^{2}s^{-1}] which describes how fast (or slow) particles diffuse. Concentration is defined as the amount of mass in a given volume represented in units of mol/cm^3 or mol/liter. The negative sign indicates that J_x is positive when the movement is down the gradient, i.e., the negative sign cancels the negative gradient along the direction of the positive net current.

Fick's law can also be written in terms of particle flux Φ, as

$$J_x = -D \frac{d\Phi}{dx} \tag{8-2}$$

where $d\Phi/dx$ $[\text{cm}^{-3}\text{s}^{-1}]$ represents the flux gradient and D [cm] is called the *diffusion coefficient*. Fick's law can be written in vector form to analyze three-dimensional space (neutron current vector is sketched in Fig. 8-3)

$$\vec{J}(\vec{r},t) = -D\nabla\Phi(\vec{r},t) \tag{8-3}$$

where $\vec{J}(\vec{r},t)$ represents the neutron current density or the net flow vector of neutrons passing through a unit area perpendicular to the direction of neutron motion per unit time (Chapter 7). Divergence of the neutron current density represents the net number of neutrons leaking from the unit volume per unit time (see next section) and the term $\nabla\Phi(\vec{r},t)$ represents the gradient of neutron flux in three-dimensional space.

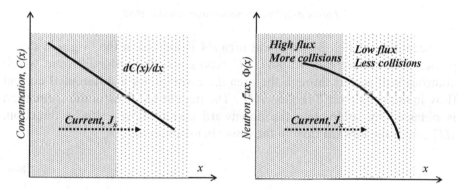

Figure 8-2. Fick's law

2.3 Diffusion Coefficient and Diffusion Length

Neutron current density, neutron flux and the diffusion coefficient are correlated variables. In order to derive the relation for the diffusion coefficient which depends on the nuclear characteristics of the medium the following assumptions are made:
1. There is no neutron sources in the medium of interest
2. The medium is homogeneous, i.e., neutron cross section is independent of spatial position
3. Angular neutron distribution in the medium is isotropic
4. Neutron flux is nearly uniform in the medium
5. The medium is considered to be infinite

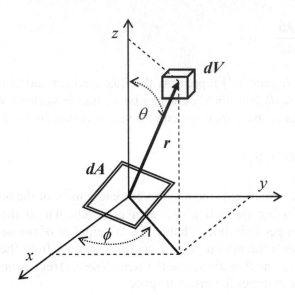

Figure 8-3. The formation of neutron current

According to Fig. 8-3, the unit area dA is located at the origin in the zy plane. The differential volume dV represents the volume from which neutrons will scatter through the area dA contributing a differential current flow in the negative z direction, dJ_z^-. The net current flow in the z direction is obtained by subtracting the downward current flow in the z direction, dJ_z^-, from the upward flow in the lower hemisphere, dJ_z^+

$$J_z = J_z^+ - J_z^- \tag{8-4}$$

The number of neutrons which are elastically scattered per unit time from the differential volume dV placed at distance $\vec{r}$ from the origin is equal to

$$\Sigma_s \Phi(\vec{r}) dV \tag{8-5}$$

where Σ_s is the elastic scattering cross section and is not a function of position due to assumption 2; however, the neutron flux is position dependent. Because it was assumed that neutrons are scattered isotropically in the medium (assumption 3) the number of neutrons which will pass through the unit area dA is

$$\Sigma_s \Phi(\vec{r}) dV \frac{dA \cos\theta}{4\pi r^2} \tag{8-6}$$

where $dA\cos\theta$ represents the projection of the unit area dA onto the plane perpendicular to r or the effective surface area as seen from dV. The number of neutrons scattered through the area dA is $dA\cos\theta / 4\pi r^2$ under the assumption that there are no interactions as neutrons travel from dV to dA. However, due to interactions between these two positions in space, the number of neutrons that reach the area dA is the fraction $\exp(-\Sigma_{tot}\, r)$ of the total neutrons, where Σ_{tot} is the total neutron cross section. The remainder of the neutrons are scattered or absorbed in the medium. In spherical coordinates, the elementary volume is defined as

$$dV \equiv dr^3 = r^2 \sin\theta\, d\theta\, d\phi\, dr \qquad (8\text{-}7)$$

Assuming the medium is only weakly absorbing ($\Sigma_{tot} \sim \Sigma_s$) the number of neutrons passing through the unit area in the plane xy in direction z is

$$J_z^- = \frac{\Sigma_s}{4\pi} \int\limits_{\phi=0}^{2\pi} \int\limits_{\theta=0}^{\pi/2} \int\limits_{r} e^{-\Sigma_{tot} r}\, \Phi\!\left(\vec{r}\right)\cos\theta \sin\theta\, d\theta\, d\phi\, dr \qquad (8\text{-}8)$$

Although flux is not known, under the assumption that it is nearly independent of spatial position (assumption 4), using the McLaurin series and by neglecting all terms except the first two, it follows that

$$\Phi\!\left(\vec{r}\right) = \Phi_0 + x\left(\frac{\partial\Phi}{\partial x}\right)_0 + y\left(\frac{\partial\Phi}{\partial y}\right)_0 + z\left(\frac{\partial\Phi}{\partial z}\right)_0 \qquad (8\text{-}9)$$

Index 0 denotes the origin point. In the spherical coordinate system

$$\begin{cases} x = r\sin\theta\cos\varphi \\ y = r\sin\theta\sin\varphi \\ z = r\cos\theta \end{cases} \qquad (8\text{-}10)$$

It therefore follows that

$$J_z^- = \frac{\Sigma_s}{4\pi} \int\limits_{\phi=0}^{2\pi} \int\limits_{\theta=0}^{\pi/2} \int\limits_{r} e^{-\Sigma_s r}\left[\Phi_0 + \left(\frac{\partial\Phi}{\partial z}\right)_0 r\cos\theta\right]\cos\theta \sin\theta\, d\theta\, d\phi\, dr \qquad (8\text{-}11)$$

Assumption 5 states that the medium is infinite and the integration over r is from 0 to infinity reducing the above integral to

$$J_z^- = \frac{\Sigma_s}{4\pi} \Phi_0 \left[\frac{-e^{\Sigma_s r}}{\Sigma_s} \right]_0^\infty \left(\frac{1}{2} \right) \left[\sin^2 \theta \right]_0^{\pi/2} \left[\phi \right]_0^{2\pi}$$

$$+ \frac{\Sigma_s}{4\pi} \left(\frac{\partial \Phi}{\partial z} \right)_0 \left[\frac{e^{-\Sigma_s r}}{\Sigma_s^2} (-\Sigma_s r - 1) \right]_0^\infty \left[-\frac{\cos^3 \theta}{3} \right]_0^{\pi/2} \left[\phi \right]_0^{2\pi} \tag{8-12}$$

Following substitution of the limits

$$J_z^- = \frac{\Sigma_s}{4\pi} \Phi_0 \left[\frac{1}{\Sigma_s} \right] \left(\frac{1}{2} \right) (2\pi) + \frac{\Sigma_s}{4\pi} \left(\frac{\partial \Phi}{\partial z} \right)_0 \left[\frac{1}{\Sigma_s^2} \right] \left(\frac{1}{3} \right) (2\pi) \tag{8-13}$$

$$J_z^- = \frac{\Phi_0}{4} + \frac{1}{6\Sigma_s} \left(\frac{\partial \Phi}{\partial z} \right)_0 \tag{8-14}$$

The upward current flow through the area dA from the lower hemisphere is obtained by integration as above with θ limits from π to $\pi/2$. Therefore, the number of neutrons passing through the unit area in direction $+z$ is

$$J_z^+ = \frac{\Phi_0}{4} - \frac{1}{6\Sigma_s} \left(\frac{\partial \Phi}{\partial z} \right)_0 \tag{8-15}$$

Thus the total net flow of neutrons in direction z is

$$J_z = J_z^+ - J_z^- = -\frac{1}{3\Sigma_s} \left(\frac{\partial \Phi}{\partial z} \right)_0 \tag{8-16}$$

The net flow of neutrons through the areas in xz and in yz planes is

$$J_y = J_y^+ - J_y^- = -\frac{1}{3\Sigma_s} \left(\frac{\partial \Phi}{\partial y} \right)_0 \tag{8-17}$$

$$J_x = J_x^+ - J_x^- = -\frac{1}{3\Sigma_s} \left(\frac{\partial \Phi}{\partial x} \right)_0 \tag{8-18}$$

The neutron current density (number of neutrons per unit time crossing unit area normal to direction of flow) is according to Fig. 7-14

$$\vec{J}(\vec{r}) = \vec{i}J_x + \vec{j}J_y + \vec{k}J_z = -\frac{1}{3\Sigma_s}\left[\vec{i}\frac{\partial\Phi}{\partial x} + \vec{j}\frac{\partial\Phi}{\partial y} + \vec{k}\frac{\partial\Phi}{\partial z}\right] \qquad (8\text{-}19)$$

In this equation the flux is valid for any point in the medium and not just at the origin as previously assumed. In comparison with Eq. (8-3)

$$D = \frac{1}{3\Sigma_s} \qquad (8\text{-}20)$$

where λ_s represents the mean free path for neutron scattering. The diffusion coefficient is corrected for anisotropic scattering using the transport mean free path. If the average cosine of the scattering angle for collision in laboratory system (as explained in Section 3.3) is

$$\bar{\mu} = \cos\psi = \frac{2}{3A} \qquad (8\text{-}21)$$

where A is the atomic mass number of the medium, the diffusion coefficient can be written as a function of the transport cross section, Σ_{tr}

$$\Sigma_{tr} = \Sigma_{tot} - \bar{\mu}\Sigma_s = \frac{1}{\lambda_{tr}} \qquad (8\text{-}22)$$

Eq. (8-20) may therefore be written as

$$D = \frac{1}{3\Sigma_{tr}} = \frac{\lambda_{tr}}{3} \qquad (8\text{-}23)$$

which, for a weakly absorbing medium, becomes

$$\Sigma_{tr} = \Sigma_s - \bar{\mu}\Sigma_s = \frac{1}{\lambda_{tr}} \qquad (8\text{-}24)$$

$$D = \frac{1}{3\Sigma_s(1-\bar{\mu})} = \frac{\lambda_s}{3(1-\bar{\mu})} \qquad (8\text{-}25)$$

Example 8.1 Diffusion coefficient
Estimate the diffusion coefficient of graphite at 1 eV(σ_s (1 eV) = 4.8 b).

For the graphite $A = 12$, thus from Eq. (8-21) it follows that

$$\overline{\mu} = \overline{\cos\psi} = \frac{2}{3A} = 0.055$$

The macroscopic scattering cross section for graphite is

$$\Sigma_s = N\sigma_s = (0.08023 \times 10^{24} \text{ at/cm}^3)(4.8 \times 10^{-24} \text{ cm}^2) = 0.385 \text{ cm}^{-1} \Rightarrow$$

$$D = \frac{1}{3\Sigma_s(1-\overline{\mu})} = \frac{1}{3 \times 0.385 (1-0.055)} = 0.916 \text{ cm}$$

Example 8.2 Neutron transport mean free path

The transport mean free path is a scattering mean free path which is corrected for the slightly larger distance traveled in the laboratory system due to preferential forward scattering. Calculate the transport mean free path for thermal neutrons in beryllium oxide (BeO), if ρ(Be) = 2.70 g/cm^3, A(BeO) = 25.01, σ_s(Be) = 7 b and σ_s(O) = 4.2 b.

The atom densities are

$$N(\text{Be}) = N(\text{O}) = \frac{\rho \times N_a}{A} = \frac{2.7 \times 6.022 \times 10^{23}}{25.01} = 6.51 \times 10^{22} \text{ at/cm}^3$$

The average cosine of the scattering angle for collision in laboratory system, from Eq. (8-21), and the transport mean free path from Eq. (8-22) are

$$\overline{\mu}(\text{Be}) = \frac{2}{3 \times 9} = 0.0741 \quad \overline{\mu}(\text{O}) = \frac{2}{3 \times 16} = 0.0417$$

$$\lambda_{tr} = \frac{1}{\Sigma_s(\text{Be})\left(1-\overline{\mu}(\text{Be})\right) + \Sigma_s(\text{O})\left(1-\overline{\mu}(\text{O})\right)}$$

$$= \frac{1}{\left(6.51 \times 10^{22}\right)\left(7 \times 10^{-24}\right)\left(1-0.0741\right) + \left(6.51 \times 10^{22}\right)\left(4.2 \times 10^{-24}\right)\left(1-0.0417\right)} =$$

$$= \frac{1}{0.422 + 0.262} = 1.46 \text{ cm}$$

The diffusion coefficient divided by the absorption cross section has the dimension of the squared length, the square root of which is called the *diffusion length, L*

$$L^2 = \frac{D}{\Sigma_a} = \frac{1}{3\Sigma_a\Sigma_s(1-\overline{\mu})}. \qquad (8-26)$$

Table 8-1. Diffusion parameters for neutrons at $T = 293$ K

Moderator	D [cm]	Σ_a [cm^{-1}]	L [cm]	ML [cm]
H_2O	0.144	0.0189	2.75	5.6
D_2O	0.810	0.00007	161.0	11.0
Be	1.85	0.00053	21.2	9.2
Graphite	1.60	0.00031	52.5	18.7

The diffusion length represents the distance a neutron passes from the point of thermalization to the point of absorption. The distance from the point where neutron is born to the point where it is thermalized is called the moderation length, ML, representing the optimum distance between adjacent fuel channels in a heterogeneous reactor (called the pitch). The moderator and diffusion lengths for few materials commonly used in thermal reactors are listed in Table 8-1. The small diffusion length of H_2O is due to high absorption cross section. If a light water reactor is over-moderated (the lattice pitch is large) it will result in increased neutron absorption. If a D_2O moderated reactor is over-moderated it will have no significant effect on neutron economy.

2.4 Neutron Diffusion Theory

2.4.1 One-Speed Neutron Diffusion Equation

The exact interpretation of neutron transport in heterogeneous domains such as a reactor core is so complex that simplified approaches are often used. Though simplified, they are accurate enough to give an estimate of the average characteristics of neutron population in a given medium. The simplest form of neutron transport equation is the *one-speed diffusion equation* developed under the following assumptions:

1. *neutrons are monoenergetic*: average neutron energy and average cross sections for neutron interactions are selected;
2. *absorption in a medium is small*: macroscopic absorption cross section is small in comparison with scattering cross section;
3. *neutron scattering is isotropic in the laboratory system*: valid for neutron scattering with heavy nuclei, and not true for thermal reactor moderators (corrections must be applied);
4. *angular neutron distribution is isotropic*: valid if neutron flux is nearly constant which is approximately satisfied far from the boundaries, neutron source or points of strong absorptions (if neutron flux gradient is large, there are preferable directions of neutron motion toward low neutron flux region).

In a reactor core, neutrons are produced and lost through capture and leakage. In general the neutron balance equation starts by defining the rate of change of neutron density (flux) that must be equal to the rate of production reduced by the rate of neutron absorption and leakage. Thus the net rate of change in neutron density per unit volume and time is

$$\frac{\partial n(\vec{r},t)}{\partial t} = S(\vec{r},t) - \Sigma_a(\vec{r})\Phi(\vec{r},t) - LE \qquad (8\text{-}27)$$

where $S(\vec{r},t)$ is neutron source rate, $\Sigma_a\Phi(\vec{r},t)$ neutron absorption rate, and LE neutron leakage per unit time and unit volume. LE represents the rate of neutrons flowing in a given direction per unit time through unit area normal to direction of flow, Fig. 8-4

Neutron leakage per unit volume $= div J(\vec{r},t) = \nabla \cdot J(\vec{r},t)$ \qquad (8-28)

Thus, Eq. (8-27) is re−written in the following way:

$$\frac{\partial n(\vec{r},t)}{\partial t} = S(\vec{r},t) - \Sigma_a(\vec{r})\Phi(\vec{r},t) - \nabla \cdot J(\vec{r},t) \qquad (8\text{-}29)$$

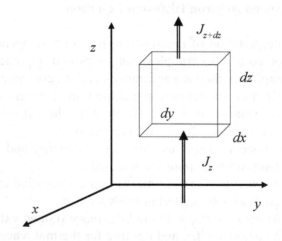

Figure 8-4. Neutron leakage from an elementary volume

Neutron diffusion through any material is the result of scattering. In reactor analysis, it is assumed that neutrons do not collide with one another. This is valid because the neutron density is much lower than the atom density of a medium. Due to nearly constant collisions, neutrons in a scattering medium travel *zigzag* trajectories. When considering a large number of neutrons assumed to be monoenergetic, there is an overall motion of neutrons from a region of higher to a region of lower neutron density (flux). This is why the Fick's law of diffusion is applied to define the net rate of neutron flow. Fick's law, Eq. (8-3), if inserted into Eq. (8-29) gives

$$\frac{\partial n(\vec{r},t)}{\partial t} = S(\vec{r},t) - \Sigma_a(\vec{r})\Phi(\vec{r},t) - \nabla \cdot [-D(\vec{r})\nabla\Phi(\vec{r},t)] \tag{8-30}$$

which can be re-written as

$$\frac{\partial n(\vec{r},t)}{\partial t} = \frac{1}{\upsilon}\frac{\partial\Phi(\vec{r},t)}{\partial t} = \nabla \cdot [D(\vec{r})\nabla\Phi(\vec{r},t)] + S(\vec{r},t) - \Sigma_a(\vec{r})\Phi(\vec{r},t) \tag{8-31}$$

Under the assumption that the medium is homogenous, the diffusion coefficient becomes independent of neutron position and the leakage term reduces to

$$LE = -D\nabla^2\Phi(\vec{r},t) \tag{8-32}$$

where ∇^2 represents the Laplacian operator which is defined for the various coordinate systems as follows:

Rectangular coordinate system: $\nabla^2 = \dfrac{\partial^2}{\partial x^2} + \dfrac{\partial^2}{\partial y^2} + \dfrac{\partial^2}{\partial z^2}$ (8-33)

Spherical coordinate system: $\nabla^2 = \dfrac{d^2}{dr^2} + \dfrac{2}{r}\dfrac{d}{dr}$ (8-34)

Cylindrical coordinate system: $\nabla^2 = \dfrac{d^2}{dr^2} + \dfrac{1}{r}\dfrac{d}{dr} + \dfrac{d^2}{dz^2}$ (8-35)

Combining Eq. (8-32) with Eq. (8-31), the one-speed diffusion equation for neutrons interacting with a homogeneous medium is

$$\frac{\partial n(\vec{r},t)}{\partial t} = \frac{1}{\upsilon}\frac{\partial\Phi(\vec{r},t)}{\partial t} = D\nabla^2\Phi(\vec{r},t) + S(\vec{r},t) - \Sigma_a(\vec{r})\Phi(\vec{r},t) \tag{8-36}$$

The following cases introduce important simplifications to Eq. (8-36)

- Steady-state condition

$$D\nabla^2 \Phi(\vec{r}) + S(\vec{r}) - \Sigma_a \Phi(\vec{r}) = 0 \tag{8-37}$$

- Steady-state, nonmultiplying medium (neutron source = 0)

$$\nabla^2 \Phi(\vec{r}) - \frac{\Phi(\vec{r})}{L^2} = 0 \tag{8-38}$$

where

$$L^2 \equiv \frac{D}{\Sigma_a} \tag{8-39}$$

2.4.2   Solution to One-Speed Neutron Diffusion Equation from a Point and Plane Source in Infinite Medium

The simplest case to demonstrate diffusion theory is diffusion of neutrons from a point source in an infinite nonmultiplying medium. Neutrons are emitted from such a source in all directions with equal probability giving a spherical symmetry in regard to the position of neutron source. If r represents the distance from the origin where point neutron source of intensity S [neutrons/sec] is located, the one-speed diffusion equation written as Eq. (8-38) but expressed in spherical coordinates becomes

$$\frac{d^2}{dr^2} \Phi(r) + \frac{2r}{r} \frac{d}{dr} \Phi(r) - \frac{\Phi(r)}{L^2} = 0 \tag{8-40}$$

Introducing $y = \Phi(r)r$ the above equation reduces to

$$\frac{d^2 y}{dr^2} - \frac{y}{L^2} = 0 \tag{8-41}$$

whose solution has the following general form

$$y = Ae^{-r/L} + Ce^{r/L} \tag{8-42}$$

Re-introducing the variable r, the solution becomes

$$\Phi(r) = A\frac{e^{-r/L}}{r} + C\frac{e^{r/L}}{r} \tag{8-43}$$

where A and C are constants determined from the two boundary conditions. First boundary condition states that far from the neutron source the neutron flux must decrease, which determines the C constant to be zero (otherwise the flux will tend to be infinitely large). Therefore the solution for the flux distribution reduces to

$$\Phi(r) = A\frac{e^{-r/L}}{r} \tag{8-44}$$

The total number of neutrons passing through the entire surface of a sphere whose center is at the point neutron source is $4\pi r^2 J$ where

$$J = -D\frac{d\Phi(r)}{dr} = -D\frac{d}{dr}\left(A\frac{e^{-r/L}}{r}\right) = DAe^{-r/L}\left(\frac{1+r/L}{r^2}\right) \tag{8-45}$$

Thus the second boundary condition states that the limiting value of the total number of neutrons passing through the surface of the sphere as distance becomes zero is equal to the source strength

$$S = \lim_{r\to 0}\left(4\pi r^2 J\right) = 4\pi DA \lim_{r\to 0}\left(e^{-r/L}\frac{1+r/L}{r^2}\right) = 4\pi DA \tag{8-46}$$

giving

$$A = \frac{S}{4\pi D} \tag{8-47}$$

and the solution for the flux is thus

$$\Phi(r) = \frac{S}{4\pi D}\frac{e^{-r/L}}{r} \tag{8-48}$$

This simple example can be used to determine the neutron mean-square distance to absorption in nonmultiplying media. The neutron mean-square distance to absorption does not represent the total neutron trajectory that is much longer due to many scattering collisions. Neutrons travel certain distance, r, during diffusion through a medium along which they collide until they are absorbed, creating a path similar to that shown in Fig. 8-5. The mean square of this distance is obtained exactly from the neutron flux distribution from a point neutron source. In a differential ring of thickness dr placed r from the neutron source, there are $4\pi r^2 dr \Sigma_a \Phi(r)$ neutrons absorbed per unit time (number of neutrons absorbed is equal to the number of neutrons created since there is no leakage of neutrons in an infinite medium). This also represents the probability that a neutron will be absorbed at a distance r from the source. Therefore,

$$\overline{r^2} = \frac{\int\limits_{r=0}^{\infty} r^2 \left(4\pi r^2 \Sigma_a \Phi(r)\right) dr}{\int\limits_{r=0}^{\infty} \left(4\pi r^2 \Sigma_a \Phi(r)\right) dr} \tag{8-49}$$

and inserting Eq. (8-48) for the neutron flux, it follows that

$$\overline{r^2} = \frac{\int\limits_{r=0}^{\infty} \left(r^3 e^{-r/L}\right) dr}{\int\limits_{r=0}^{\infty} \left(r e^{-r/L}\right) dr} = \frac{6L^4}{L^2} = 6L^2 \tag{8-50}$$

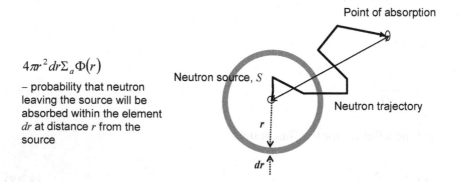

$4\pi r^2 dr \Sigma_a \Phi(r)$

– probability that neutron leaving the source will be absorbed within the element dr at distance r from the source

Figure 8-5. Distance and neutron trajectory between the point of neutron source to the point of neutron absorption

The neutron diffusion length represents $1/\sqrt{6}$ of the root of mean-square distance to absorption. In other words, the neutron diffusion length represents a measure of neutron diffusion away from the source before it is absorbed in the medium [see Problem 8.29].

An infinite planar source of intensity S in an infinite medium also produces the exponential decay of neutron flux away from the source due to neutron absorptions, Fig. 8-6. The diffusion theory solution indicates that the neutron flux decreases by a factor e for every L from the source plane. If the source is located at $x = 0$ it follows that

$$S(x)\delta(x) = \begin{cases} 0, x \neq 0 \\ \int_a^b \delta(x)dx = \begin{cases} 1, a < 0 < b \\ 0, \text{otherwise} \end{cases} \end{cases} \tag{8-51}$$

Away from the source the neutron balance diffusion equation to be solved is Eq. (8-38). The general solution for the neutron flux distribution along the x axis is

$$\Phi(x) = Ae^{-x/L} + Ce^{x/L} \tag{8-52}$$

The first boundary condition restricts the flux to zero at the distance infinitely away from the source, $\Phi(x \rightarrow \pm\infty) = 0$, which reduces the solution to

$$\Phi(x) = Ae^{-x/L} \tag{8-53}$$

The second boundary condition specifies the neutron current from the neutron source as follows

$$\lim_{x \to 0} J(x) = S/2 \tag{8-54}$$

Following the definition of the neutron current and applying the boundary condition specified with Eq. (8-54) the constant A is obtained

$$\lim_{x \to 0} J(x) = \lim_{x \to 0}\left[-D\frac{d\Phi(x)}{dx}\right] = \lim_{x \to 0}\left[DA\frac{e^{-x/L}}{L}\right] \tag{8-55}$$

$$= \frac{DA}{L} = \frac{S}{2} \quad \Rightarrow \quad A = \frac{SL}{2D}$$

giving the solution for neutron flux as follows

$$\Phi(x) = \frac{SL}{2D} e^{-x/L} \tag{8-56}$$

If the source is placed at distance b from $x = 0$, the solution changes as

$$\Phi(x) = \frac{SL}{2D} e^{-|x-b|/L}$$

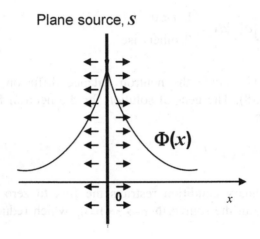

Figure 8-6. Neutron flux distribution from the infinite plane source in infinite nonmultiplying medium

2.4.3   Solution to One-Speed Neutron Diffusion Equation in Finite Medium

In order to solve the neutron diffusion equation for a medium of finite dimensions, or in the medium composed of two different materials, a set of boundary conditions are specified as follows:

(a) At the interface between the two media, A and B, with different diffusion properties (neither of which is a vacuum) the neutron flux must be the same for both media (the condition of the continuity of flux)

$$\Phi_{A_0}\left(\vec{r}\right) = \Phi_{B_0}\left(\vec{r}\right) \tag{8-57}$$

where the subscript 0 denotes the interface plane between the two media.

(b) At a plane interface between two media (neither of which is a vacuum) the neutron currents are equal. Along the x direction it follows that

$$-D_A \frac{d\Phi_{A_0}}{dx} = -D_B \frac{d\Phi_{B_0}}{dx} \qquad (8\text{-}58)$$

(c) In the case when one of the media is a vacuum (or air) the boundary conditions are different because there is no scattering from vacuum. In other words, the flow of neutrons exists only in one direction, toward vacuum. The boundary condition at the interface between the diffusion medium and the vacuum (or air) specifies that the neutron flux gradient vanishes at a certain point beyond the physical boundary, called the extrapolated distance or extrapolated boundary. The concept is hypothetical because there is no indication that neutron flux is actually zero at that particular point

$$\left(\frac{d\Phi(x)}{dx} \right)_0 = -\frac{\Phi_0}{d} \qquad (8\text{-}59)$$

where d is the distance called the linear extrapolation distance (see Fig. 8-7) and is equal to 0.71 λ_{tr} (valid for plane surfaces). Diffusion theory gives the extrapolation distance to be nearly equal to 2/3 of λ_{tr} (see Section 2.4.5).

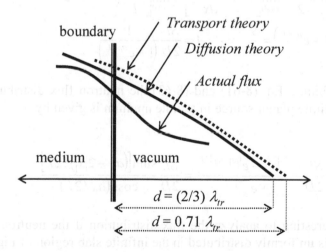

Figure 8-7. Extrapolation of neutron flux at the interface between diffusion medium and vacuum

If the planar source is placed at the center of an infinite slab of a finite thickness a with the extrapolated distance d the diffusion equation to solve for the flux distribution is

$$\nabla^2 \Phi(x) - \frac{\Phi(x)}{L^2} = \frac{S(x)\delta(x)}{D} \tag{8-60}$$

where the source is defined with Eq. (8-51). Away from the planar source the general solution for the flux distribution is given with Eq. (8-52). The first boundary condition states that the neutron flux at the extrapolated distance is zero, Fig. 8-8

$$x > 0 \quad \Phi(x = a_0/2) = Ae^{-a_0/2L} + Ce^{a_0/2L} = 0 \quad \rightarrow \quad C = -Ae^{-a_0/L} \tag{8-61}$$

reducing the general flux solution to the following

$$x > 0 \quad \Phi(x) = Ae^{-x/L} - Ae^{-a_0/L}e^{x/L} = A\left(e^{-x/L} - e^{(x-a_0)/L}\right) \tag{8-62}$$

The second boundary condition defines the magnitude of the neutron current vector and provides for the condition to determine the integration constant A as follows

$$\lim_{x \to 0} J(x) = \frac{S}{2} \quad \lim_{x \to 0}\left[-D\frac{d\Phi(x)}{dx}\right] = \lim_{x \to 0}\left[\frac{DA}{L}\left(e^{-x/L} + e^{(x-a_0)/L}\right)\right] =$$
$$= \frac{DA}{L}\left(1 + e^{a_0/L}\right) = \frac{S}{2} \quad \Rightarrow \quad A = \frac{SL}{2D}\frac{1}{\left(1 + e^{a_0/L}\right)} \tag{8-63}$$

By combining Eq. (8-61) and (8-63) the neutron flux distribution away from the infinite planar source in a finite medium is given by

$$\Phi(x) = \frac{SL}{2D}\frac{e^{-|x|/L} - e^{(|x|-a_0)/L}}{1 + e^{-a_0/L}} = \frac{SL}{2D}\frac{\sinh[(a_0 - 2|x|)/2L]}{\cosh(a_0/2L)} \tag{8-64}$$

It is interesting to analyze the flux distribution if the neutron source of intensity S is uniformly distributed in the infinite slab region of Fig. 8-8. The geometry is shown in Fig. 8-9 and the diffusion equation for this case is

$$\nabla^2 \Phi(x) - \frac{\Phi(x)}{L^2} + \frac{S}{D} = 0 \tag{8-65}$$

Eq. (8-65) is the inhomogeneous differential equation. The general solution is a sum of the homogeneous and particular solutions. The homogeneous solution (if the source is zero) is given by

$$\Phi_{hom}(x) = A\sinh\left(\frac{x}{L}\right) + C\cosh\left(\frac{x}{L}\right) \tag{8-66}$$

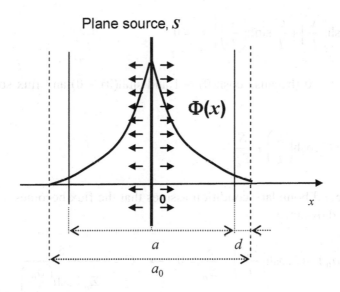

Plane source, S

$\Phi(x)$

a d

a_0

Figure 8-8. Neutron flux distribution from the infinite plane source in a finite nonmultiplying medium surrounded with vacuum

The particular solution is obtained by setting flux to be a constant and by substituting it into Eq. (8-65)

$$\Phi_p(x) = \frac{SL^2}{D} = \frac{S}{\Sigma_a} \tag{8-67}$$

Thus the complete general solution is

$$\Phi(x) = A\sinh\left(\frac{x}{L}\right) + C\cosh\left(\frac{x}{L}\right) + \frac{S}{\Sigma_a} \tag{8-68}$$

The first boundary condition is derived from the obvious symmetry along the centerline of the medium as could be seen from Fig. 8-9. The symmetry implies that there is no net current crossing the symmetry plane located at the centerline of the medium (at $x = 0$)

$$\frac{d\Phi(x)}{dx}\bigg|_{x=0} = 0 \tag{8-69}$$

By substituting Eq. (8-68) into Eq. (8-69) the following equation is obtained

$$\left[\frac{A}{L}\cosh\left(\frac{x}{L}\right) + \frac{C}{L}\sinh\left(\frac{x}{L}\right)\right]_{x=0} = 0 \tag{8-70}$$

giving $A = 0$ (because $\cosh(0) = 1$ and $\sinh(0) = 0$) and flux solution as follows

$$\Phi(x) = C\cosh\left(\frac{x}{L}\right) + \frac{S}{\Sigma_a} \tag{8-71}$$

The second boundary condition assures that the flux becomes zero at the extrapolated distance

$$\Phi(x=a_0) = C\cosh\left(\frac{a_0}{L}\right) + \frac{S}{\Sigma_a} = 0 \quad \rightarrow \quad C = -\frac{S}{\Sigma_a \cosh\left(\dfrac{a_0}{L}\right)} \tag{8-72}$$

Finally the solution for the flux distribution becomes

$$\Phi(x) = \frac{S}{\Sigma_a}\left[1 - \frac{\cosh\left(\dfrac{x}{L}\right)}{\cosh\left(\dfrac{a_0}{L}\right)}\right] \tag{8-73}$$

and is plotted in Fig. 8-9. If a medium containing homogeneous neutron source becomes infinitely large the flux becomes a constant value

$$\lim_{a_0 \to \infty} \Phi(x) = \frac{S}{\Sigma_a} \tag{8-74}$$

showing that because there is no leakage in an infinite medium, the absorption rate at any point in the medium is equal to the source intensity at that point. Thus by increasing the width of the medium the flux distribution becomes flatter as sketched in Fig. 8-9 because the second term in the brackets in Eq. (8-73) becomes smaller near the center of the medium.

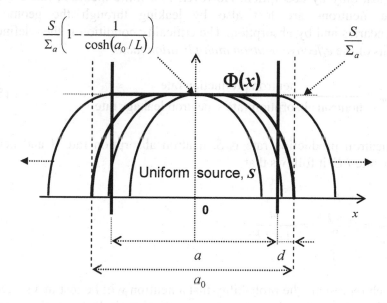

$$\frac{S}{\Sigma_a}\left(1 - \frac{1}{\cosh(a_0/L)}\right) \qquad \frac{S}{\Sigma_a}$$

Figure 8-9. Neutron flux distribution from the uniform source distributed in a finite nonmultiplying medium surrounded with vacuum

2.4.4   Neutron Diffusion in Multiplying Medium

The main interest in applying diffusion theory in neutron transport is to analyze the neutron population, neutron flux and power distribution in a reactor core. A reactor core is a finite multiplying medium with a sustaining fission chain reaction. As neutrons diffuse through the core they can be absorbed by fuel, moderator or structural materials present in the core; leak out from the geometrical boundaries of the reactor core; or act as a source for new neutrons to be born in fission reactions. In a critical (or steady-state) reactor core, the number of neutrons produced by fission is the same as the number of neutrons lost by absorption or leakage in a given unit time. Thus we can define

- *Infinite neutron multiplication factor* which represents the number of neutrons produced per fission per one neutron absorbed in a medium

$$k_\infty = \frac{\text{neutron production rate}}{\text{neutron absorption rate}} \qquad (8\text{-}75)$$

- In an infinitely large system the leakage is neglected, and the neutrons are lost only by absorption. However, in a finite medium, like a reactor core, neutrons are lost also by leaking through the geometrical boundaries and by absorption. The criticality condition is then defined in terms of the *effective neutron multiplication factor*

$$k_{eff} = \frac{\text{neutron production rate}}{\text{neutron absorption rate} + \text{neutron leakage rate}} \qquad (8\text{-}76)$$

If neutron production rate is S, neutron absorption rate A and neutron leakage rate LE, it follows that

$$\frac{k_{eff}}{k_\infty} = \frac{\dfrac{S}{A + LE}}{\dfrac{S}{A}} = \frac{A}{A + LE} \qquad (8\text{-}77)$$

which represents the probability that a neutron will be lost in a system by absorption. Since the alternative loss mechanism is leakage, this equation also represents the probability that a neutron will not be lost through leakage, i.e., it represents the non-leakage probability, $P_{non\text{-}leak}$. Therefore,

$$\frac{k_{eff}}{k_\infty} \equiv P_{non\text{-}leak} \qquad (8\text{-}78)$$

In order for a reactor to be critical the infinite multiplication factor must be greater than unity. The neutron leakage is generally proportional to the surface area, SA, and neutron production is proportional to the volume, V. If the size of a system is expressed in units of a

$$\frac{LE}{S} \prec \frac{SA}{V} \prec \frac{a^2}{a^3} \prec \frac{1}{a} \qquad (8\text{-}79)$$

the ratio between the number of neutrons leaked and number of neutrons produced is inversely proportional to the linear dimension of the finite multiplying system. Thus, by changing the size of the core, the leakage rate changes affecting the effective neutron multiplication factor to range between 0 and k_∞. The infinite multiplication factor is a function of the materials present in the core (fuel, moderator, coolant, structures). The non-leakage probability is dependent on the reactor materials and its geometry (size, shape). For thermal neutrons, exactly k_∞ new thermal neutrons are created per each neutron absorbed and thus the neutron source assuming homogeneous system is

$$S\left(\vec{r}\right) = \overline{\Sigma}_a k_\infty \Phi\left(\vec{r}\right) \tag{8-80}$$

Thus, assuming a steady-state condition and applying one-speed diffusion theory, the diffusion equation can be written in the following form

Leakage + Absorption = Production

$$-D\nabla^2\Phi\left(\vec{r}\right) + \overline{\Sigma}_a \Phi\left(\vec{r}\right) = S = \overline{\Sigma}_a k_\infty \Phi\left(\vec{r}\right)$$

$$\nabla^2\Phi\left(\vec{r}\right) + \frac{\overline{\Sigma}_a(k_\infty - 1)}{D}\Phi\left(\vec{r}\right) = 0 \tag{8-81}$$

$$\nabla^2\Phi\left(\vec{r}\right) + B^2 \Phi\left(\vec{r}\right) = 0$$

where

$$B^2 \equiv \frac{k_\infty - 1}{L^2} \tag{8-82}$$

Equation (8-81) is solved analytically for various simple reactor geometries in the following sections. The square root of Eq. (8-82) is referred to as the *material buckling* (B_m) of the reactor core because it is purely dependent on core materials. The solution to Eq. (8-81) gives the value for so-called *geometrical buckling* (B_g) because it represents a measure of the bending or the curvature of the spatial distribution of the neutron flux. When two bucklings are equal the reactor is critical and steady state. The overall neutron production must balance the neutron absorption plus leakage during the steady state operation of the reactor. The relation for material buckling can be rearranged

$$1 = k_\infty \left[\frac{1}{B_m^2 L^2 + 1}\right] \tag{8-83}$$

Comparing Eq. (8-83) with Eq. (8-78) yields an expression for the non-leakage probability in a critical reactor

$$P_{non-leak} = \frac{1}{B_m^2 L^2 + 1} = \frac{\overline{\Sigma_a \Phi(\vec{r})}}{\overline{\Sigma_a \Phi(\vec{r})} + B_m^2 \phi} = \frac{\overline{\Sigma_a \Phi(\vec{r})}}{\overline{\Sigma_a \Phi(\vec{r})} + (-D\nabla^2 \Phi(\vec{r}))} \tag{8-84}$$

Example 8.3 Material buckling

Calculate the material buckling and thermal neutron leakage probability for a critical homogeneous reactor consisting of a mixture of 200 moles of graphite per mole of 5.5% enriched uranium fuel. The overall temperature of the reactor core is $20°C$. The density of graphite and uranium are $\rho_C = 1.6$ g/cm^3 and $\rho_U = 18.9$ g/cm^3. The microscopic cross section for thermal neutron scattering at carbon is 4.8 b and at uranium is 8.3 b. The microscopic cross section for absorption in carbon is 0.0034 b, in ^{235}U is 694 b and in ^{238}U is 2.73 b. The infinite multiplication factor is 1.2. (Adapted from "*Basic Nuclear Engineering*", A. R. Foster and R. L. Wright Jr., Allyn and Bacon Inc., 1968.)

The volumes of the uranium fuel, graphite moderator and the core are

$$V_U = \frac{238 \text{ g/mole}}{18.9 \text{ g/cm}^3} = 12.6 \text{ cm}^3 \text{mole U}$$

$$V_C = \frac{200 \text{ moles C/mole U} \times 12 \text{ g C/mole C}}{1.6 \text{ g/cm}^3} = 1500 \text{ cm}^3 \text{mole C}$$

$$V = V_U + V_C = 12.6 + 1500 = 1512.6 \text{ cm}^3 \text{ mixture/mole U}$$

The atom densities of uranium, ^{235}U, ^{238}U and carbon are

$$N_U = \frac{N_a}{V} =$$

$$= \frac{6.022 \times 10^{23} \text{ atoms U/g mole U}}{1512.6 \text{ cm}^3/\text{g mole U}} = 3.98 \times 10^{20} \text{ atoms U/cm}^3 \text{ mixture}$$

$$N_{235} = 0.055 \times (3.98 \times 10^{20}) = 0.219 \times 10^{20} \text{ atoms } ^{235}\text{U/cm}^3 \text{ mixture}$$

$$N_{238} = 0.945 \times (3.98 \times 10^{20}) = 3.761 \times 10^{20} \text{ atoms } ^{238}\text{U/cm}^3 \text{ mixture}$$

$$N_C = (200 \text{ atoms C/atom U}) \times (3.98 \times 10^{20}) = 7.96 \times 10^{22} \text{ atoms C/cm}^3 \text{ mixture}$$

The transport macroscopic cross section for thermal neutrons in this mixture is

$$\Sigma_{tr}^{th} = N_C \sigma_s^C (1 - \overline{\mu_C}) + N_U \sigma_s^U (1 - \overline{\mu_U}) =$$

$$(7.96 \times 10^{22})(4.8 \times 10^{-24})\left(1 - \frac{2}{3 \times 12}\right) + (3.98 \times 10^{20})(8.3 \times 10^{-24})\left(1 - \frac{2}{3 \times 238}\right)$$

$$= 0.361 + 0.0033 = 0.364 \text{ cm}^2/\text{cm}^3 \text{mixture}$$

The average absorption cross section at the most probable neutron energy assuming $1/\upsilon$ dependence in thermal reactors (see Chapter 7, Section 4.6) is

$$\Sigma_a = \left[N_{235}\,\sigma_a^{235} + N_{238}\,\sigma_a^{238} + N_C\,\sigma_a^C\right]\frac{\sqrt{\pi}}{2}$$

$$\Sigma_a = \left[\begin{array}{l}(0.219\times10^{20})(694\times10^{-24})+(3.761\times10^{20})(2.73\times10^{-24}) \\ +(7.96\times10^{22})(0.0034\times10^{-24})\end{array}\right]\frac{\sqrt{\pi}}{2}$$

$$\Sigma_a = 0.0165\frac{\sqrt{\pi}}{2} = 0.0146 \text{ cm}^2/\text{cm}^3\text{mixture}$$

giving the material buckling and the neutron leakage probability as

$$L^2 = \frac{D}{\Sigma_a} = \frac{1}{3\Sigma_a\Sigma_s(1-\bar{\mu})} = \frac{1}{3\Sigma_a\Sigma_{tr}^{th}} = \frac{1}{3\times0.0146\times0.364} = 62.6 \text{ cm}^2$$

$$B_m^2 = \frac{k_\infty-1}{L^2} = \frac{1.2-1}{62.6} = 0.0032 \text{ cm}^{-2} \quad\rightarrow\quad B_m = 0.0566 \text{ cm}^{-1}$$

$$P_{leak} = 1 - P_{non-leak} = 1 - \frac{1}{B_m^2L^2+1} = 1 - \frac{1}{0.0032\times62.6+1}$$

$$= 1 - 0.833 = 0.167$$

2.4.5 Solution to One-Speed Neutron Diffusion Equation in Infinite Slab Bare Reactor

The diffusion equation, Eq. (8-81), is solved for an infinite slab reactor of a finite thickness in order to determine a criticality condition (directly dependent on the slab thickness). The slab is assumed to be infinite in the y or z direction, thus neutrons can leak only along the x direction through slab faces (Fig. 8-10); the neutron flow (neutron flux gradient) will exist only in the x direction. The flux falls off from the center toward either of two slab faces and falls to a zero value at the extrapolated distance.

For the half-size of the slab the distance where flux becomes zero is equal to

$$\frac{a_0}{2} = \frac{a}{2} + d \tag{8-85}$$

where d is the extrapolated distance by which the geometrical boundary of the slab reactor core is extended. In the case of a bare reactor, neutrons leave the reactor geometrical boundaries and almost none scatters back into the core. Therefore, the return current is assumed to be zero and according to Eq. (8-14)

$$J_x^- = \frac{\Phi_0}{4} + \frac{1}{6\Sigma_s}\left(\frac{d\Phi(x)}{dx}\right)_0 = \frac{\Phi_0}{4} + \frac{D}{2}\left(\frac{d\Phi(x)}{dx}\right)_0 = 0 \qquad (8\text{-}86)$$

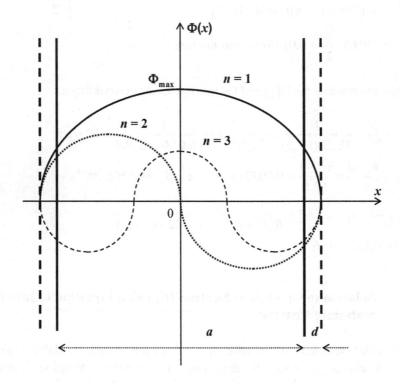

Figure 8-10. Infinite slab bare reactor

$$\left(\frac{d\Phi(x)}{dx}\right)_0 = -\frac{\Phi_0}{2D} \qquad (8\text{-}87)$$

Assuming the extrapolation of flux is a straight line the slope is equal to

$$\left(\frac{d\Phi(x)}{dx}\right)_0 = -\frac{\Phi_0}{a_0/2 - a/2} = -\frac{\Phi_0}{d} \qquad (8\text{-}88)$$

Combining Eq. (8-87) and Eq. (8-88) gives the extrapolation distance

$$d = 2D = \frac{2}{3}\lambda_{tr} \tag{8-89}$$

as mentioned in Section 2.4.3. The most sophisticated transport theory predicts the extrapolation distance to be $0.71\,\lambda_{tr}$. It is always much smaller than the size of a reactor and these two different values do not introduce significant errors into flux estimates.

The neutron flux in an infinite slab varies along the x direction. The diffusion equation reduces to an ordinary second-order linear differential equation

$$\frac{d^2\Phi(x)}{dx^2} + B^2\Phi(x) = 0 \tag{8-90}$$

If B^2 is real and positive the general solution of this equation is

$$\Phi(x) = A\cos Bx + C\sin Bx \tag{8-91}$$

The boundary conditions are

1. The neutron flux drops to zero at the extrapolated distance and is finite at the geometrical boundaries

$$\Phi\left(x = \pm\frac{a_0}{2}\right) = 0 \tag{8-92}$$

2. The neutron flux is symmetric about the origin

$$\left.\frac{d\Phi(x)}{dx}\right|_{x=0} = 0 \tag{8-93}$$

The first boundary condition when applied to the general solution given with Eq. (8-91) requires

$$\Phi\left(x = \pm\frac{a_0}{2}\right) = A\cos B\frac{a_0}{2} + C\sin B\frac{a_0}{2} = 0 \tag{8-94}$$

implying that simultaneously both terms should be equal to zero, i.e.,

$$A \cos B \frac{a_0}{2} = 0 \quad C \sin B \frac{a_0}{2} = 0 \tag{8-95}$$

If both constants, A and C, are set to zero this will give a trivial solution of a zero neutron flux which is not a true solution. This demands that B is the one to be selected such that the solution satisfies Eq. (8-94). Thus,

- $C = 0$ requires $B = B_n = n\pi/a_0$ where $n = 1, 3, 5, \ldots$ and that gives a solution for neutron flux to be

$$\Phi(x) = A_n \cos \frac{n\pi}{a_0} x \quad n = 1,3,5,\ldots \tag{8-96}$$

- $A = 0$ requires $B = B_n = n\pi/a_0$ where $n = 2, 4, 6, \ldots$ and that gives a solution for neutron flux to be

$$\Phi(x) = C_n \sin \frac{n\pi}{a_0} x \quad n = 2,4,6,\ldots \tag{8-97}$$

Therefore, the non-trivial solutions are

$$B^2 = \left(\frac{n\pi}{a_0} \right)^2 \quad n = 1,2,3,\ldots \qquad \leftarrow \quad \text{eigenvalues}$$

$$\Phi(x) = \begin{cases} A_n \cos \dfrac{n\pi}{a_0} x & n = 1,3,5,\ldots \\[2mm] C_n \sin \dfrac{n\pi}{a_0} x & n = 2,4,6,\ldots \end{cases} \qquad \leftarrow \quad \text{eigenfunctions} \tag{8-98}$$

The eigenfunctions are called harmonics. From Eq. (8-98) it can be seen that there is an infinite number of harmonics corresponding to the allowed eigenvalues of the buckling. If the reactor is critical the first harmonic is the solution and is called the *fundamental mode solution*. The higher harmonics would then represent the subcritical system. First three harmonics are sketched in Fig. 8-10.

Now, from the second boundary condition the flux gradient is obtained

$$\frac{d\Phi(x)}{dx} = -AB \sin Bx + CB \cos Bx = 0$$

At the origin, the flux gradient is zero which eliminates the sin term, i.e.,

$\sin Bx = 0$. Since B is real and positive, C must be equal to zero; hence

$$\Phi(x) = A \cos Bx \tag{8-99}$$

The first boundary condition gives Eq. (8-96) and consequently

$$B\left(\pm \frac{a_0}{2}\right) = \frac{\pi}{2}, \frac{3\pi}{2}, \frac{5\pi}{2}, \dots = \frac{n\pi}{2} \quad n = 1,3,5,\dots \tag{8-100}$$

The various values of a_0 are

$$a_0 = \frac{\pi}{B}, \frac{3\pi}{B}, \frac{5\pi}{B}, \dots \tag{8-101}$$

Only the first value is used to define the flux in critical reactors, π/B, corresponding to the fundamental mode as discussed.

For a steady-state critical infinite slab reactor therefore the final solution is

$$\Phi(x) = A_n \cos\frac{n\pi}{a_0}x \quad n = 1,3,5,\dots \tag{8-102}$$

The value π/a_0 is called the *geometric buckling*, B_g. If the reactor is critical the material buckling must be equal to the geometrical buckling and as a result

$$B_m^2 = B_g^2 \quad \Rightarrow \quad \frac{k_\infty - 1}{L^2} = \left(\frac{\pi}{a_0}\right)^2$$

The constant A in flux relation is an arbitrary value. However, at the center of a slab reactor it is equal to the maximum value of neutron flux

$$\Phi(x = 0) = \Phi_{\max} = A \tag{8-103}$$

Example 8.4 Infinite slab reactor

Calculate the thickness of a critical infinite slab homogeneous reactor consisting of a mixture of 200 moles of graphite per mole of 5.5% enriched uranium fuel. Assume that the overall temperature of the reactor core is $20°C$ and that the reactor is critical. The density of graphite and uranium are $\rho_C = 1.6$ g/cm^3 and $\rho_U = 18.9$ g/cm^3. The microscopic cross section for thermal neutron scattering at carbon is 4.8

b and at uranium is 8.3 b. The microscopic cross section for absorption in carbon is 0.0034 b, in ^{235}U is 694 b and in ^{238}U is 2.73 b. The infinite multiplication factor is assumed to be 1.2. (Adapted from "*Basic Nuclear Engineering*", A. R. Foster and R. L. Wright Jr., Allyn and Bacon Inc., 1968.)

Since the reactor is critical the material buckling is equal to the geometrical buckling, therefore

$$a_0 = \frac{\pi}{B} = \frac{\pi}{0.0566} = 55.5 \text{ cm} \quad d = 0.71\lambda_{tr} = \frac{0.71}{\Sigma_{tr}^{th}} = \frac{0.71}{0.364} = 1.95 \text{ cm}$$

$$\frac{a_0}{2} = \frac{a}{2} + d \quad \rightarrow \quad a = a_0 - 2d = 55.5 - 2 \times 1.95 = 51.6 \text{ cm}$$

2.4.6   Solution to One-Speed Neutron Diffusion Equation in Rectangular Bare Parallelepiped Reactor

For the rectangular parallelepiped reactor core shown in Fig. 8-11, Eq. (8-81) can be written as

$$\frac{\partial^2 \Phi(x, y, z)}{\partial x^2} + \frac{\partial^2 \Phi(x, y, z)}{\partial y^2} + \frac{\partial^2 \Phi(x, y, z)}{\partial z^2} + B^2 \Phi(x, y, z) = 0 \qquad (8\text{-}104)$$

This equation is solved by the method of variable separation

$$\Phi(x, y, z) = X(x)Y(y)Z(z) \equiv XYZ \qquad (8\text{-}105)$$

This expression indicates that the flux in the *x*, *y* or *z* directions is independent of that in the other two directions.

Differentiating the above equation yields

$$\frac{\partial \Phi(x, y, z)}{\partial x} = YZ \frac{dX}{dx} \qquad \frac{\partial^2 \Phi(x, y, z)}{\partial x^2} = YZ \frac{d^2 X}{dx^2}$$

$$\frac{\partial \Phi(x, y, z)}{\partial y} = XZ \frac{dY}{dy} \qquad \frac{\partial^2 \Phi(x, y, z)}{\partial y^2} = XZ \frac{d^2 Y}{dy^2} \qquad (8\text{-}106)$$

$$\frac{\partial \Phi(x, y, z)}{\partial z} = YX \frac{dZ}{dz} \qquad \frac{\partial^2 \Phi(x, y, z)}{\partial z^2} = YX \frac{d^2 Z}{dz^2}$$

Substituting these partial second order derivatives into Eq. (8-104) gives

$$YZ\frac{d^2X}{dx^2} + XZ\frac{d^2Y}{dy^2} + YX\frac{d^2Z}{dz^2} + B^2 XYZ = 0 \qquad (8\text{-}107)$$

Dividing by XYZ reduces Eq. (8-107) to

$$\frac{1}{X}\frac{d^2X}{dx^2} + \frac{1}{Y}\frac{d^2Y}{dy^2} + \frac{1}{Z}\frac{d^2Z}{dz^2} + B^2 = 0 \qquad (7\text{-}108)$$

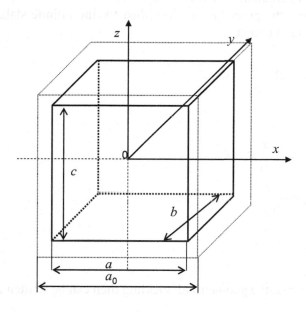

Figure 8-11. Rectangular bare parallelepiped reactor core

Since each of the terms is a function of a single variable, the above equation may be written as

$$B^2 = \alpha^2 + \beta^2 + \gamma^2 \qquad (8\text{-}109)$$

where α, β and γ are constants such that

$$\frac{d^2X}{dx^2} + X\alpha^2 = 0 \qquad \frac{d^2Y}{dy^2} + Y\beta^2 = 0 \qquad \frac{d^2Z}{dz^2} + Z\gamma^2 = 0 \qquad (8\text{-}110)$$

Since the derivatives are functions of only one variable, the partial derivative is replaced with a total derivative to give

$$X = A_x \cos \alpha X + C_x \sin \alpha x$$
$$Y = A_y \cos \beta Y + C_y \sin \beta y \tag{8-111}$$
$$Z = A_z \cos \gamma Z + C_z \sin \gamma z$$

The boundary conditions for the x direction are
1. For $x = a_0/2$, $X = 0$
2. For $x = 0$, gradient of X is zero, $dX / dx = 0$

Following the procedure as described for the infinite slab, the following equations are obtained

$$\alpha = \frac{\pi}{a_0} \quad \beta = \frac{\pi}{b_0} \quad \gamma = \frac{\pi}{c_0} \tag{8-112}$$

$$X = A_x \cos \frac{\pi}{a_0} x$$

$$Y = A_y \cos \frac{\pi}{b_0} y \tag{8-113}$$

$$Z = A_z \cos \frac{\pi}{c_0} z$$

The flux and the geometrical buckling then can be written as

$$\Phi(x, y, z) \equiv XYZ = A \cos\left(\frac{\pi}{a_0} x\right) \cos\left(\frac{\pi}{b_0} y\right) \cos\left(\frac{\pi}{c_0} z\right) \tag{8-114}$$

$$B_g^2 = \left(\frac{\pi}{a_0}\right)^2 + \left(\frac{\pi}{b_0}\right)^2 + \left(\frac{\pi}{c_0}\right)^2 \tag{8-115}$$

In case of a cubic reactor, the geometrical buckling becomes

$$B_g^2 = 3\left(\frac{\pi}{a_0}\right)^2 \quad \Leftrightarrow \quad a_0 = \frac{\pi}{B_g}\sqrt{3} \tag{8-116}$$

It can be understood that the extrapolated length of a side of a cubic core is larger than the extrapolated thickness of an infinite slab of the same material by a factor of $\sqrt{3}$.

2.4.7 Solution to One-Speed Neutron Diffusion Equation in Spherical Bare Reactor

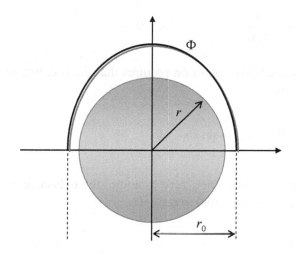

Figure 8-12. Spherical bare reactor

A spherical configuration requires the minimum amount of fuel to achieve criticality because the leakage is minimum (the area to volume ratio is minimal compared to other geometries). In a spherical reactor, neutron flux varies along the radial coordinate, Fig. 8-12. Eq. (8-81) written in spherical coordinates becomes

$$\frac{d^2\Phi(r)}{dr^2} + \frac{2}{r}\frac{d\Phi(r)}{dr} + B^2\Phi(r) = 0 \qquad (8\text{-}117)$$

This equation is solved by introducing $y = \Phi(r)r$ similar to the solution of the one-speed neutron diffusion equation shown in Section 2.4.2 for a point neutron source. With the given substitution, Eq. (8-117) reduces to

$$\frac{1}{r}\frac{d^2y}{dr^2} + B^2\frac{y}{r} = 0 \qquad (8\text{-}118)$$

with a solution of the form

$$y = A \cos Br + C \sin Br = \Phi(r)r \quad \Rightarrow$$

$$\Phi(r) = \frac{A}{r} \cos Br + \frac{C}{r} \sin Br \tag{8-119}$$

The first boundary condition specifies that the flux must be finite at the origin of the sphere resulting in

$$\lim_{r \to 0} \frac{A \cos Br}{r} = \frac{A \times 1}{0} = \infty \quad \Rightarrow \quad A = 0 \tag{8-120}$$

The second boundary condition requires that the flux becomes zero at the extrapolated radius

$$\Phi(r = r_0) = 0 \quad \Rightarrow \quad \frac{C}{r_0} \sin Br_0 = 0 \tag{8-121}$$

The constant C must be non-zero to assure the existence of neutron flux. Thus, Eq. (8-121) can be satisfied only if

$$Br_0 = 0, \pi, 2\pi, 3\pi, ... \tag{8-122}$$

The first value is a trivial solution and disregarded. The first non-zero value is a fundamental eigenvalue followed by the higher harmonic eigenvalues. The fundamental flux mode is

$$r_0 = \frac{\pi}{B} \quad \Leftrightarrow \quad B_g^2 = \left(\frac{\pi}{r_0}\right)^2 \tag{8-123}$$

$$\Phi(r) = \frac{C}{r} \sin \frac{\pi r}{r_0} \tag{8-124}$$

Example 8.5 Spherical bare reactor

Calculate the critical radius of a spherical homogeneous reactor consisting of a mixture of 200 moles of graphite per mole of 5.5% enriched uranium fuel. Determine the flux ratio between the center and the core boundary. For details see Examples 8.3 and 8.4. (Adapted from "*Basic Nuclear Engineering*", A. R. Foster and R. L. Wright Jr., Allyn and Bacon Inc., 1968.)

The extrapolated radius and macroscopic cross section are

$$r_0 = \frac{\pi}{B} = r + 0.71\lambda_{tr} \quad \rightarrow \quad r = \frac{\pi}{B} - 0.71\lambda_{tr} = \frac{\pi}{0.0566} - 1.95 = 53.55 \text{ cm}$$

$$\Sigma_f = N_{235}\sigma_f^{235}\frac{\sqrt{\pi}}{2} = (0.219 \times 10^{20})(582 \times 10^{-24})\frac{\sqrt{\pi}}{2} =$$

$$= 0.0113 \text{ cm}^2/\text{cm}^3\text{mixture}$$

Flux has its maximum value, Φ_{max}, at the center of a spherical core

$$\Phi(r) = \frac{C}{r}\sin\frac{\pi r}{r_0}$$

Constant C is determined as follows:

$$\lim_{r \to 0}\Phi(r) = \lim_{r \to 0}\frac{C}{r}\sin Br = \frac{0}{0} \quad \xrightarrow{\text{L'Hospital rule}}$$

$$\lim_{r \to 0}\Phi(r) = \lim_{r \to 0}\frac{CB\cos Br}{1} = CB = \Phi_{max}$$

$$C = \frac{\Phi_{max}}{B}$$

$$\Phi(r) = \frac{\Phi_{max}}{Br}\sin Br \quad \rightarrow \quad \frac{\Phi_{max}}{\Phi(r)} = \frac{Br}{\sin Br} = \frac{0.0566 \times 53.55}{\sin(0.0566 \times 53.55)} = 27.4$$

Notice that the angle is expressed in the units of radian.

2.4.8 Solution to One-Speed Neutron Diffusion Equation in Cylindrical Bare Reactor

The geometrical buckling of the finite one-region homogeneous cylindrical bare reactor is obtained from the neutron diffusion equation

$$\frac{d^2\Phi(r,z)}{dr^2} + \frac{1}{r}\frac{d\Phi(r,z)}{dr} + \frac{d^2\Phi(r,z)}{dz^2} + B^2\Phi(r,z) = 0 \qquad (8\text{-}125)$$

The solution is obtained by the method of separation of variables

$$\Phi(r,z) = \Phi(r)Z(z) \qquad (8\text{-}126)$$

If arbitrarily assigning that the radial part has a dimension of flux then the axial part represents the so-called form factor [Ott and Bezella, 1989].

Assuming r_0 is the extrapolated core radius and z_0 is the extrapolated core height the flux and geometrical buckling are found using the boundary conditions:

1. Flux is zero at the extrapolated boundaries; if the origin of the coordinate system is placed at the axial midplane the boundary conditions are defined as follows

$$\Phi\left(r, \pm \frac{z_0}{2}\right) = \Phi(r_0, z) = 0 \qquad (8\text{-}127)$$

2. Flux $\Phi(r, z)$ is symmetric and finite at the center of origin, i.e., at $r = 0$. With Eq. (8-126), the diffusion equation Eq. (8-125) becomes

$$\frac{1}{\Phi(r)}\left[\frac{d^2\Phi(r)}{dr^2} + \frac{1}{r}\frac{d\Phi(r)}{dr}\right] + \frac{1}{Z(z)}\frac{d^2Z(z)}{dz^2} + B^2 = 0 \qquad (8\text{-}128)$$

The first two terms must be equal to the constants for the equation to be satisfied, thus it can be written as follows

$$-B_r^2 - B_z^2 + B^2 = 0 \quad \Rightarrow \quad B^2 = B_r^2 + B_z^2 \qquad (8\text{-}129)$$

The equation

$$\frac{1}{\Phi(r)}\left[\frac{d^2\Phi(r)}{dr^2} + \frac{1}{r}\frac{d\Phi(r)}{dr}\right] + B_r^2 = 0 \qquad (8\text{-}130)$$

is the equation describing an infinite cylinder, while the equation

$$\frac{1}{Z(z)}\frac{d^2Z(z)}{dz^2} + B_z^2 = 0 \qquad (8\text{-}131)$$

describes an infinite slab (compare to Eq. (8-90)).

The general solution of Eq. (8-130) is given in the form of Bessel functions of the first kind, J_0 and Y_0

$$\Phi(r) = AJ_0(B_r r) + CY_0(B_r r) \qquad (8\text{-}132)$$

where A and C are the constants. After applying the boundary conditions and knowing the properties of the Bessel functions, the radial solution is obtained as follows

$$\Phi(r) = AJ_0(B_r r) = AJ_0\left(\frac{2.405r}{r_0}\right) \tag{8-133}$$

and the *radial buckling* is determined by the lowest root of the Bessel function J_0 (Fig. 8-13)

$$B_r^2 = \left(\frac{2.405}{r_0}\right)^2 \tag{8-134}$$

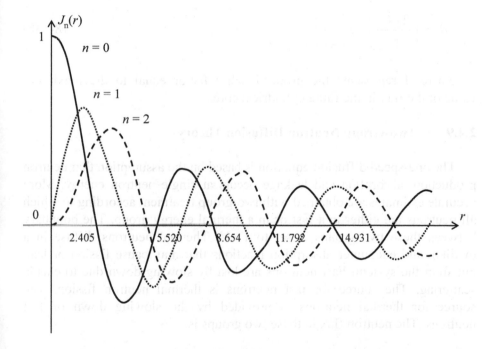

Figure 8-13. Schematic representation of the Bessel function of the first kind, $J_n(r)$, indicating the root values

The solution for the axial component of the flux and the *axial buckling* (follows the solution for the infinite slab) is

$$Z(z) = C \cos\left(\frac{\pi z}{z_0}\right) \qquad (8\text{-}135)$$

$$B_z^2 = \left(\frac{\pi}{z_0}\right)^2 \qquad (8\text{-}136)$$

Therefore, the final solution for the flux and the total bucking is

$$\Phi(r,z) = A J_0\left(\frac{2.405 r}{r_0}\right)\cos\left(\frac{\pi z}{z_0}\right) \qquad (8\text{-}137)$$

$$B^2 = \left(\frac{2.405}{r_0}\right)^2 + \left(\frac{\pi}{z_0}\right)^2 \qquad (8\text{-}138)$$

where A represents the proportionality factor equal to the maximum value of the flux in the finite cylindrical core.

2.4.9   Two-Group Neutron Diffusion Theory

The one-speed diffusion equation is based on the assumption that neutron production, absorption and leakage occurs at single neutron energy. More accurate estimates are obtained with two-group treatment according to which all neutrons are either in a fast or in a thermal energy group. The boundary between these two groups is set to 1 eV. Thermal neutrons diffuse in a medium and encounter absorption reactions that may cause fission or leak out from the system. Fast neutrons are lost by slowing down due to elastic scattering. The source for fast neutrons is thermal neutron fission. The source for thermal neutrons is provided by the slowing down of fast neutrons. The neutron flux in these two groups is

$$\text{Fast}: \qquad \Phi(\vec{r}) = \int_{1eV}^{10MeV} \Phi(E, \vec{r})\, dE$$

$$\qquad (8\text{-}139)$$

$$\text{Thermal}: \qquad \Phi(\vec{r}) = \int_{0}^{1eV} \Phi(E, \vec{r})\, dE$$

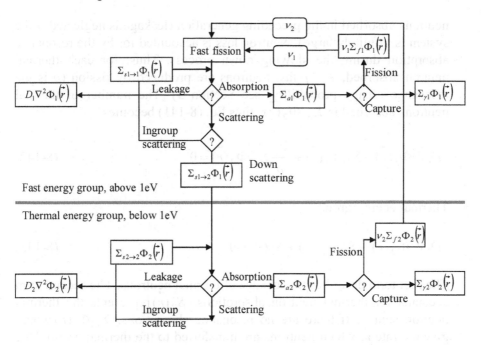

Figure 8-14. Schematic representation of the two-group diffusion equation

In the two-group approximation, neutron multiplication factor is defined as follows (see Fig. 8-14)

$$k_{eff} = \frac{v_1 \Sigma_{f1} \Phi_1(\vec{r}) + v_2 \Sigma_{f2} \Phi_2(\vec{r})}{-D_1 \nabla^2 \Phi_1(\vec{r}) - D_2 \nabla^2 \Phi_2(\vec{r}) + \Sigma_{a1} \Phi_1(\vec{r}) + \Sigma_{a2} \Phi_2(\vec{r})} \qquad (8\text{-}140)$$

The diffusion equations for the two energy groups become
- Fast energy group

$$D_1 \nabla^2 \Phi_1(\vec{r}) - \Sigma_{a1} \Phi_1(\vec{r}) + S_1(\vec{r}) = 0 \qquad (8\text{-}141)$$

First term describes the fast neutron leakage which involves a fast diffusion coefficient. The second term represents the removal of fast neutrons by thermalization. Σ_{a1} is the sum of the fission, capture and scattering (from group 1 to group 2) cross sections and is called the *removal cross section*. Fast neutron production is described by $S_1(\vec{r})$ representing the fast neutron source which depends on thermal neutron flux at the spatial positions where thermal neutrons cause fission (Fig. 8-14). The infinite multiplication factor for a thermal reactor represents the ratio of neutrons produced in any generation to the neutrons absorbed in the preceding generation (leakage is neglected if the system is infinitely large). Neutron loss is accounted for by the resonance absorption during the slowing down process. Thus, for each thermal neutron absorbed, k_∞ / p fast neutrons are produced by fission (p is the resonance escape probability, see Section 3). The number of thermal neutrons absorbed is $\Sigma_{a2}\Phi_2(r)$, thus Eq. (8-141) becomes

$$D_1\nabla^2\Phi_1(\vec{r}) - \Sigma_{a1}\Phi_1(\vec{r}) + \frac{k_\infty}{p}\Sigma_{a2}\Phi_2(\vec{r}) = 0 \qquad (8\text{-}142)$$

- Thermal energy group

$$D_2\nabla^2\Phi_2(\vec{r}) - \Sigma_{a2}\Phi_2(\vec{r}) + S_2(\vec{r}) = 0 \qquad (8\text{-}143)$$

The first term describes the leakage of thermal neutrons. The second term accounts for thermal neutron absorptions. $S_2(r)$ represents the thermal neutron source. If there are no resonance absorptions, $\Sigma_{a1}\Phi_1(r)$ would give the rate at which neutrons are transferred to the thermal group. Due to the resonance capture, the probability that a fast neutron will be thermalized is p. Thus, the thermal neutron source depends on fast neutron flux, and Eq. (8-143) becomes

$$D_2\nabla^2\Phi_2(\vec{r}) - \Sigma_{a2}\Phi_2(\vec{r}) + p\Sigma_{a1}\Phi_1(\vec{r}) = 0 \qquad (8\text{-}144)$$

Both equations are dependent on fast and thermal flux and thus represent a coupled system of equations. In addition, for a critical steady-state system the following equations from the diffusion theory are applied

$$\nabla^2\Phi_1(\vec{r}) + B^2\Phi_1(\vec{r}) = 0$$
$$\nabla^2\Phi_2(\vec{r}) + B^2\Phi_2(\vec{r}) = 0 \qquad (8\text{-}145)$$

Note that the buckling is same for both energy groups because it depends only on the core geometry for the critical system. By substituting Eq. (8-145) into Eq. (8-143) and Eq. (8-144), the following system is obtained

$$-\left(D_1 B^2 + \Sigma_{a1}\right)\Phi_1(\vec{r}) + \frac{k_\infty}{p}\Sigma_{a2}\Phi_2(\vec{r}) = 0$$
$$-\left(D_2 B^2 + \Sigma_{a2}\right)\Phi_2(\vec{r}) + p\Sigma_{a1}\Phi_1(\vec{r}) = 0 \qquad (8\text{-}146)$$

The solution of these coupled equations is found by setting the determinant of the coefficients to zero (the Cramer's rule)

$$
\begin{vmatrix}
D_1 B^2 + \Sigma_{a1} & -\dfrac{k_\infty \Sigma_{a2}}{p} \\
-p\Sigma_{a1} & D_2 B^2 + \Sigma_{a2}
\end{vmatrix} = 0
\tag{8-147}
$$

or

$$
\left(D_1 B^2 + \Sigma_{a1} \right)\left(D_2 B^2 + \Sigma_{a2} \right) - k_\infty \Sigma_{a1}\Sigma_{a2} = 0
\tag{8-148}
$$

giving

$$
\frac{k_\infty}{\left(1 + L_1^2 B^2\right)\left(1 + L_{th}^2 B^2\right)} = k_{eff} = 1
\tag{8-149}
$$

$$
L_1^2 \equiv D_1 / \Sigma_{a1} \qquad L_{th}^2 \equiv D_2 / \Sigma_{a2}
\tag{8-150}
$$

Equation (8-149) represents the two-group diffusion approximation for the critical bare reactor. In comparison with the one-speed diffusion equation there is one additional leakage term.

Equation (8-149) can be rewritten as

$$
\frac{k_\infty}{1 + B^2\left(L_{th}^2 + L_1^2\right) + B^4\left(L_{th}^2 L_1^2\right)} = 1
\tag{8-151}
$$

For large reactors for which $B^2 \ll 1$, Eq. (8-151) reduces to

$$
\frac{k_\infty}{1 + B^2\left(L_{th}^2 + L_1^2\right)} = 1 \quad \rightarrow \quad B^2 = \frac{k_\infty - 1}{M^2}
\tag{8-152}
$$

where M is the migration length and is explained in Section 3. The effective multiplication factor for a finite system thus becomes

$$
k_{eff} = k_\infty P_{non-leak}^{th} P_{non-leak}^{fast}
\tag{8-153}
$$

$$P_{non-leak}^{fast} = \frac{1}{\left(1 + L_1^2 B^2\right)} \quad P_{non-leak}^{th} = \frac{1}{\left(1 + L_{th}^2 B^2\right)} \tag{8-154}$$

2.4.10   Multi-Group Neutron Diffusion Theory

Neutrons born in fission are always fast. In thermal reactors, they slow down to lower energies due to scattering with the medium:
-   If a medium consists of dominantly heavy nuclei, neutrons scatter through inelastic processes creating energy spectrum shifted toward the lower energies.
-   If a medium consists of dominantly light nuclei, neutrons scatter through the elastic processes resulting in thermal energy spectrum.
-   In both cases, neutrons possess a wide spectrum of energies that require more than one or two energy groups for accurate estimates.

In a multi-group approach, neutrons are divided into a number of groups such that to every group corresponds an average energy and velocity with which neutrons diffuse through a medium until they are absorbed or due to slowing down removed to a lower energy group. The n-group diffusion theory is represented by the following series of equations

$$D_1 \nabla^2 \Phi_1(\vec{r}) - \Sigma_{a1} \Phi_1(\vec{r}) - \sum_{h=2}^{n} \Sigma_{s1 \to h} \Phi_1(\vec{r}) + S_1(\vec{r}) = 0$$

$$D_2 \nabla^2 \Phi_2(\vec{r}) - \Sigma_{a2} \Phi_2(\vec{r}) - \sum_{h=3}^{n} \Sigma_{s2 \to h} \Phi_2(\vec{r}) + S_2(\vec{r}) + \Sigma_{s1 \to 2} \Phi_1(\vec{r}) = 0$$

$$D_3 \nabla^2 \Phi_3(\vec{r}) - \Sigma_{a3} \Phi_3(\vec{r}) - \sum_{h=4}^{n} \Sigma_{s3 \to h} \Phi_3(\vec{r}) + S_3(\vec{r}) + \sum_{h=1}^{2} \Sigma_{sh \to 3} \Phi_h(\vec{r}) = 0$$

$$\vdots \tag{8-155}$$

$$D_i \nabla^2 \Phi_i(\vec{r}) - \Sigma_{ai} \Phi_i(\vec{r}) - \sum_{h=i+1}^{n} \Sigma_{si \to h} \Phi_3(\vec{r}) + S_i(\vec{r}) + \sum_{h=1}^{i-1} \Sigma_{sh \to i} \Phi_h(\vec{r}) = 0$$

$$\vdots$$

$$D_n \nabla^2 \Phi_n(\vec{r}) - \Sigma_{an} \Phi_n(\vec{r}) + S_n(\vec{r}) + \sum_{h=1}^{n-1} \Sigma_{sh \to n} \Phi_h(\vec{r}) = 0$$

For each of these groups, the diffusion equation is written such that for energy group i, $\Sigma \Phi$ describes the neutron losses (absorptions, $\Sigma_{ai} \Phi_i$, and removal to a lower energy group, $\Sigma_i \Phi_i$). The source term in first energy group, S_1, takes into account neutrons which are emitted with energies corresponding to that interval. In all other energy groups the source, S_i, is defined as a sum of neutrons emitted from that source plus all neutrons that come from other energy groups.

3. SLOWING DOWN OF NEUTRONS

Neutrons are slowed down in both elastic and inelastic scattering collisions with the nuclei of the atoms in a medium. In each collision, the neutron transfers a portion of its kinetic energy to the target nucleus in the form of kinetic energy if the collision is elastic or excitation energy if the collision is inelastic. Inelastic scattering is dominant with heavy nuclei, while elastic scattering is dominant with light nuclei. Moderator materials have low mass numbers and remove a large amount of energy from neutrons in a single collision and are also weak absorbers. This is why in a fast reactor materials of low mass number are avoided thus keeping the neutron population at a high average energy (the range where inelastic scattering by uranium or plutonium nuclei plays an important role). The one-group neutron diffusion equation provides a basic understanding of neutron transport.

Most of the neutrons produced in fission have energies in range of 1–2 MeV. In collisions with the materials in a reactor core, neutron energies range from small fractions of eV to a few MeV. Thermal reactors incorporate moderator materials in order to reduce the neutron energies to the thermal region where fission is most likely to occur. In an accurate reactor analysis both elastic and inelastic scattering are analyzed. A simple mathematical description of the elastic scattering processes can be developed under the following assumptions:

- target nuclei are at rest relative to the neutrons and
- the nuclei are not bound in a solid, liquid or gaseous molecule.

However, in the thermal region, the energies of the nuclei cannot be neglected in comparison with the neutron kinetic energies, and the scattering nuclei should be considered bound. In this condition, low-energy inelastic scattering cannot be neglected. Also, neutrons can gain or lose energy in a collision. An increase in energy is called *up-scattering*, and a decrease is called *down-scattering*.

The slowing down of a neutron from fission energies to roughly 1 eV is called *moderation* and the slowing down below 1 eV is called the *thermalization*. The following description of the neutron slowing down process refers to the moderation process for which the two assumptions stated above are acceptable.

3.1   Elastic Scattering in the Moderating Region

Elastic scattering in the moderator is described by assuming that the colliding particles behave as elastic spheres and that the target nuclei are stationary. In considering the scattering collision processes, two frames of references (Fig. 8-15) are used:

(a) *The laboratory system* (*LS*): scattering nucleus is at rest before the collision, and the neutron is moving toward the nucleus; after the collision, the neutron changes its direction of motion and velocity, and the nucleus moves from the rest position with some velocity. The viewpoint is that of a stationary external observer.

(b) *The center of mass system* (*COM*): neutron and nucleus are stationary in the collision. The observer is located at the center of mass of neutron plus the nucleus (compound nucleus) and travels with the velocity of the compound nucleus. The center of mass is an imaginary point where the system is balanced.

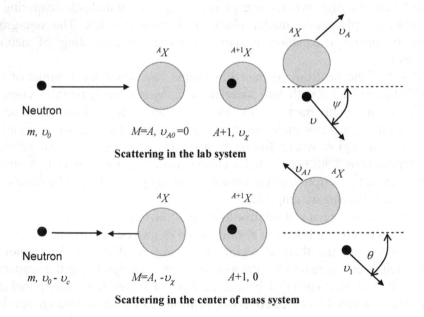

Figure 8-15. Scatter in lab (*LS*) and center of mass systems (*COM*)

Actual measurements are made in *LS* system, while the theoretical treatment is easier in the *COM* system. Since data are measured in the *LS* reference frame and the theoretical predictions are made in the *COM* reference system, a coordinate transformation is needed to compare theory to the experiment. Both systems are shown in Fig. 8-15 with

v_0: initial neutron velocity in *LS*

$v_{A0} = 0$: nucleus velocity in *LS*

v_c: compound nucleus velocity in *LS*

v_A: recoil nucleus velocity in the *LS*

v: scattered neutron velocity in the *LS*

ψ: neutron scattering angle in *LS* with respect to original neutron direction

v_{A1}: recoil nucleus velocity in the *COM* system

v_1: scattered neutron velocity in the *COM* system

θ: neutron scattering angle in the *COM* system

Laboratory system: Since the nucleus is stationary, its velocity is equal to zero and the momentum of a compound nucleus in *LS* is equal to the momentum of the incoming neutron

$$mv_0 + Mv_{A0} = (m + M)v_c \tag{8-156}$$

giving the velocity of the compound nucleus to be

$$v_c = \frac{v_0}{1+A} \quad A = \frac{M}{m} \tag{8-157}$$

Center of mass system: In order to follow the splitting of the compound nucleus it is convenient to transfer to the *COM*. In this system, the observer travels at the velocity and direction of the compound nucleus after the collision. Thus, the velocity of the neutron and nucleus before the collision must be reduced by the velocity of the compound nucleus v_C. The velocity of a compound nucleus itself will become zero as it will appear stationary after the collision (Fig. 8-15). Thus

• The velocity of incident neutron:

$$v_0 - v_c = v_0 - \frac{v_0}{(1+A)} = \frac{Av_0}{(1+A)} \tag{8-158}$$

• The velocity of a nucleus: $-v_c$

According to the conservation of energy law, the kinetic energy before the collision must equal the kinetic energy of the particles after the collision. The binding energy to form and break up the compound nucleus is the same and thus cancels out. The only energy to be conserved is the kinetic energy. The kinetic energy before the collision and available to the compound nucleus is the sum of the kinetic energies of a neutron and a nucleus

$$E(COM) = \frac{1}{2}(v_0 - v_c)^2 + \frac{1}{2}A(-v_c)^2 \qquad (8\text{-}159)$$

Eliminating the target nucleus velocity from the above equation gives

$$E(COM) = \frac{A v_0^2}{2(1 + A)} = \frac{A}{1 + A} E(LS)_0 \qquad (8\text{-}160)$$

where $E(LS)_0$ represents kinetic energy of the incident neutron in *LS*. From the above equation it can be seen that the kinetic energy before the collision in the *COM* system for light nuclei is half that of the incident neutron energy in *LS*, while for an interaction with ^{235}U (that creates ^{236}U as a compound nucleus) it is the fraction 235/236 of the incident neutron energy. Thus, the difference between these two systems is more evident for light nuclei. According to Fig. 8-15 the kinetic energy in the *COM* system is shared between the scattered neutron and scattered nucleus flying away in opposite directions. Thus, the conservation energy law in *COM* gives

$$\frac{A v_0^2}{2(1 + A)} = \frac{A v_{A1}^2}{2} + \frac{v_1^2}{2} \qquad (8\text{-}161)$$

The conservation of momentum equation gives

$$v_1 = A v_{A1} \qquad (8\text{-}162)$$

By combining the last two equations it follows that

$$\frac{A v_0^2}{2(1 + A)} = \frac{A v_{A1}^2}{2} + \frac{A^2 v_{A1}^2}{2} \qquad (8\text{-}163)$$

$$v_{A1} = \frac{v_0}{1 + A} \qquad v_1 = \frac{A v_0}{1 + A} \qquad (8\text{-}164)$$

Laboratory system (*LS*): It is useful to now convert back to the *LS* in order to compare the kinetic energy of the scattered neutron with the kinetic energy of the incoming neutron. Conversion from the *COM* system to the *LS* system is depicted in Fig. 8-16 and shows the transfer of velocities from one system to another using the Pythagorean theorem

$$v^2 = (v_1 \sin \theta)^2 + (v_1 \cos \theta + v_C)^2 \tag{8-165}$$

or

$$v^2 = \left(\frac{Av_0}{1+A} \sin \theta\right)^2 + \left(\frac{Av_0}{1+A} \cos \theta + \frac{v_0}{1+A}\right)^2 \tag{8-166}$$

which gives

$$v^2 = \frac{A^2 + 2A \cos \theta + 1}{(1+A)^2} v_0^2 \tag{8-167}$$

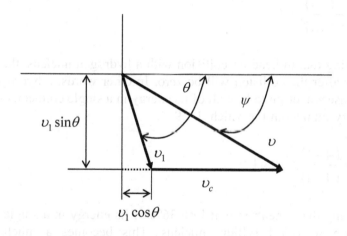

Figure 8-16. Diagram of velocities for conversion from *COM* to LS

From this equation it is possible to obtain the ratio of kinetic energy of the neutron after collision to that before the collision

$$\frac{E(LS)}{E(LS)_0} = \frac{v^2/2}{v_0^2/2} = \frac{A^2 + 2A \cos \theta + 1}{(1+A)^2} = \frac{1}{2}[(1+\alpha) + (1-\alpha)\cos \theta] \tag{8-168}$$

where α is defined with Eq. (8-169). Very often $E(LS)$ is denoted E' (as it will be used later in this chapter) and $E(LS)_0$ is denoted as E (for simplicity). This equation leads to the following conclusions:

- This ratio reaches its maximum when $\theta = 0$, or a glancing collision. Therefore, in forward scattering, neutron energy is not changed

$$\left(\frac{E(LS)}{E(LS)_0}\right)_{max} = \frac{A^2 + 2A + 1}{(1+A)^2} = 1$$

- The minimum ratio of energies is obtained for a head-on collision in which the neutron does not change its direction, or $\theta = \pi$.

$$\left(\frac{E(LS)}{E(LS)_0}\right)_{min} = \frac{A^2 + 2A(-1) + 1}{(1+A)^2} = \frac{(A-1)^2}{(A+1)^2} \equiv \alpha \qquad (8\text{-}169)$$

In the example of hydrogen ($A = 1$), the value of the defined parameter α becomes

$$\alpha_H = \frac{(A-1)^2}{(A+1)^2} = 0$$

indicating that in head-on collision with a hydrogen nucleus, the neutron energy after the collision will be zero. In other words, a hydrogen atom can cause a neutron to lose all of its energy in a single collision event. For beryllium atom for which $A = 9$

$$\alpha_{Be} = \frac{(A-1)^2}{(A+1)^2} = 0.64$$

indicating that a neutron will lose 36% of its energy in a single head-on collision with a beryllium nucleus. This becomes a much smaller percentage for a heavy nucleus like ^{235}U

$$\alpha_{235_U} = \frac{(A-1)^2}{(A+1)^2} = 0.98$$

giving that only 2% of the initial neutron energy will be lost in a single head-on collision. Thus, for heavy nuclei in which $A \gg 1$, it is expected that $\alpha \sim 1$ indicating, as shown in the example of uranium atom, that neutron energy after the collision is nearly equal to its energy before the collision.

Example 8.6 Scattering of a neutron in *COM* and *LS*

A neutron traveling through a medium is scattered by ^{9}Be. If the initial neutron energy is 0.1 MeV and the scattering angle 45° in the *COM* system calculate the fraction of energy that the neutron will lose as well as the scattering angle in the *LS*. From

$$\frac{E(LS)}{E(LS)_0} = \frac{A^2 + 2A\cos\theta + 1}{(1+A)^2} = \frac{9^2 + 2\times 9 \times \cos 45 + 1}{(1+9)^2} = 0.937$$

it follows that the scattered neutron energy is 0.937x0.1 MeV = 93.7 keV. The fraction of energy that neutron has lost in this collision is

$$\frac{0.1\,\text{MeV} - 93.7\,\text{keV}}{0.1\,\text{MeV}} = 0.063 \quad \rightarrow \quad 63\%$$

From Fig. 8-16 it follows that

$$\upsilon_1 \sin\theta = \upsilon \sin\psi \qquad \frac{A\upsilon_0}{A+1}\sin\theta = \upsilon\sin\psi \qquad \frac{\upsilon}{\upsilon_0} = \sqrt{\frac{E(LS)}{E(LS)_0}} = \frac{A}{A+1}\frac{\sin\theta}{\sin\psi}$$

$$\sin\psi = \frac{9}{9+1}\frac{\sin 45}{\sqrt{0.937}} = 0.657 \quad \rightarrow \quad \psi = 41.1°$$

The probability that neutrons will scatter within a certain angle while changing the energy is defined with the scattering kernel, $F(E \rightarrow E')$ introduced in Chapter 7, Section 4.1. In the *COM* the neutron scattering is isotropic, Fig. 8-17. All neutrons that scatter into the ring area as shown in Fig. 8-17 will have the energy that changes from E to E' within the energy interval of dE. Since the scattering energy E' decreases as the scattering angle increases the probability of scattering into a given angle or within the energy interval is defined as follows

$$F(\vec{\Omega} \rightarrow \vec{\Omega'})d\Omega' = -F(E \rightarrow E')\,dE'$$

where both probabilities are normalized to unity over the range of the corresponding variable. The angular probability is defined as the ratio of the angular cross section in the *COM*, $\sigma_s(\theta)$, to the total scattering cross section, σ_s. The angular cross section in the *COM* is equal to $\sigma_s/4\pi$ because the scattering is isotropic. Thus,

$$F(E \rightarrow E')dE' = -\frac{\sigma_s(\theta)}{\sigma_s}2\pi\sin\theta\,d\theta$$

From Eq. (8-168) we can express $\sin\theta\,d\theta$ as follows

$$dE' = -\frac{E}{2}(1-\alpha)\sin\theta\, d\theta$$

to obtain the general expression for the elastic neutron scattering kernel

$$F(E \to E') = \begin{cases} \dfrac{4\pi\sigma_s(\theta)}{\sigma_s(1-\alpha)E}; & \alpha E < E' < E \\ 0; & 0 < E' < \alpha E \end{cases} \qquad (8\text{-}170a)$$

For the isotropic scattering

$$F(E \to E') = \begin{cases} \dfrac{1}{(1-\alpha)E}; & \alpha E < E' < E \\ 0; & 0 < E' < \alpha E \end{cases} \qquad (8\text{-}170b)$$

This equation indicates that the probability to obtain any energy between E and αE is uniform and independent of scattering energy.

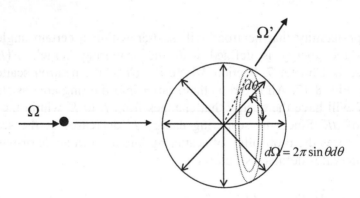

Figure 8-17. Isotropic scattering of neutrons in *COM*

3.2   Energy Distribution in Elastic Scattering – Logarithmic Energy Decrement

The energy that a neutron loses in an elastic collision with the nuclei of a medium is a function of medium atomic number and the scattering angle. The logarithmic energy decrement is defined as the logarithm of neutron energy per collision

$$\xi = \overline{\ln E_0 - \ln E} = -\overline{\ln \frac{E}{E_0}} \qquad (8\text{-}171)$$

In the *COM* system the breakup of a compound nucleus does not depend on the mode of its creation and neutrons scatter in random manner having equal probability for all directions (isotropic scattering, Fig. 8-17). The probability that a neutron will scatter into an angle between θ and $\theta + d\theta$ is the ratio of the area of the differential ring to the total area of the unit sphere

$$\frac{2\pi \sin\theta \, d\theta}{4\pi} = \frac{\sin\theta \, d\theta}{2} \tag{8-172}$$

The differential number of neutrons, dn, scattered into a differential angle is the product of the total number of neutrons, n, and the probability that neutrons will scatter into a differential angle between θ and $\theta + d\theta$

$$dn = n\frac{\sin\theta \, d\theta}{2} \tag{8-173}$$

The differential number of neutrons, dn, multiplied by the logarithmic decrement

$$\xi dn = -\ln\frac{E}{E_0}dn = \frac{n}{2}\sin\theta \, d\theta\left[-\ln\frac{A^2 + 2A\cos\theta + 1}{(A+1)^2}\right] \tag{8-174}$$

and integrated from 0 to π will give the total logarithmic decrement for all n neutrons of the system

$$\xi = -\ln\frac{E}{E_0} = \frac{1}{n}\int_0^\pi \frac{n}{2}\sin\theta \, d\theta\left[-\ln\frac{A^2 + 2A\cos\theta + 1}{(A+1)^2}\right] \tag{8-175}$$

This integral can be solved introducing the following change of variables

$$x = \frac{A^2 + 2A\cos\theta + 1}{(A+1)^2} \Rightarrow dx = -\frac{2A\sin\theta \, d\theta}{(A+1)^2} \tag{8-176}$$

with the appropriate adjustment of the limits, Eq. (8-174) becomes

$$\xi = -\ln\frac{\overline{E}}{E_0} = \int\limits_{1}^{\alpha}\frac{\ln x}{2}\left[\frac{(A+1)^2}{2A}\right]dx = \frac{(A+1)^2}{4A}\int\limits_{1}^{\alpha}\ln x\, dx \qquad (8\text{-}177)$$

The constant term can be rearranged in the following way

$$\frac{(A+1)^2}{4A} = \frac{(A+1)^2}{(A+1)^2-(A-1)^2} = \frac{1}{1-\dfrac{(A-1)^2}{(A+1)^2}} = \frac{1}{1-\alpha} \qquad (8\text{-}178)$$

and the logarithmic decrement becomes

$$\xi = 1+\frac{\alpha}{1-\alpha}\ln\alpha \qquad\qquad \text{or} \qquad\qquad (8\text{-}179a)$$

$$\xi = 1+\frac{(A-1)^2}{2A}\ln\left(\frac{A-1}{A+1}\right) \qquad (8\text{-}179b)$$

The average logarithmic energy loss per collision is only a function of the target nucleus mass and is not dependent on neutron energy; it is usually approximated with

$$\xi = \frac{2}{A+2/3} \qquad (8\text{-}180)$$

Since ξ represents the average logarithmic energy loss per collision, the total number of collisions necessary for a neutron to lose a given amount of energy may be determined by expanding ξ into a difference of natural logarithms of the energy range in question. The number of collisions (N) to travel from any energy, E_{high}, to any lower energy, E_{low}, is

$$N = \frac{\ln E_{high} - \ln E_{low}}{\xi} \qquad (8\text{-}181)$$

For a non-homogeneous medium, it follows that

$$\xi = -\ln\frac{\overline{E}}{E_0} = \frac{\xi_1\sigma_{s1}+\xi_2\sigma_{s2}+\cdots}{\sigma_{s1}+\sigma_{s2}+\cdots} \qquad (8\text{-}182)$$

Example 8.7 Average number of neutron elastic collisions

Calculate the number of collisions in ^{9}Be and ^{238}U required to reduce neutron energy from 2 MeV to thermal energies (0.025 eV).

- ^{9}Be:

$$\xi = 1 + \frac{(9-1)^2}{2 \times 9} \ln\left(\frac{9-1}{9+1}\right) = 0.207 \quad N = \frac{\ln E_{high} - \ln E_{low}}{\xi} = \frac{\ln\left(2 \times 10^6 / 0.025\right)}{0.207} = 88$$

- ^{238}U:

$$\xi = 1 + \frac{(238-1)^2}{2 \times 238} \ln\left(\frac{238-1}{238+1}\right) = 0.0084 \quad N = \frac{\ln\left(2 \times 10^6 / 0.025\right)}{0.0084} = 2446$$

Although the logarithmic energy decrement is a convenient measure of the ability of a material to slow neutrons, it does not measure all necessary properties of a moderator. How rapidly slowing down will occur in a material is measured by the *macroscopic slowing down power* (*MSDP*) which is defined as the product of the logarithmic energy decrement and the macroscopic scattering cross section for the material

$$MSDP = \xi \, \Sigma_s$$

MSDP thus represents the slowing down power of all nuclei in a unit volume of a moderator and does not give full information about material properties such as probability of scattering or absorption of neutrons. For example (Table 8-2), helium gas would have a good logarithmic energy decrement but very poor slowing down power due to the small probability of scattering neutrons due to its low density. Another example is boron that has a high logarithmic energy decrement and a good slowing down power, but it is a poor moderator because of very high probability of absorbing neutrons.

The most complete measure of the effectiveness of a moderator is the *moderating ratio* (*MR*) which is defined as the ratio of the *MSDP* to the macroscopic cross section for absorption. The higher the *MR*, the more effectively the material performs as a moderator

$$MR = \xi \frac{\Sigma_s}{\Sigma_a} \tag{8-183}$$

For a single element this reduces to

$$MR = \xi \frac{\sigma_s}{\sigma_a} \qquad (8\text{-}184)$$

while for a mixture of two elements it becomes

$$MR = \frac{\xi_1 \Sigma_{s1} + \xi_2 \Sigma_{s2}}{\Sigma_{a1} + \Sigma_{a2}} \qquad (8\text{-}185)$$

Table 8-2. Characteristics of moderators

Moderator	ξ	N to thermalized	*MSDP*	*MR*
Water	0.927	19	1.425	62
Heavy water	0.510	35	0.177	4,830
Helium	0.427	42	8.87×10^{-6}	51
Beryllium	0.207	86	0.724	126
Boron	0.171	105	0.092	0.00086
Carbide	0.258	114	0.083	216

Relative merits of some moderator materials used in current thermal reactors are given in Table 8-2. Ordinary water has high ξ and a good *MSDP*. However, because of 0.332 b absorption cross section it has the lowest *MR* of all moderators. The use of enriched fuel is thus required for a reactor to be critical. But the low cost and high availability are crucial factors in the wide use in the majority of nuclear power plant designs. Graphite is also widely used due to good moderation parameters and low cost. Heavy water has superior characteristics as a moderator, but is very expensive and therefore used in only a small number of reactor configurations. Helium is not used because of its low density while beryllium is avoided due to its high toxicity.

3.3 Average Cosine of the Scattering Angle

As described in Section 3.1, actual physical measurements are made in the *LS* system, while theoretical treatment is usually done in the *COM* system because it is simpler. In the *COM* system the scattering of neutrons is considered to be isotropic while in *LS* there is a preferential forward scattering and scattering is therefore anisotropic. This can be shown by deriving the relation for the average cosine of the scattering angle.

In the *COM* system the average value of the cosine of the scattering angle is calculated as a product of a number of neutrons scattering into an angle between θ and $\theta + d\theta$ and $\cos\theta$ integrated from 0 to π divided by the total number of scattered neutrons

$$\overline{\cos\theta} = \frac{1}{n}\int_0^\pi \left(\frac{n}{2}\right)\cos\theta\sin\theta \, d\theta = 0 \qquad (8\text{-}186)$$

This gives a value for the scattering angle in *COM* system of $90°$; this means that an equal number of neutrons scatter forward and backward therefore proving that scattering is isotropic in the *COM* system. Back to the *LS* system, the scattering angle becomes (Fig. 8-15 and 8-16)

$$\cos\psi = \frac{v_1\cos\theta + v_c}{\sqrt{(v_1\cos\theta + v_c)^2 + (v_1\sin\theta)^2}} = \frac{A\cos\theta + 1}{\sqrt{A^2 + 2A\cos\theta + 1}} \qquad (8\text{-}187)$$

The average cosine of scattering angle is

$$\overline{\mu} = \overline{\cos\psi} = \frac{1}{2}\int_0^\pi \cos\psi\sin\theta \, d\theta = \int_0^\pi \frac{(A\cos\theta + 1)\sin\theta \, d\theta}{2\sqrt{A^2 + 2A\cos\theta + 1}} = \frac{2}{3A} \qquad (8\text{-}188)$$

For example, the average cosine of the scattering angle for graphite indicates that the scattering is nearly isotropic in *LS*

$$\overline{\mu} \equiv \overline{\cos\psi} = \frac{2}{3A} = \frac{2}{3\times 12} = 0.056$$

while scattering on hydrogen indicates strong forward scattering

$$\overline{\mu} \equiv \overline{\cos\psi} = \frac{2}{3A} = \frac{2}{3\times 1} = \frac{2}{3}$$

3.4 Slowing Down of Neutrons in Infinite Medium

3.4.1 Slowing Down Density (Neutron Moderation) Without Absorption

All analyses presented in this section are valid for a steady-state reactor, under the assumptions that there is no loss of neutrons by absorptions or leakage since the medium is assumed infinite during the slowing down process. It is also assumed that the energy-dependent relations are already integrated over the spatial coordinates (spatial dependence is given in

Section 3.5). The no-absorption assumption requires a moderator that does not absorb neutrons with energies greater than thermal energies. Another assumption is that the neutron source is provided inside the moderator to produce neutrons at a uniform rate and at a definite energy, $S(E_1)$. A sink is provided to absorb only neutrons which have slowed down to thermal energies. Therefore, at steady state there will be no accumulation of neutrons and the number of neutrons that enter any energy increment, dE, at given energy E will be exactly equal to the number of neutrons leaving it. The slowing down process is shown schematically in Fig. 8-18.

The *slowing down density*, $q(E)$, is defined as the number of neutrons per unit volume that pass a given energy E per unit time. The derivations which follow are given for energies far from the source energy. These solutions are called *asymptotic solutions*.

Solutions applicable near the source are complex (with the exception of hydrogen moderator) and are called the *transient solutions*. Each neutron generated at E_1 will be either scattered or absorbed. The scattering collisions will distribute neutrons uniformly over the energy range from E_1 to αE_1. The slowing down density at E_1 is defined as the number of neutrons that slows down from E_1 per unit volume and in unit time

$$q(E_1) = S(E_1) \frac{\Sigma_s(E_1)}{\Sigma_{tot}(E_1)} \tag{8-189}$$

According to the assumption of no absorption in the system, the above relation reduces to

$$\Sigma_{tot}(E_1) = \Sigma_s(E_1) \quad \rightarrow \quad q(E_1) = S(E_1) \tag{8-190}$$

In an energy increment dE' (see Fig. 8-18) lying in energy interval between E/α and E, the number of collisions per unit volume in unit time is

$$\Sigma_s(E')\Phi(E')dE' \tag{8-191}$$

If the fraction of neutrons that will have energy less than E after scattering from dE' is

$$\frac{E - \alpha E'}{E' - \alpha E'} \tag{8-192}$$

then the number of neutrons passing an energy level E in unit volume per unit time that originate from energy increment dE' is

$$\Sigma_s(E')\Phi(E')\frac{E-\alpha E'}{E'-\alpha E'}dE' \qquad (8\text{-}193)$$

Therefore, the slowing down density becomes

$$q(E)=\int_E^{E/\alpha}\Sigma_s(E')\Phi(E')\frac{E-\alpha E'}{E'-\alpha E'}dE' \qquad (8\text{-}194)$$

A more explicit relation can be obtained for the slowing down density by recognizing that at steady state the number of neutrons scattered into the increment dE at E must be equal to those scattered out (because there is no absorption and no leakage of neutrons)

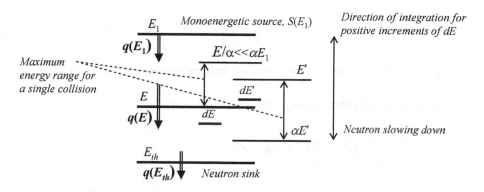

Figure 8-18. Neutron slowing down in energy space

$$\Phi(E)=\frac{q(E)}{E\xi\,\Sigma_s(E)}=\frac{S(E)}{E\xi\,\Sigma_s(E)} \qquad (8\text{-}195)$$

Since no neutrons are lost in an infinite non-absorbing medium at steady state, the number of neutrons slowing down past any energy is constant, or in other words the slowing down density is constant. The scattering cross section does not vary greatly in moderating energy region (see examples in Fig. 8-19) and the flux is proportional to $1/E$.

Example 8.8 Slowing down of neutrons

Neutrons of 1.5 MeV are introduced at the rate of 2×10^{15} n/cm^{3}s in an infinite slab of graphite. Calculate the number of elastic scattering collisions occurring per second in cm^3 in the energy interval from 0.5 MeV to 0.3 MeV.

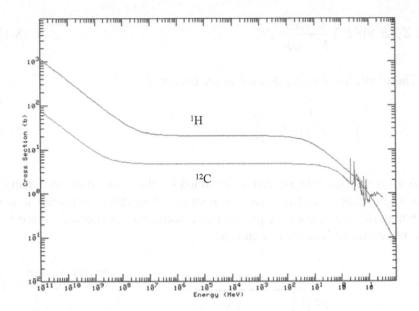

Figure 8-19. Elastic scattering cross section in moderating energy region for ^{1}H and ^{12}C

The average logarithmic energy loss per collision is

$$\xi = 1 + \frac{(A-1)^2}{2A} \ln\left(\frac{A-1}{A+1}\right) = 0.158 \quad A = 12$$

Because no absorptions or leakage are assumed, the only interaction neutrons may undergo is scattering with the nuclei in the graphite slab. Thus, the slowing down density equals the neutron source, i.e., the neutron rate at energy 1.5 MeV. Hence, we may write

$$\Phi(E) = \frac{S(E)}{E\xi\Sigma_s(E)} \quad \rightarrow$$

$$\Phi(E)\Sigma_s(E) = \int_{0.3}^{0.5} \frac{S(E)}{E\xi} dE = \frac{S}{\xi} \int_{0.3}^{0.5} \frac{dE}{E} = \frac{S}{\xi} \ln E\Big|_{0.3}^{0.5}$$

$$= \frac{2\times 10^{15}}{0.158} \ln\frac{0.5}{0.3} = 6.466\times 10^{15} \text{ n/cm}^3\text{s}$$

3.4.2   Lethargy

The equations involving energy and energy changes may be expressed in terms of a quantity called *lethargy*. By definition, the lethargy is

$$u = \ln\frac{E_0}{E} \tag{8-196}$$

where E_0 is an arbitrary starting energy usually taken to be 10 MeV. As neutron energy decreases the lethargy increases (Fig. 8-20). Low lethargy media are such that the energy change after a collision is small. This is true for high mass nuclei. If E_1 represents the neutron initial energy and E_2 neutron energy after the collision, the corresponding lethargies are u_1 and u_2, respectively, and then the lethargy change is given by

$$\Delta u = u_1 - u_2 = \ln\frac{E_1}{E_2} \tag{8-197}$$

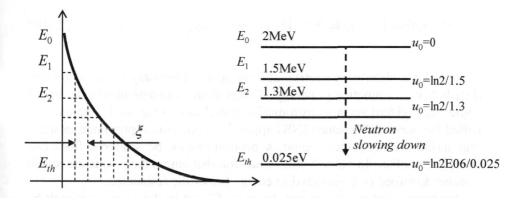

Figure 8-20. Neutron lethargy distribution

The average value of $\ln E_1/E_2$ represents the average logarithmic energy loss, ξ

$$\overline{\Delta u} = \xi \tag{8-198}$$

which can be also regarded as the average change in lethargy. As stated before, for the isotropic scattering in the moderating energy region of the *COM* system, ξ is independent of energy. This means that neutrons, regardless of their initial energy, must undergo on average the same number of collisions in a given medium to increase their lethargy by a specific amount (because the reciprocal value $1/\xi$ represents the average number of collisions).

3.4.3   Slowing Down Density (Neutron Moderation) with Absorption

In any actual situation neutrons are absorbed during the slowing down process. The slowing down density with absorption can be expressed as

$$q(E) = q(E')p(E) = S(E')p(E) \tag{8-199}$$

where $q(E')$ is the slowing down density without absorption and $p(E)$ is the fraction of neutrons that escape capture while slowing down from E' (energy of the source neutrons) to E and is called the *resonance escape probability*.

For a homogeneous system assumed to be infinite the neutron flux is independent of position. The neutron absorption rate is then

$$\text{Absorption from } E' \text{ to } E = \int_{E}^{E'} \Sigma_a(E'')\Phi(E'')dE'' \tag{8-200}$$

In order to determine neutron absorption it is necessary to know the flux distribution as a function of energy. That is difficult to determine exactly and some approximations are customarily introduced. One such approach is called the *narrow resonance* (NR) approximation. This approximation states that inside the resonance region, a neutron cannot be scattered from one energy to another. In other words, a neutron that enters the resonance region is either absorbed or is scattered to energy below the resonance.

Neutrons reaching the energy interval dE within the resonances will be only those scattered from higher energies. In the NR approximation, the number of neutrons entering this energy interval is independent of whether or not neutrons are absorbed in this region and it is equal to $q(E)\,dE/E\xi$, or $S(E)\,dE/E\xi$. Neutrons are lost from the energy interval dE by absorption and scattering. The loss rate is $(\Sigma_{aF} + \Sigma_s)\,\Phi(E)\,dE$, where Σ_{aF} is the absorption cross section in the fuel (absorber) and Σ_s is the total scattering cross section of the fuel and moderator, all of which are functions of energy.

In steady state, the number of neutrons entering an energy interval dE is equal to the number of neutrons which are lost

$$S(E)\frac{dE}{E\xi} = \left(\Sigma_{aF} + \Sigma_s\right)\Phi(E)dE \tag{8-201}$$

giving the flux to be

$$\Phi(E) = \frac{S(E)}{E\xi(\Sigma_{aF} + \Sigma_s)} \tag{8-202}$$

The presence of the absorption cross section in the denominator means that the neutron flux decreases in the resonance region.

The resonance escape probability then becomes

$$p(E) = \exp\left[-\frac{1}{\xi}\int\frac{\Sigma_{aF}}{\Sigma_{aF} + \Sigma_s}\frac{dE}{E}\right] \tag{8-203}$$

where the integration is over the resonance region energies.

A further approximation is called the narrow resonance infinite mass, NRIM, approximation in which the mass number of the absorber is assumed to be infinitely large. In such a case, the scattering cross section is that for the moderator only. An alternative expression for the resonance escape probability may be written in the following way

$$p(E) = \exp\left[-\frac{N_F}{\xi\Sigma_s}\int\frac{\Sigma_s}{\Sigma_{aF} + \Sigma_s}\sigma_{aF}\frac{dE}{E}\right] \tag{8-204}$$

where the scattering cross section and the average energy loss are assumed to be independent of energy. The integral in the last equation is called the *effective resonance integral, I,* and has the same dimension as the microscopic cross section. Thus, $N_F I$ has the dimension of macroscopic cross section.

Actual reactors are heterogeneous systems where fuel and moderator are physically separated. The fuel is present in distinct units called fuel rods that are spaced in a lattice array with the moderator region in between. If the neutron mean free path at given energy is less than or equal to the rod diameter, the probability that neutron will be absorbed in the fuel rod is large. This means that the flux at that given energy in the fuel rod will be lower than the flux in the moderator region. Resonance neutrons are largely absorbed in the outer regions of the fuel rods, especially if the resonance peak is narrow and high. As a result, nuclei in the interior are exposed to a very low neutron flux and the amount of absorptions is small. This effect is called *self-shielding.* The net result is that the probability of resonance capture is less than in the case of a uniform distribution of the fuel within the moderator. Therefore, the resonance escape probability is larger in heterogeneous systems. Also, it increases with fuel radius. Another factor that increases the escape probability in a fuel region is that some neutrons are slowed down in the moderator region to energies below the resonance region and therefore they escape capture. Neutrons absorbed by resonance capture in a thermal reactor fuel region (i.e., ^{238}U) are lost from the fission chain reaction. Thus, most thermal reactors are designed to maximize the resonance escape probability. For fuel rods placed far enough that resonance neutrons cannot pass directly between the rods, the rods are said to be "isolated". The resonance escape probability is then found to be

$$p(E) = \exp\left[-\frac{N_F V_F}{\xi_F \Sigma_F V_F + \xi_M \Sigma_M V_M} I\right] \qquad (8\text{-}205)$$

In closely packed or so-called "tight" lattices, like in water-moderated reactors, some resonance neutrons that would normally enter a fuel rod will be intercepted by adjacent fuel rods. The resonance flux is then less, on average, than it would be if the rods were well separated. Thus, each fuel rod in a tight lattice configuration is said to be partially "shadowed" by the other rods. The effective resonance integral is smaller, and thus the escape probability is larger. Corrections are made by introducing the Dancoff factor which depends on the spacing and radius of fuel rods and the fuel material cross sections.

3.5   Spatial Distribution of the Slowing Down Neutrons

3.5.1   Fermi Model

The preceding models were developed for an infinite medium in which the neutron flux distribution was not a function of spatial coordinates. However, in reality the system has finite dimensions in which the neutron flux distribution is a function of energy as well as spatial position $\Phi(E, \vec{r})$. A useful approach in studying the spatial distribution of neutrons is to consider the slowing down density in the moderating region. Slowing down density can be expressed analytically only under certain approximations. A fairly simple analytical approach is the so-called continuous slowing down model or the Fermi model. In the Fermi model the following assumptions are made:
1. The scattering of neutrons is isotropic in the *COM* system, thus the *average logarithmic energy decrement*, ξ, is independent of neutron energy. *This* also represents the *average increase in lethargy* per collision, i.e., after n collisions the neutron lethargy will be increased by $n\xi$ units (see Section 3.4.2).
2. Every neutron gains exactly ξ units of lethargy in every collision, i.e., each neutron is supposed to behave as an average neutron. Therefore, the

only lethargy values possible in the moderating region are discrete values of $n\xi$, where $n = 1, 2, 3$, etc.

3. The lethargy is a continuous function, i.e., the steps in lethargy change are approximated by continuous change, see Fig. 8-21.

The Fermi model is reasonably good for describing neutron slowing down process in a material with a large mass number because the average logarithmic energy loss is small (the spread of neutron energies after scattering is relatively small). Thus, the assumption that each neutron behaves like an average neutron is nearly accurate. In addition, since ξ is small, the steps shown in Fig. 8-21 are small in height but large in number. Therefore, it is acceptable to approximate the steps with the continuous curve. If neutrons slow down in materials of low mass number (like hydrogenous materials), the energy spread after collision is large and the average lethargy change is large. For example, in hydrogen it is possible that a neutron would lose all of its energy in a single collision. In this case, the Fermi model is inapplicable.

The neutron conservation equation in a reactor for the energy range E and $E + dE$, assuming

- continuous slowing down of neutrons
- weak neutron absorptions in the moderator
- finite size of the reactor (leakage cannot be neglected)

  may be written as

$$-\left[-D\nabla^2\Phi(E,\vec{r})\right]dE - \Sigma_a(E)\Phi(E,\vec{r})dE + S(E,\vec{r}) = 0 \qquad (8\text{-}206)$$

with the terms defined as follows:

$-\left[-D\nabla^2\phi(E,\vec{r})\right]dE$: number of neutrons with energy dE leaking out of the system

$\Sigma_a(E)\Phi(E,\vec{r})dE$: number of neutrons with energy dE being absorbed in the medium

$S(E,\vec{r})$: neutron source (number of neutrons slowing down) out of dE as shown in Fig. 8-22.

The source term can thus be expressed in terms of slowing down density

$$S(E,\vec{r}) = q(E + dE,\vec{r}) - q(E,\vec{r}) = \frac{\partial}{\partial E}q(E,\vec{r})dE \qquad (8\text{-}207)$$

which may be inserted into Eq. (8-206) to give

$$-\left[-D\nabla^2\Phi(E,\vec{r})\right]dE - \Sigma_a(E)\Phi(E,\vec{r})dE + \frac{\partial}{\partial E}q(E,\vec{r})dE = 0 \qquad (8\text{-}208)$$

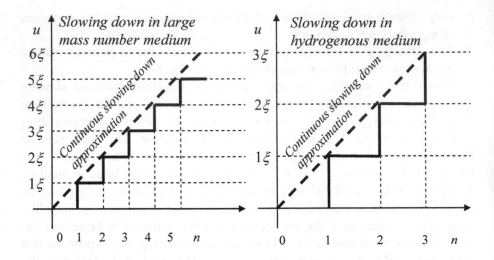

Figure 8-21. Continuous slowing down approximation

If the medium is a weak absorber, Eq. (8-208) reduces to

$$DV^2\phi\left(E,\vec{r}\right)+\frac{\partial}{\partial E}\,q\left(E,\vec{r}\right)=0 \qquad (8\text{-}209)$$

In the absence of absorption, the change in neutron slowing down density is due to leakage.

Combining Eq (8-209) with Eq. (8-195) gives

$$DV^2\left[\frac{1}{\xi\Sigma_s\left(E\right)}\frac{q\left(E,\vec{r}\right)}{E}\right]=-\frac{\partial}{\partial E}\,q\left(E,\vec{r}\right) \qquad (8\text{-}210)$$

or,

$$\nabla^2 q\left(E,\vec{r}\right)=-\frac{1}{\left(\dfrac{D}{\xi\Sigma_s\left(E\right)E}\right)}\frac{\partial}{\partial E}\,q\left(E,\vec{r}\right) \qquad (8\text{-}211)$$

Equation (8-211) can be simplified by introducing the variable, τ, called the *Fermi age*:

$$dt = -\frac{D}{\xi \Sigma_s(E)E} \qquad \tau = -\int_{E_0}^{E} \frac{D}{\xi \Sigma_s(E)E} dE = \int_{E}^{E_0} \frac{D}{\xi \Sigma_s(E)E} dE \qquad (8\text{-}212)$$

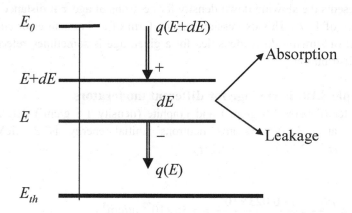

Figure 8-22. Definition of the neutron source term

Fermi age does not have units of time but the units of distance squared and represents the *chronological age of neutrons*. In other words, it indicates the time elapsing as neutrons travel away from their source (with energy E_0) to the point where its energy has been reduced to E. For neutrons of source energy ($E = E_0$) the Fermi age is zero, $\tau = 0$. The Fermi age increases as energy decreases (as a neutron slows down its age increases).

The slowing down density can be now expressed in terms of Fermi age as

$$\nabla^2 q(\tau,\vec{r}) = \frac{\partial}{\partial \tau} q(\tau,\vec{r}) \qquad (8\text{-}213)$$

and is valid for a medium with no absorption of neutrons. However, the age equation can be modified for weakly absorbing medium in the following way: if $q(E)$ is the neutron density in non-absorbing medium, then $q(E)p(E)$ is the slowing down density in a medium in which there is weak absorption of neutrons, $p(E)$ being the resonance escape probability.

Example 8.9: Fermi age equation
Find and interpret the solution for the Fermi age equation for a point source of monoenergetic fast neutrons (10 MeV) undergoing continuous slowing down in non-absorbing medium.
The solution of the equation

$$\nabla^2 q(\tau,\vec{r}) = \frac{\partial}{\partial \tau} q\left(\tau,\vec{r}\right) \qquad \text{is} \qquad q(\tau,\vec{r}) = \frac{e^{-\frac{r^2}{4\tau}}}{\left(4\pi\tau\right)^{3/2}} \qquad (8\text{-}214)$$

It represents the slowing down density for neutrons of age τ at distance r from a point source of 1 *n/s*. This expression has the form of a Gaussian error curve, thus distribution of slowing down densities for a given age is sometimes referred to as Gaussian distribution.

Example 8.10: Fermi age for different moderators

For water (density 1.0 g/cm^3) and graphite (density 1.6 g/cm^3) determine the Fermi age at 1 MeV. Assume neutrons' initial energy is 2 MeV. Data: $\sigma_s^H = 38b$, $\sigma_s^O = 3.76b$, $\sigma_s^C = 4.75b$.

Water

$$N^{H_2O} = \frac{\rho N_a}{M} = \frac{1 \times 6.023 \times 10^{23}}{18} = 3.35 \times 10^{22} \text{ at/cm}^3$$

$$\Sigma_s^{H_2O} = 0.0335\,(2 \times 38 - 3.76) = 2.66 \text{ cm}^{-1}$$

$$\Sigma_{tr}^{H_2O} = \Sigma_s^{H_2O}\left(1 - \frac{2}{3A}\right) =$$

$$0.0335\left[2 \times 38 \times \left(1 - \frac{2}{3 \times 1}\right) + 3.76 \times \left(1 - \frac{2}{3 \times 16}\right)\right] = 0.969 \text{ cm}^{-1}$$

$$D^{H_2O} = \frac{1}{3\Sigma_{tr}} = 0.344 \text{ cm} \qquad \xi^{H_2O} = \frac{2\sigma_s^H \xi^H + \sigma_s^O \xi^O}{2\sigma_s^H + \sigma_s^O} = 0.958$$

$$\tau^{H_2O} = -\frac{D}{\xi\Sigma_s}\ln\frac{E}{E_0} = -\frac{0.344}{0.958 \times 2.66}\ln\frac{1}{2} = 0.093 \text{ cm}^2$$

Carbon

$$N^C = \frac{\rho N_a}{M} = \frac{1.6 \times 6.023 \times 10^{23}}{12} = 8.03 \times 10^{22} \text{ at/cm}^3$$

$$\Sigma_s^C = 0.0803 \times 4.75 = 0.381 \text{ cm}^{-1}$$

$$\Sigma_{tr}^C = \Sigma_s^C\left(1 - \frac{2}{3A}\right) = 0.381\left(1 - \frac{2}{3 \times 12}\right) = 0.360 \text{ cm}^{-1}$$

$$D^C = \frac{1}{3\Sigma_{tr}} = 0.926 \text{ cm} \qquad \xi^C = 0.158$$

$$\tau^C = -\frac{D}{\xi\Sigma_s}\ln\frac{E}{E_0} = -\frac{0.926}{0.158 \times 0.381}\ln\frac{1}{2} = 10.66 \text{ cm}^2$$

3.5.2   Migration Length

The Fermi age of neutrons is related to the mean square distance traveled while slowing down. For thermal neutrons of age τ_{th} the $\sqrt{\tau_{th}}$ represents a measure of net vector distance traveled from the formation as fission neutrons to their appearance as thermal neutrons. The mean square distance which corresponds to the Fermi age is calculated as

$$\overline{r^2} = \frac{\int\limits_0^\infty r^2 q(E,\vec{r})4\pi r^2 dr}{\int\limits_0^\infty q(E,\vec{r})4\pi r^2 dr} = \frac{\int\limits_0^\infty r^4 e^{-\frac{r^2}{4\tau}} dr}{\int\limits_0^\infty r^2 e^{-\frac{r^2}{4\tau}} dr} = 6\tau \qquad (8\text{-}215)$$

The neutron age is analogous to the square of diffusion length. The above equation means that neutron travels 1/6th the mean square distance in going from the lethargy level before collision ($u_0 = 0$) to lethargy level after the collision (u). It also represents the slowing down length. The sum of the square of the diffusion length and the age is called the *migration area*

$$M^2 = l_{th}^2 + \tau_{th} \qquad (8\text{-}216)$$

and its square root the *migration length* (see Table 8-3). The criticality equation in the slowing down approximation for a large reactor is

$$k_\infty = 1 + M^2 B^2 \qquad (8\text{-}217)$$

Table 8-3. Migration lengths for most common moderators and thermal neutrons

Moderator	Diffusion length (cm)	Slowing down length (cm)	Migration length (cm)
Water	0.027	0.052	0.059
Heavy water	1.000	0.114	1.010
Beryllium	0.210	0.100	0.233
Graphite	0.540	0.192	0.575

Example 8.11: Critical core dimensions

Calculate the migration length, critical core radius and critical mass of a spherical reactor moderated by unit density water. The core contains ^{235}U at concentration of 0.0145 g/cm^3. *Data*: Fermi age is 27 cm^2, thermal diffusion area 3.84 cm^2 and buckling 2.8×10^{-3}cm^{-2}.

From

$$M = \sqrt{L_{th}^2 + \tau_{th}} = \sqrt{3.84 + 27} = 5.55 \, \text{cm}$$

the geometrical buckling for the spherical core will give the critical radius

$$B_g^2 = \left(\frac{\pi}{R_c}\right)^2 \quad \rightarrow \quad R_c = 59.4 \, \text{cm}$$

Thus the critical reactor core mass is

$$m_c = 0.0145 \times \frac{4\pi}{3} R_c^3 = 12.7 \, \text{kg}$$

4. NEUTRON TRANSPORT IN THERMAL REACTORS

4.1 Neutron Lifetime in Thermal Reactors

The neutron lifetime in a reactor is characterized through the fast fission factor, fast non-leakage probability, resonance escape probability, thermal non-leakage probability, thermal fuel utilization factor and reproduction factor used to define the six-factor formula.

Fast Fission Factor, ε

In a thermal reactor some fast neutrons before they slow down will cause fission of both ^{235}U and ^{238}U. At neutron energies above 1 MeV, most of the fissions will be in ^{238}U because of its large proportion in the fuel. Since each single fission event produces more than one neutron, there will be an increase in the number of neutrons available. This effect is described by the *fast fission factor* (Fig. 8-23) which represents the ratio of the total number of neutrons ($k_1 + k_2$), to the number of neutrons produced by thermal fissions (k_1). The fast fission factor is fixed once the fuel is fabricated. As the fuel ages (due to fuel burnup), the number of ^{238}U atoms is depleted by fast fissions (and consequently converted into ^{239}Pu). ^{239}Pu is fissionable with the epithermal neutrons. In further considerations, these fissions are included in the fast fission factor as fast fissions. Thus, the change of fast fission factor over the reactor core lifetime can be assumed to be insignificant.

Fast Non-leakage Probability, P_f

In thermal reactor, where there is a significant amount of moderator material, fast neutrons will slow down. They also may leak out of the reactor core or may proceed to slow down through interactions with the nuclei in media. The ratio of the number of fast neutrons which begin to slow down to the number of fast neutrons from all fissions is called the *fast neutron non-leakage probability* (Fig. 8-23).

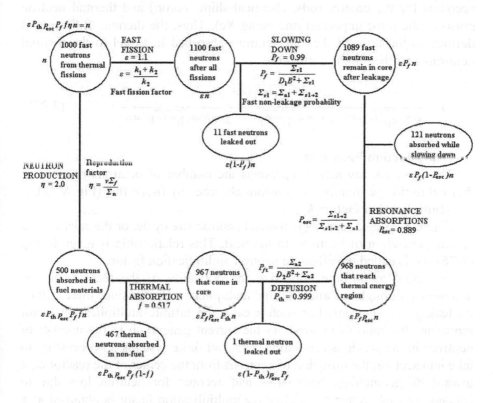

Figure 8-23. Full neutron life cycle, $k_1 = \nu_1 \Sigma_{f1}/(D_1 B^2 + \Sigma_{a1})$; $k_2 = \Sigma_{s1 \rightarrow 2} \nu_2 \Sigma_{f2}/(D_1 B^2 + \Sigma_{a1})(D_2 B^2 + \Sigma_{a2})$
[Refer to Appendix 7]

Resonance Escape Probability, p_{esc}

During the slowing down process neutrons may escape or may be captured in the resonance region. The number of neutrons which become thermalized to the number of neutrons that started to slow down represents the so-called *resonance escape probability* (Fig. 8-23).

Thermal Non-leakage Probability, P_{th}

Of the total number of neutrons which are thermalized, a certain number will leak out of the core. The ratio of the number of thermal neutrons that are absorbed in the core to the number of neutrons that are thermalized is called the *thermal non-leakage probability* (Fig. 8-23). Like the fast non-leakage probability, the thermal non-leakage probability also strongly depends on the core size. As the core is smaller, the leakage is larger.

Thermal Utilization Factor, f

One of the most important factors in the life cycle of neutrons is the *thermal utilization factor* (Fig. 8-23). This factor takes into account absorption of thermal neutrons in materials other than the fissile fuel. It accounts for the control rods, chemical shim (boron) and thermal neutron poisons (the most important one being Xe). Thus, the thermal utilization is defined as the ratio of thermal neutrons absorbed in a fuel to the thermal neutrons absorbed in the entire core

$$f = \frac{\Sigma_a^{\text{fuel}} \overline{\Phi^{\text{fuel}}} V^{\text{fuel}}}{\Sigma_a^{\text{fuel}} \overline{\Phi^{\text{fuel}}} V^{\text{fuel}} + \Sigma_a^{\text{mod}} \overline{\Phi^{\text{mod}}} V^{\text{mod}} + \Sigma_a^{\text{other}} \overline{\Phi^{\text{other}}} V^{\text{other}}} \qquad (8\text{-}218)$$

Reproduction Factor, η

The *reproduction factor* represents the number of neutrons released in thermal fission per number of neutrons absorbed by fissile fuel (Fig. 8-23).

Multiplication factor: k_{eff}

Figure 8-23 also represents the full neutron life cycle, or the relationship of one generation of neutrons to the next. This relationship is given in Eq. (8-76) and is called the effective neutron multiplication factor.

Assuming an infinite core size the criticality of the reactor will be determined through so-called infinite multiplication factor since there will be no leakage of neutrons. For such a case, the infinite multiplication factor represents the ratio of neutrons in the current generation to the number of neutrons in the previous generation. In actual finite systems it is necessary to take into account the diffusion of neutrons from the center of the reactor core toward its geometrical boundaries and account for neutron loss due to leakage. In such systems the effective multiplication factor is obtained as a product of the infinite multiplication factor and neutron non-leakage probability as defined in Eq. (8-76).

For a reactor to be critical the effective multiplication factor must be equal to unity. This means that the number of neutrons is constant in each generation and that the fission rate, and thus the reactor power, is maintained at the constant rate. With the k_{eff} greater than one the reactor power will raise exponentially and the reactor becomes supercritical. With the k_{eff} below unity reactor becomes subcritical and the number of neutrons in every coming generation decreases causing the reactor power to drop.

The infinite multiplication factor must be greater than unity for the reactor to be critical to allow for

- Loss of neutrons due to leakage
- Build-up of fission fragments with time as some of them have very large absorption cross sections that toward the end of fuel cycle will reduce the neutron population, and thus the reactor power

- Consumption of fissionable nuclei that is depleted by time and thus neutron population decreases toward the end of fuel cycle
- Changes in temperature and pressure in the core that may cause change in fission rates.

Example 8.12 Infinite multiplication factor

A bare spherical reactor is made of a homogeneous mixture of heavy water and ^{235}U, with the composition that for every uranium atom there are 2,000 heavy water atoms. Using the one-speed diffusion theory, calculate the total absorption cross section, the thermal utilization factor and the infinite multiplication factor if

$$\eta = 2.06 \quad \sigma_a^{235} = 678\,\text{b} \quad D = 0.87\,\text{cm} \quad p_{esc} = 0.6$$
$$\Sigma_a^{D_2O} = 3.3 \times 10^{-5}\,\text{cm}^{-1} \quad \sigma_a^{D_2O} = 0.001\,\text{b}$$

For the homogeneous reactor the neutron flux is the same in the core regardless of the material type, thus

$$f = \frac{\Sigma_a^{fuel}\overline{\Phi}^{fuel}V^{fuel}}{\Sigma_a^{fuel}\overline{\Phi}^{fuel}V^{fuel} + \Sigma_a^{mod}\overline{\Phi}^{mod}V^{mod} + \Sigma_a^{other}\overline{\Phi}^{other}V^{other}} = \frac{\Sigma_a^{fuel}}{\Sigma_a^{tot}}$$

$$\Sigma_a^{tot} = \Sigma_a^{fuel} + \Sigma_a^{D_2O} \qquad \Sigma_a^{D_2O} = N^{D_2O}\sigma_a^{D_2O} \qquad \Sigma_a^{fuel} = N^{fuel}\sigma_a^{fuel}$$

$$\frac{N^{D_2O}}{N^{fuel}} = 2000 \qquad N^{D_2O} = \frac{\Sigma_a^{D_2O}}{\sigma_a^{D_2O}}$$

$$\Sigma_a^{tot} = \Sigma_a^{fuel} + \Sigma_a^{D_2O} = \Sigma_a^{D_2O}\left[1 + \frac{N^{fuel}}{N^{D_2O}}\frac{\sigma_a^{fuel}}{\sigma_a^{D_2O}}\right] = 0.01122\,\text{cm}^{-1}$$

$$f = \frac{\Sigma_a^{fuel}}{\Sigma_a^{tot}} = \frac{678}{678 + 2000 \times 0.001} = 0.997$$

For the homogeneous mixture $\varepsilon = 1$.

$$k_\infty = \varepsilon p_{esc} f\eta = 1 \times 0.6 \times 0.997 \times 2.06 = 1.232$$

Example 8.13 Neutron generation doubling time

If the effective multiplication factor is 1.1 how many generations of neutrons are required to double neutron population? If there are 1,000 neutrons at the beginning (Fig. 8-23) how many neutrons will produce 50 generations?

After n generations there will be k_{eff}^n neutrons produced. In order to double the number of neutrons

$$k_{eff}^n = 2 \quad \rightarrow \quad n = \frac{\ln 2}{\ln k_{eff}} = \frac{\ln 2}{\ln 1.1} = 7$$

The number of neutrons generated after 50 generations is

$$k_{eff}^n = N \quad \rightarrow \quad N = (1.1)^{50} = 117$$

Meaning, the initial number of neutrons is increased 117 times, therefore the total number of neutrons after 50 generations is $1,000 \times 117 = 117,000$.

4.2   Homogeneous and Heterogeneous Reactors

The models used in the previous sections to describe the neutron transport and parameters of thermal nuclear reactors were related to a homogeneous mixture of fuel and moderator. In a homogeneous reactor core the nuclear properties like neutron flux and average cross sections are spatially uniform (see Fig. 7-14 in Chapter 7). Although homogeneous systems are practical to use for theoretical analysis, in practice most reactor concepts are based on heterogeneous configurations. In heterogeneous cores the fuel and moderator are separated as all other structural and reactor control components. Thus, nuclear properties change from one region to another such that for example neutron flux can vary drastically over a very short distance. The basic reason for the spatial variation of neutron flux in heterogeneous reactors is because adjacent material regions can have different absorption cross sections or some zones can have materials with strong resonance peaks. The neutron flux is always depressed in a material region of high absorption cross section, like the control rods consisting of strong absorber materials (see Chapter 8). Materials with high resonance absorptions (i.e., ^{238}U or ^{232}Th) cause the neutron flux to be reduced in the resonance energy region. The neutron resonance absorption rate in ^{238}U is smaller in heterogeneous than in homogeneous reactors. The following is a brief summary of how some main reactor parameters which influence criticality conditions change in heterogeneous as opposed to homogeneous systems.

Homogeneous reactors

In homogeneous reactors the fission neutrons are in immediate contact with the atoms of the moderator. The neutrons are moderated through elastic scattering before they are absorbed by the nuclei of the fuel. As a consequence, the neutron will not have the energy necessary to cause fission in ^{238}U (fast fissions), thus fast fission factor is nearly equal to unity, $\varepsilon \cong 1$ (the ratio of the total number of (fission, fast) neutrons slowing down past the fission threshold of ^{238}U to the number of neutrons produced by thermal fission).

The value of η does not vary since it depends on the composition of the fuel alone. In the case of natural uranium for example $\eta = 1.34$. Assuming an infinite reactor in its critical state, it follows that

- $\varepsilon = 1$
- $\eta = 1.34$ $\varepsilon \cdot \eta$
- $p_{esc} \cdot f = k_\infty / (\varepsilon \cdot \eta)$

Thus the value for $p_{esc} \cdot f = 1/(1 \cdot 1.34) = 0.746$ assures for chain reaction to be maintained. This value can be varied by changing the ratio of a moderator to fuel in a homogeneous mixture. Examples of homogeneous systems include natural uranium and graphite, natural uranium and D_2O, natural uranium and H_2O and natural uranium and beryllium.

Heterogeneous reactors

In heterogeneous reactors the fuel rods are surrounded by moderator material (see numerical example at the end of this chapter as well as Figs. 7-7 and 7-12 in Chapter 7). Fission takes place within the fuel and the neutrons are partially moderated through inelastic scattering until they escape from the fuel and initiate the principal process of moderation via elastic scattering with the nuclei of the moderator. The separation which exists between two fuel rods determines how many elastic scattering collisions can take place. Since neutrons travel through a fuel region before they enter moderator region there is a slight gain in fast fissions given that the neutrons emitted within the fuel rod can cause fast fission with ^{238}U before escaping the rod. This is why the value of the fast fission factor, ε, increases to some extent. The ε value ranges between 1.02 and 1.03.

The resonance escape probability p_{esc} increases significantly in heterogeneous systems as a result of two effects:

- *Pitch* (distance between the fuel rods): if the pitch is large, the majority of the neutrons will be moderated below resonant energies before entering a fuel element.
- *Fuel Self-Shielding*: fast neutrons born in the fuel region are mainly slowed down in the moderator region. After being thermalized, neutrons may diffuse back into the fuel region. Those with energies that correspond to the peak resonance region of the fertile nuclei in the fuel

region will immediately be absorbed. The most significant resonance in the case of ^{238}U is at energy of 6.7 eV with a peak cross section of 8,000 b. Therefore, absorptions at this energy level arise on the surface of the fuel rods, permitting the interior of the fuel to "see" no neutrons of epithermal energies consequently reducing the number of ^{238}U atoms available for resonant capture. The result is an improvement in the resonance escape probability since only a small fraction of fuel volume is involved in resonance capture.

As a result of these two reasons, the usual value for resonance escape probability is about 0.9.

The thermal utilization factor, f, decreases in heterogeneous cores because of the fuel self-shielding for the absorption of thermal neutrons. The general expression for thermal utilization factor can be written as

$$f = \frac{\Sigma_a^{fuel}}{\Sigma_a^{fuel} + \Sigma_a^{mod} \dfrac{\Phi^{mod}}{\Phi^{fuel}} + \Sigma_a^{other} \dfrac{\Phi^{other}}{\Phi^{fuel}}} \qquad (8\text{-}219)$$

If the flux in the moderator region and other components of the reactor core is larger than the flux in the fuel region, the thermal utilization factor will be reduced. The flux ratios in the denominator are called the *thermal disadvantage factors*.

The flux distribution of fast and thermal neutrons in a heterogeneous lattice is sketched in Fig. 8-24 (a) (numerical example is given at the end of this chapter and some other examples in Chapter 7). Fast neutrons are born in the fuel region from fission events. Once they reach the moderator region they are lost from the fast group because they slow down in elastic collisions and become thermal neutrons. The slow neutrons are therefore born in moderator region where as fast neutrons they lose the energy. Fuel elements represent a strong sink (absorber) for the thermal neutrons and thus the thermal flux drops in fuel region and peaks in the moderator region. Conversely, the fast flux peaks in the fuel region and dips in the moderator region. Figures 8-24(b) and (c) show the fast and thermal neutron flux distribution, respectively, in a more complex geometry of the research reactor assemblies. The research reactors usually have thin fuel plates as opposite to the power reactors that usually consist of cylindrical fuel rods arranged in a grid of rectangular or hexagonal shape.

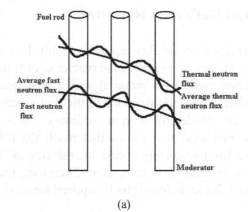

(a)

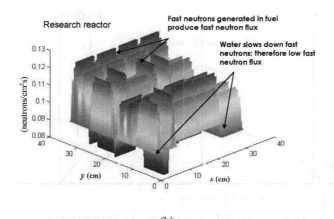

(b)

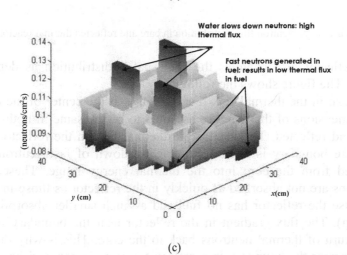

(c)

Figure 8-24. (**a**) Variation of thermal and fast neutron flux in a heterogeneous reactor fuel
lattice. (**b**) Fast and (**c**) thermal neutron flux profile in typical research reactor assemblies
(note the very thin rectangular fuel plates)

4.3   Bare and Reflected Reactors

The theory developed so far has referred to only bare reactors (reactors without a reflector). However, in reality the reactor core is usually surrounded by a neutron reflector made of a material which possesses good scattering properties. As a general rule, the reflector in thermal reactors is made of the same material as the moderator region; ordinary water, heavy water or graphite. Since the majority of neutrons that reach the reflector region are returned to the core from scattering collisions, the size of the critical reactor core is smaller than in case of bare reactor. Therefore, the use of reflector decreases the mass of fissile material (fuel) required for a critical system.

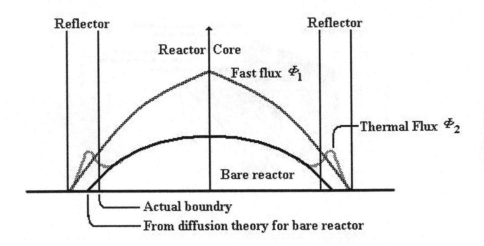

Figure 8-25. Neutron flux distribution in bare and reflected thermal reactors

The reflector also changes the flux spatial distribution as depicted in Fig. 8-25. The figure shows the following:
- The peak in the thermal flux distribution is at the center of the core and the dimensions of the core are assumed to be the same in both reactors, bare and reflected. The increase in thermal flux in the reflector beyond the core boundary is due to the slowing down of fast neutrons which escaped from the core into the thermal energy range. These thermal neutrons are not absorbed as quickly in the reflector as those in the core (because the reflector has no fuel and a much smaller absorption cross section). The flux gradient in the reflector near the boundary results in the return of thermal neutrons back to the core. This is why the flux is greater near the boundary in a reflected reactor in comparison to a bare reactor. Therefore, the reflector acts as a source of thermal neutrons due to the slowing down of fast neutrons (in thermal reactors).

- The average flux over the entire reactor core is increased in a reflected reactor. Since the power of a reactor is proportional to the neutron flux, the addition of a reflector increases power output.

A reflector reduces the critical size of the reactor and therefore the fissile mass needed to produce a critical reactor. The decrease in critical dimension of a reactor with the reflector is called the *reflector savings, δ*

$$\delta = R_0 - R \qquad (8\text{-}220)$$

where R_0 is the bare core radius, and R is the core radius of a reactor with the reflector. The reflector savings depend on the thickness of the reflector. The maximum reflector savings is obtained for the reflector thickness of about two migration lengths (assuming the reflector and moderator are of the same material).

5. CONCEPT OF THE TIME-DEPENDENT NEUTRON TRANSPORT

All previous chapters were concerned with a reactor in which the flux (or neutron population) varied only with spatial position assuming a steady-state reactor (reactor that operates at constant power). Analysis of how the neutron population varies with time is also very important and is called the transient behavior of the reactor.

Issues of the time-dependent reactor can be grouped as follows:

1. Behavior of the reactor in the non-critical regime (for example, at startup of a reactor or when its power is to be raised a reactor has to be supercritical; also in order to shut down a reactor it must be subcritical). The study of the behavior of a neutron population in a non-critical reactor is called reactor kinetics. It assumes the analysis of the prompt neutron lifetime, the reactor without delayed neutrons, the reactor with delayed neutrons, the prompt critical stage and the prompt jump approximation (all to be described in this chapter).

2. Regulation of the degree of reactor criticality (reactor is usually regulated by the use of control rods or chemical shim, where control rods are parts of fuel assemblies, and chemical shim is usually a boric acid mixed with the water moderator or coolant). Insertion of control rods makes the reactor subcritical (more neutrons are absorbed), while withdraw causes the neutron multiplication factor to increase. In the case of chemical shim, the reactor is controlled by changing the concentration of a neutron absorbing chemical in the moderator or coolant region. The basics of reactor control are described in Chapter 8.

3. Temperature effects on neutron population (several of the factors defining the multiplication factor are temperature dependent) as described in Chapter 8.
4. Fission product poisoning (accumulation of fission products takes place during the operation of reactor). Some fission products have very large absorption cross sections and their presence in a reactor can have a profound effect on the neutron population. Xenon-155 and Samarium-149 are particularly important in analyzing reactor fuel consumption as well as the condition of the reactor after shutdown. This aspect is described in Chapter 8.
5. Reactor core properties during the lifetime of the core (fuel burnup and fission product formation affect the power level over time and thus power costs). Analysis related to this issue is called fuel management and is not addressed in this book.

The departure from the steady-state neutron population or the percent change in multiplication factor is called *reactivity*

$$\text{Reactivity for a finite reactor: } \rho = \frac{k_{eff} - 1}{k_{eff}} = \frac{\Delta k_{eff}}{k_{eff}} \tag{8-221}$$

$$\text{Reactivity for an infinite reactor: } \rho = \frac{k_\infty - 1}{k_\infty} = \frac{\Delta k_\infty}{k_\infty} \tag{8-222}$$

From these equations it can be understood that the reactivity changes according to

$$\rho = 1 - \frac{1}{k_{eff}} = \begin{bmatrix} k_{eff} = 1 & \rightarrow & \rho = 0 & \text{critical} \\ k_{eff} > 1 & \rightarrow & \rho > 0 & \text{supercritical} \\ k_{eff} < 1 & \rightarrow & \rho < 0 & \text{subcritical} \end{bmatrix} \tag{8-223}$$

Thus, reactivity is restricted to the following ranges from $-\infty < \rho < 1$.

5.1   Neutron Lifetime and Reactor Period Without Delayed Neutrons

The total neutron lifetime accounts for the average time that a neutron spends in a reactor before it is absorbed or leaks out. In a thermal reactor it represents the sum of the slowing down time and the thermal (diffusion) time

$$l = l_s + l_{th} \tag{8-224}$$

The slowing down lifetime, l_s, is much shorter than the thermal neutron lifetime, l_{th}. It represents the time that a neutron spends while slowing down from fission energies to thermal energies. The thermal or diffusion lifetime corresponds to the time that neutrons spend diffusing before they are absorbed. In an infinite thermal reactor the neutron lifetime is obtained as the ratio of the absorption mean free path and the average neutron velocity

$$l_\infty \approx l_{th\infty} = \frac{\lambda_{th}}{\overline{\upsilon}} \tag{8-225}$$

The neutron lifetime in a finite thermal reactor is shorter than that in an infinite reactor. If N represents the number of neutrons per generation and $P_{non-leak}$ represents the non-leakage probability then $N \times P_{non-leak}$ neutrons remain in the core to contribute to the effective neutron lifetime

$$Nl_{th} = Nl_{th\infty} P_{non-leak} \quad \rightarrow$$

$$l \approx l_{th} = l_{th\infty} P_{non-leak} = \frac{l_{th\infty}}{1 + B^2 L^2} = \frac{1}{\Sigma_a \overline{\upsilon} \left(1 + B^2 L^2\right)} \tag{8-226}$$

The slowing and thermal lifetimes in thermal reactors are shown in Table 8-4 for a few most common moderator materials. The slowing down lifetime in fast reactors has no practical meaning. The total neutron lifetime in fast reactors is on the order of 10^{-7} s. The neutron generation time is defined as the integral time until a neutron is produced

$$\Lambda = \frac{\lambda_p}{\overline{\upsilon}} = \frac{1}{\upsilon \Sigma_f} \tag{8-227}$$

where λ_p represents the mean free path for neutron production.

If G represents the number of neutron generations, or the number of neutron lifetimes, between 0 and time t the effective multiplication factor is very close to unity. The neutron density and neutron flux in a thermal infinite reactor will change as follows:

$$n(t) = n(0)\left(k_{eff}\right)^G \quad \leftrightarrow \quad \Phi(t) = \Phi(0)\left(k_{eff}\right)^G \tag{8-228}$$

where
$\Phi(0)$ = initial (steady-state) neutron flux

$$k_{eff} = \Delta k_{eff} + 1$$

$$G = \frac{t}{l_\infty}$$

Thus, the flux change can be expressed as

$$\ln\frac{\Phi(t)}{\Phi(0)} = \frac{t}{l_\infty}\ln\left(1 + \Delta k_{eff}\right) = \frac{t}{l_\infty}\left[\Delta k_{eff} - \frac{1}{2}\left(\Delta k_{eff}\right)^2 + \cdots\right] \qquad (8\text{-}229)$$

For very small change in multiplication factor, Eq. (8-229) reduces to

$$\ln\frac{\Phi(t)}{\Phi(0)} = \frac{t}{l_\infty}\left[\Delta k_{eff}\right] \quad \rightarrow \quad \Phi(t) = \Phi(0)e^{\frac{\Delta k_{eff}}{l_\infty}t} \qquad (8\text{-}230)$$

Table 8-4. Neutron lifetime in thermal reactors

Moderator	Slowing down time (s)	Thermal time (s)
Carbon	1.5×10^{-4}	1.8×10^{-2}
Water	5.6×10^{-6}	2.1×10^{-4}
Heavy water	4.3×10^{-5}	1.4×10^{-1}
Beryllium	5.7×10^{-5}	3.7×10^{-3}

The reactor period taking into account only prompt neutrons (or *e* folding time), T, is defined as the time needed for flux to change by a factor *e*

$$T \equiv \frac{l_\infty}{\Delta k_{eff}} \qquad (8\text{-}231)$$

The reactor period must be long enough to prevent a dangerous excursion of reactor power. All reactors employ automatic safety systems to suddenly shutdown a reactor if the period becomes too short. The following example illustrates the importance of this concept.

Example 8.14 Reactor period in the absence of delayed neutrons

For the reactor described in Example 8.2 calculate how the reactor power changes if $\Delta k_{eff} = 0.01$?
From the data calculated in Example 8.2

$$l_\infty \approx l_{th\infty} = \frac{\overline{\lambda_{th}}}{\overline{\upsilon}} = \frac{1}{\Sigma_a \overline{\upsilon}} = \frac{1}{0.0146\,\text{cm}^{-1} \times 2.2 \times 10^5\,\text{cm/s}} = 3.1 \times 10^{-4}\,\text{s}$$

$$T = \frac{l_\infty}{\Delta k_{eff}} = \frac{3.1 \times 10^{-4}}{0.01} = 3.1 \times 10^{-2} \quad \rightarrow \quad \Phi(t) = \Phi(0)e^{32t}$$

This means every second the power will increase by a factor of e^{32}.

5.2 Delayed Neutrons and Average Neutron Lifetime

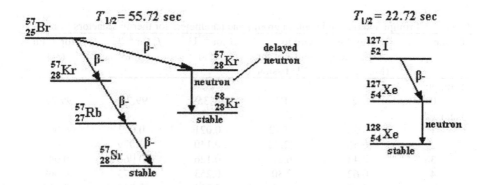

Figure 8-26. Delayed neutron precursors

Not all neutrons are released at the same time following a fission event. Nearly 99% of all neutrons are released virtually instantaneously (within about 10^{-13} s) after the actual fission event. These neutrons are called prompt neutrons. The remainder of neutrons are released after the decay of fission products. These neutrons are called delayed neutrons (with respect to the fission event). They are emitted immediately following the first β decay of a fission fragment, known as a delayed neutron precursor. Although delayed neutrons represent a very small fraction of the total number of neutrons, they play an extremely important role in the control of the reactor. Beta delayed neutron emission is improved when the emitted neutron binding energy is minimum. This is true when the neutron emitter has an odd neutron number, just above neutron shell closure. In particular, β decaying nuclei with neutron numbers equal to 52 ($N = 50$ closed shell) and 84 ($N = 82$ closed shell) are very important delayed neutron precursors as shown for [87]Br and [137]I in Fig. 8-26. The delay is determined by the β decay constant. Delays vary from fraction of seconds to tens of seconds. Probabilities for delayed neutron emission are on the order of less than 1% per fission or per prompt fission neutron. For example, the decay time of 55.72 s corresponds to the half-life of ^{87}Br (Fig. 8-26) and defines the first decay group. Similarly, the decay of ^{131}I followed by the emission of neutron after 2.72 s specifies the s group. In total, there are six groups of delayed neutrons (see Table 8-5).

Beta delayed neutrons are characterized by their *yields* β_i, relative to the total neutron number per fission, and their *decay constants* τ_i. The *total delayed neutron yield* per fission depends on the actual nuclear fuel that is used in a reactor and is defined as

$$\beta = \sum_{i=1}^{6} \beta_i \tag{8-232}$$

Table 8-5. Prompt and delayed neutron groups and parameters for thermal reactors

Group	Energy for ^{235}U fission (MeV)	Group half-life for ^{235}U fission	β_i for ^{235}U (%)	β_i for ^{233}U (%)	β_i for ^{239}Pu (%)
Prompt:					
0	~2	~10^{-3}	99.359	99.736	99.790
Delayed:					
1	0.25	55.72	0.021	0.023	0.007
2	0.56	22.72	0.140	0.079	0.063
3	0.43	6.22	0.126	0.066	0.044
4	0.62	2.30	0.253	0.073	0.069
5	0.42	0.61	0.074	0.014	0.018
6		0.23	0.027	0.009	0.009
β			0.641	0.264	0.210

Delayed neutrons do not have the same properties as the prompt neutrons released directly from fission. The average energy of prompt neutrons is about 2 MeV which is much greater than the average energy of delayed neutrons, ~0.5 MeV (see Table 8-5). The fact that delayed neutrons are born at lower energies has two significant impacts on the way they proceed through the neutron life cycle:

- Delayed neutrons have a much lower probability of causing fast fissions than prompt neutrons because their average energy is less than the minimum required for fast fission to take place.
- Delayed neutrons have a lower probability of leaking out of the core while they are at fast energies, because they are born at lower energies and subsequently travel a shorter distance as fast neutrons.

The average neutron lifetime in a thermal reactor is defined as

$$\bar{l} = \frac{\sum\limits_{i=0}^{6} \beta_i \bar{l}_i}{\sum\limits_{i=0}^{6} \beta_i} \tag{8-233}$$

where $\bar{l}_i$ represents the mean lifetime of a delay group defined as reciprocal of the decay constant of the delayed group. For example, for ^{235}U (from Table 8-5) the average neutron lifetime for all neutrons (prompt and delayed) is 0.0843 s. This means that 0.641% of the total number of neutrons increases the effective neutron generation time by a factor of 84.

Example 8.15 Reactor period including all neutrons

For the reactor described in Example 8.2 and Example 8.14 calculate how the reactor power changes if $\Delta k_{eff} = 0.01$ and all neutrons are considered. Use Table 8-5 to estimate the delayed neutron contribution.

In Table 8-5 the prompt neutron lifetime was assumed to be 0.001 s. In Example 8.14 that time was calculated to be 0.00031 s. Thus, the average neutron lifetime for all neutrons is 0.0836 s.

$$T = \frac{\bar{l}}{\Delta k_{eff}} = \frac{0.836}{0.01} = 83.6\ s \quad \rightarrow \quad \Phi(t) = \Phi(0)e^{t/83.6}$$

Thus every second the reactor power will change by factor $e^{0.012} = 1.012$.

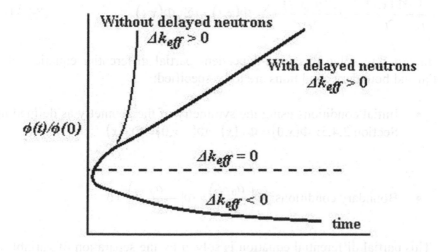

Figure 8-27. Effect of delayed neutron on power change in thermal reactors

The effect of delayed neutrons on reactor power changes as shown in Fig. 8-27:

- Without delayed neutrons the power will rise exponentially and in a very short time (see Example 8.14).
- When the effect of delayed neutrons is taken into account, the power of a thermal reactor changes as shown in Fig. 8-27 for $\Delta k_{eff} > 0$. At the very beginning the reactor behaves as if all neutrons were prompt. This is because the delayed neutrons are not yet effective. However, after a few seconds when delayed neutrons start to appear the rate of neutron flux and reactor power starts to level off. The rate of flux increase approaches the constant value determined by the stable reactor period.
- When $\Delta k_{eff} < 0$ the rate at which neutron power decreases is very fast and as soon as delayed neutrons appear the curve tends to flatten out. Since the flux is dying out, the short-lived delayed neutrons disappear completely and the curve approaches a slope with the value determined by the longest lived neutron group.

5.3   Diffusion Equation for Transient Reactor

5.3.1   Time-Dependent Infinite Slab Reactor

The criticality condition for the infinite slab reactor based on the one-group diffusion equation was derived in Section 2.4.5. The time-dependent one-group diffusion equation for this reactor is written as follows:

$$\frac{1}{\upsilon}\frac{\partial \Phi(x,t)}{\partial t} - D\frac{\partial^2 \Phi(x,t)}{\partial x^2} + \Sigma_a \Phi(x,t) = \nu\Sigma_f \Phi(x,t) \qquad (8\text{-}234)$$

In order to solve this time-dependent partial differential equation, the initial and boundary conditions are to be specified:

- Initial conditions using the symmetry of the geometry as defined in Section 2.4.5: $\Phi(x,0) = \Phi_0(x) \quad \Phi(-x,0) = \Phi_0(x)$

- Boundary conditions: $\Phi\left(\dfrac{a_0}{2},t\right) = \Phi\left(\dfrac{-a_0}{2},t\right) = 0$

This partial differential equation is solved by the separation of variables. Thus, introducing

$$\Phi(x,t)=\varphi(x)\phi(t) \tag{8-235}$$

and replacing into Eq. (8-234) it follows that

$$\frac{1}{\upsilon}\frac{\partial\varphi(x)\phi(t)}{\partial t}-D\frac{\partial^2\varphi(x)\phi(t)}{\partial x^2}+\Sigma_a\varphi(x)\phi(t)=\nu\Sigma_f\varphi(x)\phi(t) \tag{8-236}$$

Dividing by Eq. (8-235) gives

$$\frac{1}{\phi(t)}\frac{d\phi(t)}{\partial t}=\frac{\upsilon}{\varphi(x)}\left[D\frac{d^2\varphi(x)}{\partial x^2}+\left(\Sigma_a-\nu\Sigma_f\right)\varphi(x)\right]=-\lambda \tag{8-237}$$

Since the left-hand side of Eq. (8-237) depends only on time and the right-hand side depends only on spatial coordinates, both sides of this equation must be equal to some constant; the constant is set to be $-\lambda$. Thus Eq. (8-237) reduces to two ordinary differential equations

$$\frac{d\phi(t)}{\partial t}=-\lambda\phi(t)$$

$$D\frac{d^2\varphi(x)}{\partial x^2}+\left(\Sigma_a-\nu\Sigma_f\right)\varphi(x)=-\frac{\lambda}{\upsilon}\varphi(x) \tag{8-238}$$

The time-dependent differential equation has a solution

$$\phi(t)=\phi(0)e^{-\lambda t} \tag{8-239}$$

In order to solve the spatially dependent differential equation the boundary conditions as discussed in Section 2.4.5 are used to obtain the values for symmetric eigenfunctions and eigenvalues

$$\varphi_n(x)=A_n\cos B_n x$$

$$B_n^2=\left(\frac{n\pi}{a_0}\right)^2 \qquad n=1,3,5,\ldots \tag{8-240}$$

that is valid for the eigenvalue equation specified as follows (see Section 2.4.5)

$$\frac{d^2\varphi_n(x)}{dx^2} + B_n^2\varphi_n(x) = 0 \tag{8-241}$$

Thus, simply by comparing Eq. (8-241) and spatial differential equation in Eq. (8-238) it is clear that the constant must be the time eigenvalue as follows

$$B_n^2 = \frac{(\Sigma_a - \nu\Sigma_f)}{D} + \frac{\lambda_n}{D\upsilon} \tag{8-242}$$

$$\lambda_n = \upsilon D B_n^2 + \upsilon(\Sigma_a - \nu\Sigma_f)$$

Thus the general solution of Eq. (8-234) is

$$\Phi_n(x,t) = \varphi(x)\phi(t) = \sum_{\substack{n \\ \text{odd}}} A_n e^{-\lambda_n t}\cos\left(\frac{n\pi}{a_0}x\right) =$$

$$\sum_{\substack{n \\ \text{odd}}} A_n e^{-\lambda_n t}\cos B_n x \tag{8-243}$$

The initial conditions are used to determine the constant A_n

$$\Phi(x,0) = \Phi_0(x) = \sum_{\substack{n \\ \text{odd}}} A_n \cos\left(\frac{n\pi}{a_0}x\right)$$

If the time is long enough, i.e., $t \to \infty$, Eq. (8-243) reduces to the fundamental mode flux only

$$\Phi_1(x,t) = A_1 e^{-\lambda_1 t}\cos B_1 x \tag{8-244}$$

Based on this derivation there are three aspects that can be connected to the previous explanations: the criticality condition for the steady-state infinite slab reactor; the multiplication factor for the steady-state core; and the reactor period for time-dependent condition.

    (a) *Criticality condition*: the requirement for the time-independent (steady-state) flux is that the fundamental time eigenvalue becomes zero:

$$\lambda_1 = \upsilon D B_1^2 + \upsilon \left(\Sigma_a - \nu \Sigma_f \right) = 0$$

$$B_1^2 \equiv B_g^2 = \left(\frac{\pi}{a_0} \right)^2 \tag{8-245}$$

giving the condition for criticality in which geometrical buckling is equal to the material buckling and is specified with

$$B_g^2 \equiv B_m^2 = \left(\frac{\pi}{a_0} \right)^2 = \frac{\nu \Sigma_f - \Sigma_a}{D} \tag{8-246}$$

Thus, from Eq. (8-246) it follows that

- $B_m^2 > B_g^2 \rightarrow \dfrac{\nu \Sigma_f - \Sigma_a}{D} > \left(\dfrac{\pi}{a_0} \right)^2$; $\lambda_1 < 0$: supercritical reactor

- $B_m^2 = B_g^2 \rightarrow \dfrac{\nu \Sigma_f - \Sigma_a}{D} = \left(\dfrac{\pi}{a_0} \right)^2$; $\lambda_1 = 0$: critical reactor

- $B_m^2 < B_g^2 \rightarrow \dfrac{\nu \Sigma_f - \Sigma_a}{D} < \left(\dfrac{\pi}{a_0} \right)^2$; $\lambda_1 > 0$: subcritical reactor

(b) *Multiplication factor and neutron lifetime*: since $L = \sqrt{D / \Sigma_a}$ Eq. (8-245) can be re-written in the form

$$\lambda_1 = \upsilon \Sigma_a \left(1 + L^2 B_g^2 \right) \left(1 - \frac{\nu \Sigma_f / \Sigma_a}{1 + L^2 B_g^2} \right) \tag{8-247}$$

In an infinite system,

$$\frac{\nu \Sigma_f}{\Sigma_a} = \frac{\nu \Sigma_f}{\Sigma_a^{fuel}} \frac{\Sigma_a^{fuel}}{\Sigma_a} = \eta f = k_\infty \tag{8-248}$$

In Section 2.4.9 we showed that the non-leakage probability is defined with

$$P_{non-leakage} = \frac{1}{1 + L^2 B_g^2} \tag{8-249}$$

giving the multiplication factor in a finite reactor

$$k_{eff} = \eta f P_{non-leakage} = k_\infty P_{non-leakage} \tag{8-250}$$

Since in an infinite reactor $1/\upsilon\Sigma$ would always represent the mean free path for the neutron absorption, its product with the non-leakage probability would define the neutron mean lifetime in a finite reactor

$$P_{non-leakage} \frac{1}{\upsilon\Sigma_a} = \frac{1}{1 + L^2 B_g^2} \frac{1}{\upsilon\Sigma_a} \equiv l \tag{8-251}$$

By comparing Eq. (8-249) and (8-251) it follows that the fundamental time eigenvalue is the inverse of a reactor period (see also Eq. (8-231))

$$\lambda_1 = \upsilon\Sigma_a \left(1 + L^2 B_g^2\right)\left(1 - k_{eff}\right) = \frac{1 - k_{eff}}{l}$$

$$-\lambda_1 = \frac{k_{eff} - 1}{l} = \frac{1}{T} \tag{8-252}$$

5.3.2 Derivation of the Point Kinetics Equations

In an accurate reactor kinetics analysis all six groups of delayed neutrons are considered in detail (their production and decay). In order to simplify the complex calculation procedure, these six groups are often considered as one group of delayed neutrons that appear from the decay of a single hypothetical precursor. The time-dependent diffusion equation is given by Eq. (8-234).

The neutron source in a transient reactor takes into account both prompt and delayed neutrons

$$S(\vec{r},t) = S_p(\vec{r},t) + S_d(\vec{r},t) \tag{8-253}$$

The fraction of prompt neutrons that slows down to thermal energies is $1-\beta$ (see Fig. 8-28). This fraction contributes to the neutron source as

$$S_p\left(\vec{r},t\right)=(1-\beta)k_\infty\Phi\left(\vec{r},t\right)\Sigma_a=(1-\beta)\nu\Sigma_f\Phi\left(\vec{r},t\right) \tag{8-254}$$

where $k_\infty\Phi\left(\vec{r},t\right)\Sigma_a$ means that k_∞ thermal neutrons will appear for each neutron absorbed.

The delayed neutron source is defined with six delayed neutron groups that are for simplicity in this derivation assumed to all belong to one group. The contribution to the delayed neutron source is equal to the rate of decay for all precursors

$$S_d\left(\vec{r},t\right)=\lambda C\left(\vec{r},t\right) \tag{8-255}$$

where λ is the decay constant of the precursor and $C\left(\vec{r},t\right)$ is the precursor concentration or the number of delayed neutrons reaching thermal energies. In case all six groups of delayed neutrons are considered the delayed neutron source will be written as follows:

$$S_d\left(\vec{r},t\right)=\sum_{i=1}^{6}\lambda_i C_i\left(\vec{r},t\right) \tag{8-256}$$

The balance equation describing the precursor concentration change in time depends on the rate by which the precursors are produced (see Fig. 8-28) and decayed

$$\frac{\partial C\left(\vec{r},t\right)}{\partial t}=\beta\nu\Sigma_f\Phi\left(\vec{r},t\right)-\lambda C\left(\vec{r},t\right) \tag{8-257}$$

Thus, the system of equations for the time-dependent reactor is

$$\frac{1}{\upsilon}\frac{\partial\Phi\left(\vec{r},t\right)}{\partial t}-D\nabla^2\Phi\left(\vec{r},t\right)+\Sigma_a\Phi\left(\vec{r},t\right)=(1-\beta)\nu\Sigma_f\Phi\left(\vec{r},t\right)+\lambda C\left(\vec{r},t\right)$$

$$\frac{\partial C\left(\vec{r},t\right)}{\partial t}=\beta\nu\Sigma_f\Phi\left(\vec{r},t\right)-\lambda C\left(\vec{r},t\right) \tag{8-258}$$

and is solved assuming asymptotic situation that allows to separate both the neutron flux and the precursor concentration into the space- and time-dependent components (such as Eq. (8-259)):

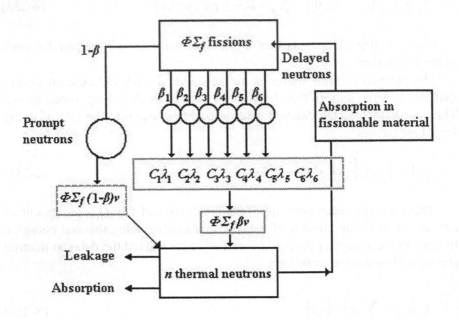

Figure 8-28. Neutron cycle chart in transient thermal reactor

$$\Phi\!\left(\vec{r},t\right)=\varphi_1\!\left(\vec{r}\right)\!\phi(t)=\upsilon n(t)\varphi_1\!\left(\vec{r}\right)$$
$$C\!\left(\vec{r},t\right)=C(t)\varphi_1\!\left(\vec{r}\right)$$

(8-259)

where $\varphi_1\!\left(\vec{r}\right)$ represents the fundamental mode defined by

$$\nabla^2\varphi_n\!\left(\vec{r}\right)+B_n^2\varphi_n\!\left(\vec{r}\right)=0$$

(8-260)

After substituting Eq. (8-259) into the set of equations defined by Eq. (8-238) and applying the definitions derived in Section 5.3.1 assuming a bare homogeneous reactor we obtain the point reactor kinetics equations

$$\frac{dn(t)}{dt}=\frac{k_{eff}\left(1-\beta\right)-1}{l}n(t)+\lambda C(t)$$

$$\frac{dC(t)}{dt} = \beta \frac{k_{eff}}{l} n(t) - \lambda C(t) \tag{8-261}$$

These equations are however usually written as a function of reactivity, $\rho(t)$, and mean neutron generation time. The mean neutron generation time, Λ, represents the time between the neutron birth by fission and its absorption inducing fission

$$\Lambda = \frac{l}{k_{eff}} \tag{8-262}$$

Thus,

$$\frac{dn(t)}{dt} = \frac{\rho(t) - \beta}{\Lambda} n(t) + \lambda C(t)$$

$$\frac{dC(t)}{dt} = \frac{\beta}{\Lambda} n(t) - \lambda C(t) \tag{8-263}$$

In case of all six delayed neutron groups the above point reactor kinetics equations become

$$\frac{dn(t)}{dt} = \frac{\rho(t) - \beta}{\Lambda} n(t) + \sum_{i=1}^{6} \lambda_i C_i(t)$$

$$\frac{dC_i(t)}{dt} = \frac{\beta}{\Lambda} n(t) - \lambda_i C_i(t) \tag{8-264}$$

5.3.3 Solution of the Point Kinetics Equations

The one-group delayed neutron approximation will be used to show the solutions for the point reactor kinetics equations. In addition, we will assume that the reactor was operating at the constant power $P(t = 0) = P_0$ prior to time $t = 0$ when the reactivity changes to ρ_0 (non-zero value). During the steady-state condition ($t < 0$) the power and precursor concentration were constant, giving the initial conditions

$$\frac{dP(t)}{dt} = 0 \rightarrow P(t=0) \equiv P_0$$

$$\frac{dC(t)}{dt} = 0 \rightarrow C(t=0) \equiv C_0 = \frac{\beta}{\lambda \Lambda} P_0$$

The yield fraction and the average decay constant for all six groups of delayed neutrons represented with only one group are defined as

$$\beta = \sum_{i=1}^{6} \beta_i \quad \lambda = \left[\frac{1}{\beta} \sum_{i=1}^{6} \frac{\beta_i}{\lambda_i} \right]^{-1} \tag{8-265}$$

Since the reactor power is directly proportional to neutron density the one-group point kinetics equations can be written as follows:

$$\frac{dP(t)}{dt} = \frac{\rho_0 - \beta}{\Lambda} P(t) + \lambda C(t)$$

$$\frac{dC(t)}{dt} = \frac{\beta}{\Lambda} P(t) - \lambda C(t) \tag{8-266}$$

The solution to the above equations could be assumed to be given as

$$P(t) = P e^{\omega t}$$
$$C(t) = C e^{\omega t} \tag{8-267}$$

where P, C and ω are the constants to be determined. Substituting these solutions into the point reactor kinetics equations it follows that

$$\omega P = \frac{\rho_0 - \beta}{\Lambda} P + \lambda C \quad \rightarrow \quad \left[\omega - \frac{\rho_0 - \beta}{\Lambda} \right] P - \lambda C = 0$$

$$\omega C = \frac{\beta}{\Lambda} P - \lambda C \quad \rightarrow \quad \frac{\beta}{\Lambda} P - (\lambda + \omega) C = 0 \tag{8-268}$$

This set of homogeneous equations are solved by equalizing the determinant to zero

$$\left[\omega - \frac{\rho_0 - \beta}{\Lambda}\right](\lambda + \omega) - \frac{\lambda\beta}{\Lambda} = 0$$

$$\Lambda\omega^2 + (\lambda\Lambda + \beta - \rho_0)\omega - \rho_0\lambda = 0 \tag{8-269}$$

It follows therefore that the general solution will be given in the form

$$P(t) = P_1 e^{\omega_1 t} + P_2 e^{\omega_2 t}$$

$$C(t) = C_1 e^{\omega_1 t} + C_2 e^{\omega_2 t} \tag{8-270}$$

The unknowns are then determined applying the initial conditions. However, the simplified solution is usually discussed for which it is assumed that

$$(\lambda\Lambda + \beta - \rho_0)^2 \gg 4\Lambda\lambda\rho_0$$

where the two roots are approximately given as

$$\omega_1 \approx \frac{\lambda\rho_0}{\beta - \rho_0} \qquad \omega_2 \approx -\frac{\beta - \rho_0}{\Lambda}$$

and the solution for the power change as

$$P(t) \approx P_0\left[\left(\frac{\beta}{\beta - \rho_0}\right)\exp\left(\frac{\lambda\rho_0}{\beta - \rho_0}\right)t - \left(\frac{\rho_0}{\beta - \rho_0}\right)\exp\left(-\frac{\beta - \rho_0}{\Lambda}\right)t\right] \tag{8-271}$$

5.3.4 The Inhour Equation

The solution to the point kinetics equation can be discussed in a somewhat different way. The equation

$$\left[\omega - \frac{\rho_0 - \beta}{\Lambda}\right](\lambda + \omega) - \frac{\lambda\beta}{\Lambda} = 0$$

$$\Lambda\omega^2 + (\lambda\Lambda + \beta - \rho_0)\omega - \rho_0\lambda = 0$$

can be re-written such as to give a solution for the reactivity (using the definition of mean neutron generation time and expressing the multiplication factor as a function of reactivity) as follows:

$$\rho_0 = \frac{\omega l}{1 + \omega l} + \frac{1}{1 + \omega l}\frac{\beta\omega}{\omega + \lambda} \qquad (8\text{-}272)$$

This is known as the reactivity equation for one group of delayed neutrons or the inhour equation. The right-hand side (RHS) of this equation can be plotted as a function of parameter ω:

- If $\omega = 0$, the RHS $= 0$ (the solution curve will pass through the origin as shown in Fig. 8-29).
- For $\omega \to \pm\infty$, the RHS $\to 1$.
- When $\omega = -\lambda$ or $\omega = -1/l$, the RHS $\to \infty$.
- Since the reactivity can be positive or negative, there are two roots ω_1 and ω_2.

  Therefore, the flux or the power time behavior can be represented with

$$\phi(t) = A_1 e^{\omega_1 t} + A_2 e^{\omega_2 t} \qquad (8\text{-}273)$$

From Fig. 8-29 it can be observed that

- When reactivity is positive ($\rho > 0$) then ω_1 is positive and ω_2 is negative. Thus, as time increases the second term in the flux equation dies out and the flux increases as $e^{\omega_1 t}$.
- When the reactivity is negative ($\rho < 0$) then both roots are negative. With time, the second term will die out faster than the first term because ω_2 is more negative than ω_1. Thus, the flux will decrease as $e^{\omega_1 t}$.

From these considerations it can be concluded that in either case, positive or negative reactivity, the flux will approach the term $e^{\omega_1 t}$. The reciprocal of ω_1 is called the reactor period or the stable period.

The one-group delayed neutron reactivity equation can be generalized to include all six delayed neutron groups:

$$\rho_0 = \frac{l\omega}{1 + l\omega} + \frac{\omega}{1 + l\omega}\sum_{i=1}^{6}\frac{\beta_i}{\lambda_i + \omega} \qquad (8\text{-}274)$$

Using the same analysis as for the one-group delayed neutron reactivity equation a similar plot can be obtained (see Fig. 8-30). In this case, however, there are seven roots for either positive or negative reactivity. The flux is given as a sum of exponentials

$$\phi(t) = A_1 e^{\omega_1 t} + A_2 e^{\omega_2 t} + \cdots + A_7 e^{\omega_7 t} \qquad (8\text{-}275)$$

With increasing time flux again approaches $e^{\omega_1 t}$ since all other exponents die out fast.

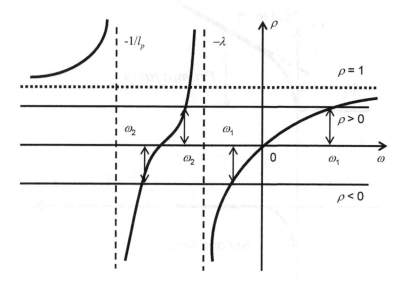

Figure 8-29. Reactivity equation for one-group delayed neutrons

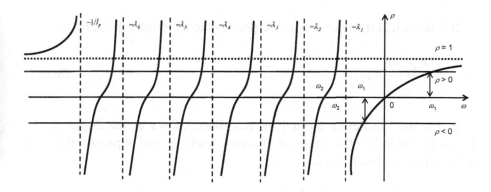

Figure 8-30. Reactivity equation for six-group delayed neutrons

5.4   The Prompt Jump Approximation and Inhour Formula

The amount of reactivity necessary to make a reactor prompt critical corresponds to the prompt neutrons' multiplication factor (Fig. 8-31)

$$(1 - \beta)k_{eff} = 1 \tag{8-276}$$

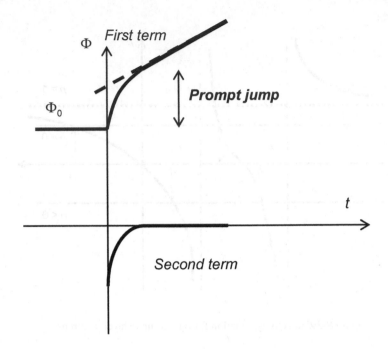

Figure 8-31. Flux change with the step reactivity insertion.

It follows that the reactivity corresponding to a prompt critical reactor is

$$\rho = \frac{k_{eff} - 1}{k_{eff}} = \beta \tag{8-277}$$

and is used to define the unit of reactivity known as the dollar, $. As shown in Table 8-5 the value of β varies with the fuel type and thus the dollar is not an absolute unit.

Example 8.16 Reactivity in dollars

Calculate the reactivity of a homogeneous ^{235}U reactor if it suddenly becomes supercritical with $k_{eff} = 1.005$.

$$\rho = \frac{k_{eff} - 1}{k_{eff}} = \frac{1.005 - 1}{1.005} = 0.00497 \rightarrow \rho(\$) = \frac{\rho}{\beta} = \frac{0.00497}{0.0065} = 0.765\ \$ = 76.5\ \text{cents}$$

The exact computation of the early response of the reactor to a sudden change in reactivity is complex. However, under certain assumptions it can be significantly simplified. One such approach is called the *prompt jump approximation* and is based on the assumption that the concentration of the delayed neutron precursors does not change over the time following a sudden decrease or increase in neutron flux.

With time, as explained in the previous section, the second term in Eq. (8-273) will die out quickly and the flux will decrease or increase with the reactor period T. Exact calculations predict that the constant A_2 is negative for positive reactivity and positive for negative reactivity. Therefore, the fast die out of a negative term will give a sudden rise in flux following the insertion of positive reactivity (Fig. 8-31). On the other hand, the fast die out of a positive term will give a sudden drop in flux for negative reactivity insertion. With the assumption that the delayed neutron precursor concentration does not change over the time during the sudden decrease or increase of the neutron flux, it follows that

$$\frac{dC(t)}{dt} = \beta k_\infty \Sigma_a \phi(t) \frac{1}{P_{esc}} - \lambda C(t) \rightarrow C(t) = \frac{\beta \Sigma_a \phi_0}{P_{esc} \lambda} \tag{8-278}$$

It is also assumed that the infinite reactor was originally critical. The flux value in the above equation is the flux prior to a sudden change in reactivity. Therefore,

$$l_{th\infty} \frac{d\phi(t)}{dt} = \phi(t) + (1-\beta)k_\infty \phi + \beta \phi_0 \tag{8-279}$$

where k_∞ is the multiplication factor after the reactivity change. The solution is

$$\phi(t) = \phi_0 e^{\frac{(1-\beta)k_\infty - 1}{l_{th\infty}}t} + \frac{\beta \phi_0}{1-(1-\beta)k_\infty}\left[1 - e^{\frac{(1-\beta)k_\infty - 1}{l_{th\infty}}t}\right] \tag{8-280}$$

or introducing the reactor stable period

$$T = \frac{l_{th\infty}}{(1-\beta)k_\infty - 1} \tag{8-281}$$

Eq. (8-280) becomes

$$\phi(t) = \phi_0 e^{t/T} + \frac{\beta \phi_0}{1-(1-\beta)k_\infty}\left[1 - e^{t/T}\right] \tag{8-282}$$

The condition for a reactor to be less than prompt critical is that Eq. (8-223) is less than one. The two exponential terms in Eq. (8-282) will die out with a reactor period as given by Eq. (8-231) which is the reactor period taking into account only prompt neutrons (see Section 5.1). Thus,

$$\phi = \frac{\beta\phi_0}{1-(1-\beta)k_\infty} = \frac{\beta(1-\rho)\phi_0}{\beta-\rho} \tag{8-283}$$

where $k_\infty = 1/(1-\rho)$.

The above equation can be analyzed for the following two cases:

(a) *Positive reactivity change*: this is an example of the reactivity required to increase the reactor power. This increase is usually small and takes a short period of time. For the example of a thermal reactor fueled with ^{235}U the time is less than ~2 min for a reactivity insertion of 0.0006. Thus, the flux will change according to

$$\phi(t) = \frac{\beta(1-\rho)\phi_0}{\beta-\rho} = \frac{\beta(1-0.0006)}{\beta-0.0006}\phi_0 \approx \phi_0$$

This result indicates that the prompt jump in flux is usually negligible and can be assumed to rise from the initial value with the stable period.

(b) *Negative reactivity change*: introduced whenever a reactor needs to be shutdown. The negative reactivity insertion can thus be very large. For example, if 20% in negative reactivity is suddenly introduced into a reactor fueled with ^{235}U ($\beta = 0.0065$), the flux will drop by ~4% of its initial value:

$$\phi(t) = \frac{\beta(1-\rho)\phi_0}{\beta-\rho} = \frac{0.0065(1-(-0.2))}{0.0065-(-0.2)}\phi_0 = 0.038\phi_0$$

The reactivity can be also expressed in terms of the inverse hour, or "inhour" unit. The inhour reactivity is defined as the reactivity necessary to make the reactor stable period equal to 1 h. The general inhour formula for a finite reactor including all six groups of delayed neutrons is

$$\rho(\text{inhours}) = \frac{l_{th}}{3600k_{eff}} + \sum_{i=1}^{6} \frac{\beta_i}{1+3600\lambda_i} \tag{8-284}$$

NUMERICAL EXAMPLE

Method of Characteristics Solution to Neutron Transport in Nuclear Reactor Assembly Geometry

This numerical example illustrates the computational method of characteristics solution to the neutron transport equation. A representative geometry of a complex reactor assembly is selected to show the distribution of neutron flux and reaction rates as a function of neutron energy group and spatial coordinates. The method of characteristics solves an integro-differential form of the transport equation along straight lines throughout the geometric domain in a discrete number of spatial directions and for discrete number of energy groups. These straight lines are interpreted as neutron trajectories similar to the Monte Carlo neutron trajectories. The method itself requires fine spatial subdivision of the geometrical domain into so-called flat flux zones, where the material properties are assumed to be constant. The following example is based on the methodology developed in the AGENT code and the list of references is provided for further reading for those interested in computational neutron transport modeling.

The selected example is a two-dimensional assembly consisting of 17×17 lattice with the square fuel pin cells, as shown in Fig. 8-32. The side length of every fuel pin cell is 1.26 cm and every cylinder is of radius 0.54 cm. The spatial fine-mesh flux distribution for each of energy regions is shown in Figs. 8-33, 8-34, 8-35, 8-36, 8-37, 8-38, and 8-39, while neutron absorption rate is shown in Figs. 8-40, 8-41, 8-42, 8-43, 8-44, 8-45, and 8-46. Fig. 8-47 shows the absorption rate integrated over all energies.

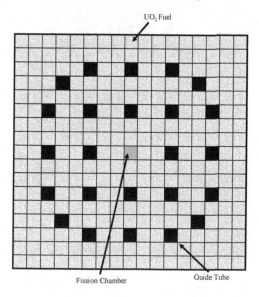

Figure 8-32. Fuel assembly geometry modeled with AGENT code

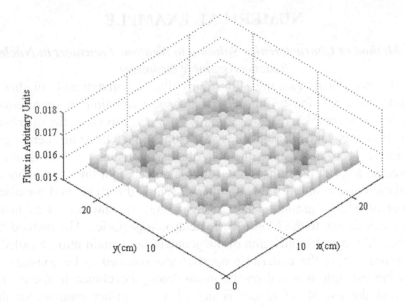

Figure 8-33. Neutron flux distribution for energies from 13.53 MeV to 20.00 MeV

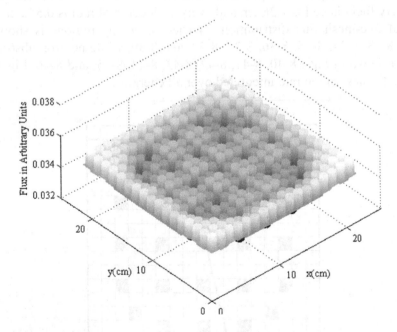

Figure 8-34. Neutron flux distribution for energies from 9.12 keV to 13.53 MeV

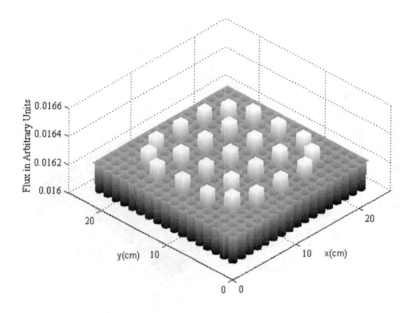

Figure 8-35. Neutron flux distribution for energies from 3.93 eV to 9.12 keV

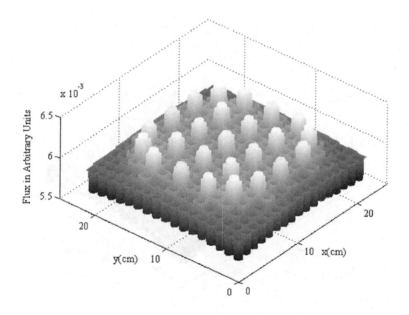

Figure 8-36. Neutron flux distribution for energies from 0.63 eV to 3.93 eV

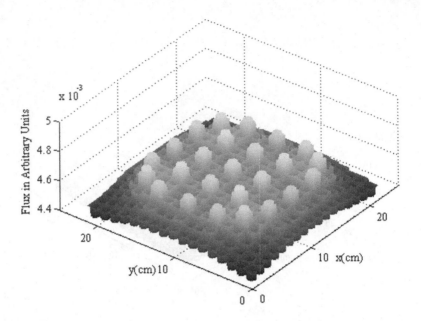

Figure 8-37. Neutron flux distribution for energies from 0.15 eV to 0.63 eV

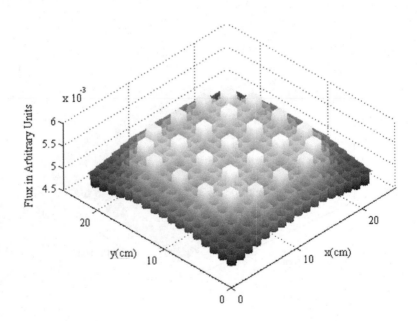

Figure 8-38. Neutron flux distribution for energies from 0.057 eV to 0.15 eV

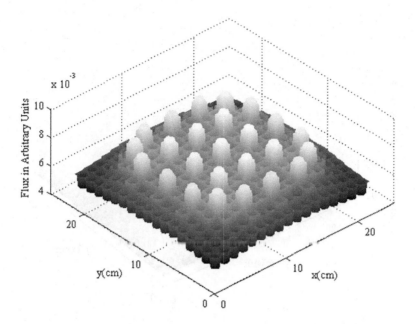

Figure 8-39. Neutron flux distribution for energies from 0.00 eV to 0.057 eV

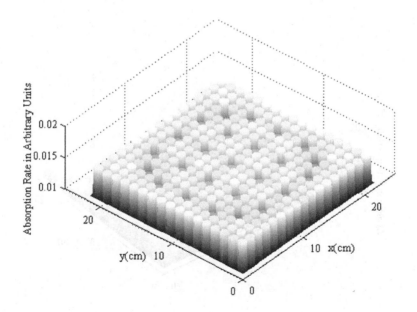

Figure 8-40. Absorption rate distribution for energies from 13.53 MeV to 20.00 MeV

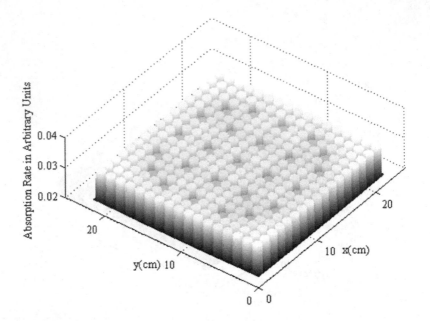

Figure 8-41. Absorption rate distribution for energies from 9.12 keV to 13.53 MeV

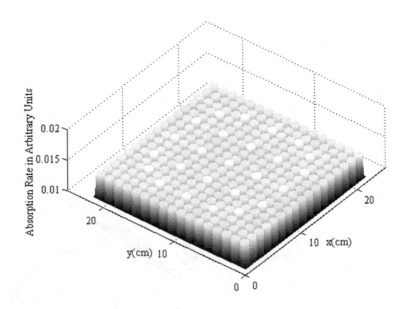

Figure 8-42. Absorption rate distribution for energies from 3.93 eV to 9.12 keV

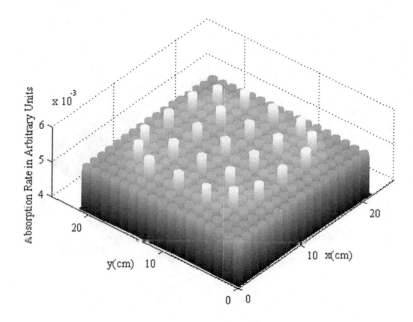

Figure 8-43. Absorption rate distribution for energies from 0.63 eV to 3.93 eV

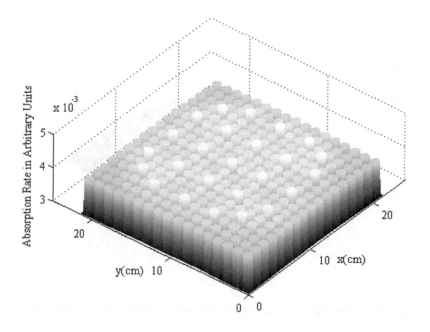

Figure 8-44. Absorption rate distribution for energies from 0.15 eV to 0.63 eV

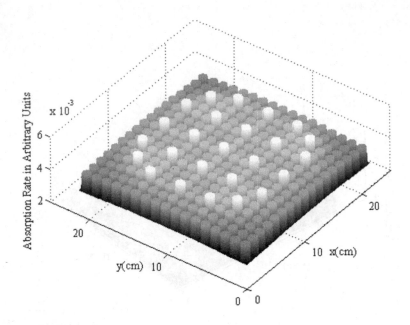

Figure 8-45. Absorption rate distribution for energies from 0.057 eV to 0.15 eV

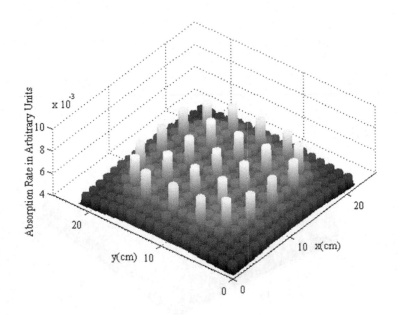

Figure 8-46. Absorption rate distribution for energies from 0.00 eV to 0.057 eV

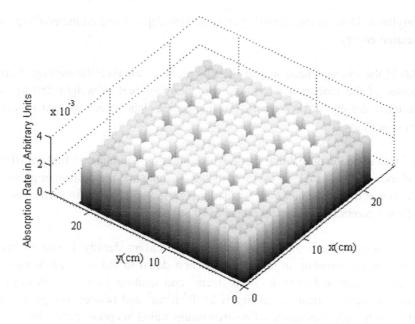

Figure 8-47. Absorption rate integrated over all energies

PROBLEMS

8.1 Calculate how many kilograms of ^{235}U are in 150 kg of U_3O_8.

8.2 Calculate the energy in eV for hydrogen atom moving at a speed of 2,200 m/s. Compare it to the energy of thermal neutron traveling at the same speed at the room temperature (293 K).

8.3 How much power will be produced from the spontaneous fission decay of 1 mg of ^{256}Fm (half-life is 158 min)? Assume that each fission event would release 220 MeV. How much ^{235}U would be needed to produce 6 MW of power?

8.4 Calculate the neutron density in a thermal reactor with the neutron flux of 10^{12} n/cm^{2}s. How does this value compare with the particle density in a volume of 1 cm^3 at standard conditions and with the number of hydrogen atoms in water?

8.5 Using on-line data for neutron cross sections (http://atom.kaeri.re.kr) calculate the scattering mean free path for thermal neutrons in graphite, lead, and beryllium. Discuss the scattering cross section dependence on material type and neutron energy.

8.6 If the average neutron flux is 10^{13} n/cm^{2}s, calculate the average thermal power of the reactor with 5% enriched uranium fuel of weight 150 kg. The uranium density is 18.7 g/cm^3. Use the online data library to read necessary cross sections.

8.7 Calculate the probability that a 2 MeV neutron will undergo first collision in 3/9 inch dia. UO$_2$ fuel rod enriched to 4%. Assume that the neutron originated in the center of the rod and travels radially. The fuel rod has density which is 94% of the theoretical fuel density (equal to 10.96 g/cm^3).

8.8 Calculate the neutron flux and neutron current density if two beams of neutrons are traveling in the same direction down to the same guide tube: (a) beam 1: neuron density is 5×10^7 n/cm^2 and neutron energy is 10 keV; (b) neutron beam 2: neutron density is 2×10^7 n/cm^2 and neutron energy is 1 eV. How do these values change if neutron beams travel in opposite direction?

8.9 Show all steps in deriving the solution to the diffusion equation for a point neutron source placed in an infinite large medium.

8.10 A large bare reactor has the infinite multiplication factor of 1.022. The neutron diffusion length is 35 cm. Determine and compare the critical volumes of the following reactor shapes: sphere, cube, cylinder with height twice its radius and rectangular parallelepiped having $a = b = c/4$.

8.11 For the homogeneous one-speed reactor of cylindrical configuration, derive the formula to obtain its minimum volume (mass). Discuss the values in terms of reactor buckling.

8.12 Calculate the non-leakage probability for the bare cubic homogeneous reactor with diffusion length of 10 cm and $a = b = c = 100$ cm. Assume the absorption cross section of 0.1 cm^{-1}.

8.13 Determine the number of elastic scattering events occurring per 1 cm^3 in the energy interval from 0.5 MeV to 0.3 MeV for neutrons of 1.5 MeV passing through an infinite slab of graphite at the rate of 2×10^{15} n/cm^{3}sec.

8.14 How many collisions are needed to slow neutrons from 2 MeV down to the thermal energy region in Be and D moderators?

8.15 Calculate the critical core radius and the critical mass of a spherical reactor moderated and reflected by water. The ^{235}U fuel density in a core is 0.0145 g/cm^3. How does critical mass of the bare reactor compare to the one with the reflector?

8.16 A homogeneous, spherical, bare reactor of volume 250 m^3 is composed of 5% enrichment ^{235}U and graphite. Using the six-factor formula, calculate k_{eff} for the given data at a thermal energy:
Uranium-to-moderator ratio: 5:1
Graphite density: 2,267 kg/m^3
Graphite molar weight: 12.0107 g/m
Uranium density: 19,050 kg/m^3
Graphite microscopic absorption cross section: 0.009 b
Graphite microscopic scattering cross section: 10 b
Uranium-238 microscopic absorption cross section: 90 b
Uranium-238 molar weight: 238.0507847 g/m
Uranium-235 microscopic total absorption cross section: 360 b
Uranium-235 microscopic fission cross section: 270 b
Uranium-235 molar weight: 235.0439242 g/m
$v = 2.2; p_{esc} = 1; \varepsilon = 1; P_f = 1$

8.17 In a thermal nuclear reactor at the beginning of its life for every 1,000 neutrons,
      500 neutrons are absorbed in ^{235}U
      225 neutrons are absorbed in ^{238}U
      125 neutrons are absorbed in coolant and cladding
      150 neutrons leak out from the geometrical core boundaries.
Calculate the multiplication factor for this reactor if $v = 2.43$. By definition the conversion factor represents the ratio of number of fissile nuclei produced to the number of fissile nuclei lost. What is the conversion ratio value for this reactor?

8.18 Inserting the control rods into the thermal reactor from problem 8.17, the absorption in other materials increases such that
      450 neutrons are absorbed in ^{235}U
      215 neutrons are absorbed in ^{238}U
      185 neutrons are absorbed in coolant, control rods and cladding and
      150 neutrons leak out from the geometrical core boundaries.
Calculate the multiplication factor for this reactor.

8.19 The simplest form of neutron diffusion equation for thermal neutrons is one-speed theory. What are the assumptions upon which this theory is valid?

8.20 Write a computer program to follow the histories of 100 neutrons starting with energy 100 keV and slowing down to 10 eV in graphite (density 1.6 g/cm^3): absorption cross section is zero, scattering cross section is 4.8 b.

8.21 Repeat problem 8.20 but use water instead with scattering cross section on hydrogen equal to 20 b and oxygen equal to 4 b.

8.22 Repeat previous two problems by including the absorption of neutrons. Assume that the cross section for absorption in carbon is 0.004 b, in hydrogen is 0.335 b and in oxygen is 0.002 b.

8.23 A reactor is critical at a power level of 400 MW. How long will it take to reach the power level of 3300 MW on a stable period of 100 s?

8.24 Using the one-group delayed neutron equation calculate how long it would take to increase the power of a reactor by 10% with the reactivity addition of 0.02% $\delta k/k$? Assume that the reactor is critical before the addition of reactivity with thermal neutron lifetime of 5×10^{-5} s.

8.25 Calculate the effective multiplication factor for the ^{235}U reactor having reactivity of $-1\$$. If fuel is replaced with ^{239}Pu, what is the multiplication factor value?

8.26 Calculate the new stable period if control rods inserted into a supercritical reactor with the stable period of 20 s add -0.01% $\delta k/k$ to the reactivity. Assume thermal neutron lifetime is 0.0001 s.

8.27 Calculate the size of a thermal bare spherical reactor containing ^{235}U and water in the atom ratio of $N(\text{water})/N(^{235}\text{U}) = 198$ if the neutron spectrum follows Maxwellian distribution at 20°C. How does the result change if temperature is changed to 300°C?

8.28 A slab of graphite contains a plane neutron source in the center. The slab is in a large pool of water. The albedo (reflection coefficient) of water is defined as $J_{\text{out}}/J_{\text{in}}$ where J represents the neutron current. Evaluate the albedo if the slab is 60 cm thick. Assume that the source produces thermal neutrons.

8.29 For thermal neutrons in graphite, determine the diffusion length. Compare the mean-square distance to absorption with the total neutron trajectory length. Show all steps of calculation and list the sources you used to read nuclear data.

8.30 Derive and sketch the flux distribution if the infinite planar source is placed at the center of an infinite slab made of two regions where the second outer region is of infinite thickness.

8.31 In an infinite medium with two planar infinite neutron sources located at a and b from $x = 0$, show that the solution for neutron flux is

$$\Phi(x) = \frac{S_a L}{2D} e^{-|x-a|/L} + \frac{S_b L}{2D} e^{-|x-b|/L}$$

Plot the neutron flux.

8.32 Discuss the scattering kernel for hydrogen. Use Eq. (8-170) and plot the kernel distribution.

8.33. Discuss the scattering kernel for $A > 1$. Plot the distribution.

8.34. Derive the equation to show that for the critical reactor the material buckling is equal to the geometrical buckling. [*Hint*: Start from the one-group time-dependent diffusion equation of the form,

$$\frac{1}{\upsilon} \frac{\partial \Phi(\vec{r},t)}{\partial t} = D\nabla^2 \Phi(\vec{r},t) - \Sigma_a \Phi(\vec{r},t) + S(\vec{r},t) .]$$

$S(\vec{r},t) \rightarrow$ source of thermal neutrons

8.35. What is the neutron velocity for which the Maxwellian thermal flux distribution peaks? [*Hint*: Flux is a product of neutron density and neutron velocity.]

8.36. For a nucleus that can interact with a neutron only through the radiative capture and elastic scattering show that the total cross section at the resonance can be estimated as follows:

$$\sigma_{tot} = \frac{\lambda^2}{\pi} \frac{\Gamma_n}{\Gamma} .$$

Chapter 9

NUCLEAR REACTOR CONTROL

Methods of reactor control, Fission product poisoning and Reactivity coefficients

There are two possible outcomes: If the result confirms the hypothesis, then you've made a measurement. If the result is contrary to the hypothesis, then you've made a discovery. *Enrico Fermi* (1901–1954)

1. METHODS OF REACTOR CONTROL

In a reactor of given volume in which fission is caused by neutrons of specified energy, the thermal power is proportional to the neutron flux and macroscopic fission cross section. As the reactor operates, the macroscopic cross section decreases as number of fissile nuclides decreases. However, over an essentially short period of time, the cross section remains constant, and the power is assumed to change only with neutron flux.

In most situations a reactor is controlled by varying the neutron flux. Among the general methods available, the insertion and withdrawal of a neutron absorber are most commonly used in power reactors. Materials used as a control absorber have large absorption cross sections, like boron, cadmium or hafnium. Strong absorbers in a core compete with fissile material for neutrons. In other words, neutrons which are absorbed by the controller are no longer available to induce fission, thus reducing the power.

1.1 Control Rods

The change in reactivity caused by control rod motion is referred to as *control rod worth*. The maximum effect (insertion of the most negative

T. Jevremovic, *Nuclear Principles in Engineering*,
DOI 10.1007/978-0-387-85608-7_9, © Springer Science+Business Media, LLC 2009

reactivity) of a control rod is at the location in the reactor where the flux has its maximum value. Control rods are used to

- *Change reactivity in order to lower or elevate the reactor power placing it on a stable period* – rod worth is defined as the magnitude of reactivity required to give the observed period.
- *Keep reactor critical by compensating for changes over reactor operating time* – rod worth is measured in terms of change in neutron multiplication factor for which the rod can compensate.

Control rods can be inserted fully or partially. In either of these two cases the neutron flux is perturbed and reactor power changed. The following two sections address the effects of control rod insertion and withdrawal on fission rate, reactor flux distribution and the resulting power change.

1.1.1 Effect of Fully Inserted Control Rod on Neutron Flux in Thermal Reactors

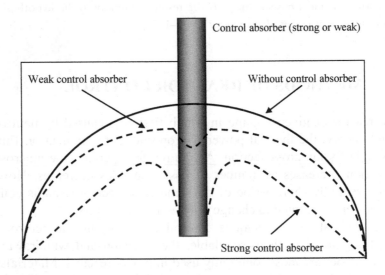

Figure 9-1. Effect of a control rod on flux perturbation

The material used for the control rods varies depending on reactor design. Generally, the control rod material should have a high-absorption cross section as well as a long lifetime in the reactor (not burn out too rapidly). A control rod which absorbs essentially all incident neutrons is referred to as a "black" absorber and generates large flux depression (Fig. 9-1). A "gray" absorber absorbs only a fraction of incident neutrons. While it takes more gray rods than black rods for a given reactivity effect, the gray rods are often preferred because they cause smaller flux depressions in the vicinity of the rod. This leads to a flatter neutron flux profile resulting in a more even power distribution across the core. Since the thermal neutron flux density generally peaks in the center of reactor core, this is where high-efficiency control rods are generally placed.

A bare cylindrical reactor which is critical with control rods removed may be described by the one-speed neutron diffusion equation

$$\nabla^2 \phi + B_{out}^2 \phi = 0 \tag{9-1}$$

The multiplication factor, which is equal to unity, is given by

$$k_{out} = \frac{k_\infty}{1 + B_{out}^2 M^2} = 1 \tag{9-2}$$

If a strongly absorbing control rod is fully inserted into the core the neutron flux will change as shown in Fig. 9-1 due to high neutron absorption in the rod. The flux distribution can be described as

$$\nabla^2 \phi + B_{in}^2 \phi = 0 \tag{9-3}$$

When the control rod is inserted the multiplication factor changes as

$$k_{in} = \frac{k_\infty}{1 + B_{in}^2 M^2} \tag{9-4}$$

Notice that the core buckling changes with control rod insertion and the change in multiplication factor will give the reactivity

$$\rho = \frac{k_{out} - k_{in}}{k_{in}} \tag{9-5}$$

The control rod worth, ρ_w, by definition, is equal to the magnitude of this reactivity change

$$\rho_w = |\rho| = \frac{\left(B_{in}^2 - B_{out}^2\right)M^2}{1 + B_{out}^2 M^2} \tag{9-6}$$

In order to obtain the control rod worth, Eqs. (9-1) and (9-3) must be solved to obtain the buckling for both cases. In initially critical reactor without control rods the buckling is given with

$$B_{out}^2 = \left(\frac{2.405}{R}\right)^2 + \left(\frac{\pi}{H}\right)^2 \tag{9-7}$$

However, the calculation of buckling when the control rod is inserted is difficult because the geometry is complicated and because the presence of strong absorber tends to deform the flux such that the diffusion approximation is not valid in its vicinity. In this case, a solution can be obtained by assuming that d represents the extrapolated distance and that the flux satisfies the following boundary condition at the surface of the control rod

$$\frac{1}{\phi}\frac{d\phi}{dr} = \frac{1}{d} \tag{9-8}$$

The final result for the extrapolation distance and control rod worth is (detailed derivation can be found elsewhere)

$$d = 2.131\overline{D}\frac{a\Sigma_{tot} + 0.9354}{a\Sigma_{tot} + 0.5098} \tag{9-9}$$

$$\rho_w = \frac{7.43M^2}{\left(1 + B_{out}^2 M^2\right)R^2}\left[0.116 + \ln\left(\frac{R}{2.405a}\right) + \frac{d}{a}\right]^{-1} \tag{9-10}$$

where a is the radius of a control rod, R is extrapolated radius of the bare cylindrical core and H is its extrapolated height, $\overline{D}$ is the diffusion coefficient and Σ_{tot} is the macroscopic cross section.

The cross section and diffusion coefficient are those for the materials surrounding the control rod which is assumed to be a black absorber.

1.1.2   Control Rod Worth in Fast Reactors

The most promising material to be used as the control absorber in fast reactors is boron–carbide (B_4C) enriched in ^{10}B, because unlike other materials, absorption cross section for the boron is still significant at high neutron energies. Although considerably higher than for other materials, the boron absorption cross section at energies of importance in fast reactors (0.1–0.4 MeV) is only 0.27 b (see Chapter 7). Therefore, the absorption neutron mean free path in a medium containing boron is large the atom density of boron is 0.087×10^{24} atoms/cm^3 at a B_4C density of 2 g/cm^3 giving $\lambda_a = 42.6$ cm. This is considerably larger than the diameter of any control rod size used in fast reactors which means that the neutron flux inside the control rod is more or less the same as in the surrounding medium. Therefore the boron contained in the rod can be assumed to be uniformly distributed in the reactor. This assumption will only affect the calculation of the fuel utilization factor in determining the control rod worth.

In actual reactor design, control rod worth is calculated using computer codes and a multigroup approach. The following is a simplified one group estimate of control rod worth in fast reactor. The multiplication factor for a fast reactor is given by

$$k_{eff} = k_\infty P_{non-leakage} = \eta f P_{non-leakage} \tag{9-11}$$

Since the uniformly distributed poison in fast reactors has an effect only on the fuel utilization factor, the control rod worth reduces to

$$\rho_w = \frac{k_{out} - k_{in}}{k_{in}} = \frac{f_{out} - f_{in}}{f_{in}} \tag{9-12}$$

$$f_{in} = \frac{\Sigma_a^{fuel}}{\Sigma_a^{fuel} + \Sigma_a^{coolant} + \Sigma_a^{boron}} \tag{9-13}$$

$$f_{out} = \frac{\Sigma_a^{fuel}}{\Sigma_a^{fuel} + \Sigma_a^{coolant}} \tag{9-14}$$

giving

$$\rho_w = \frac{\Sigma_a^{boron}}{\Sigma_a^{fuel} + \Sigma_a^{coolant}} \tag{9-15}$$

1.1.3 Effect of Partially Inserted Control Rod on Neutron Flux in Thermal Reactors

At the time of reactor start-up, all or most of, the control rods are fully inserted. After the start-up, they are slowly withdrawn in order to keep the reactor critical as the fuel is consumed and fission products accumulate. Therefore, it is necessary to know the control rod worth as a function of its insertion distance. The one group approximation is used to illustrate the computation of control rod worth for partially inserted rods in a thermal reactor.

For a cylindrical reactor let

- $\rho_w(x)$: the worth of one or more control rods inserted at the distance x parallel to the axis of the reactor core with total height H
- $\rho_w(H)$: the worth of fully inserted control rods.

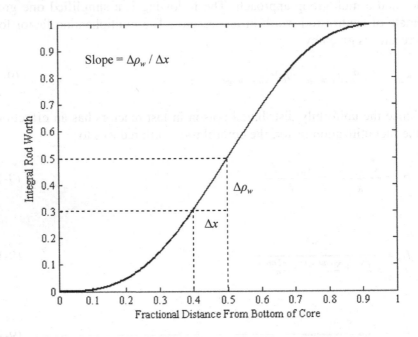

Figure 9-2. Integral control rod worth as given by Eq (9-16)

The exact effect of control rods on reactivity may be determined experimentally. For example, a control rod can be withdrawn in small increments, and the change in reactivity determined for each increment of withdrawal. By plotting the resulting reactivity versus rod position, a graph similar to that shown in Fig. 9-2 is obtained. The graph depicts *integral control rod worth* over the full range of rod withdrawal. Integral control rod worth represents the total reactivity worth of the rod at that particular degree of withdrawal

$$\rho_w(x) = \rho_w(H)\left[\frac{x}{H} - \frac{1}{2\pi}\sin\left(\frac{2\pi x}{H}\right)\right]$$

(9-16)

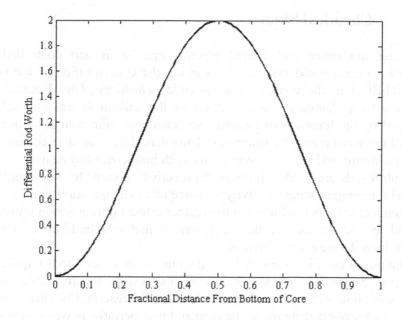

Figure 9-3. Differential control rod worth as given by Eq. (9-17)

The slope of the curve, and therefore the amount of reactivity inserted per unit of withdrawal, is greatest when the control rod is midway out of the core. This is because the neutron flux is maximum near the center of the core, thus the neutron absorption rate is also greatest in this area. If the slope of the curve for integral rod worth in Fig. 9-2 is plotted, the result is a value for the rate of change of control rod worth as a function of position. Such a plot is referred to as the *differential control rod worth* and is shown in Fig. 9-3. At the bottom of the core there are few neutrons so rod movement has little effect; therefore the change in rod worth over distance is nearly constant. As the rod approaches the center of the core its effect becomes greater, and the change in rod worth per distance becomes significant. At the center of the core, the differential rod worth is greatest and varies little with rod motion. From the center of the core to the top, the rod worth per distance is the opposite of the rod worth per distance from the center to the bottom.

The integral rod worth at a given withdrawal is the summation of the entire differential rod worth up to that point of withdrawal and is also the area under the differential rod worth curve at any given withdrawal position. The differential control rod worth is obtained as a derivative of $\rho_w(x)/\rho_w(H)$

$$\frac{1}{\rho_w(H)}\frac{\rho_w(x)}{dx} = \frac{1}{H}\left[1 - \cos\left(\frac{2\pi x}{H}\right)\right]$$

(9-17)

1.2   Chemical Shim

Water moderated and cooled reactors can be in part controlled, in addition to control rod systems, by varying the concentration of the boric acid (H_3BO_3) in the coolant. This is called *chemical shim*. Because the response to a change in concentration of the solvent is not as quick as obtained by the insertion of control rods, chemical shim cannot be used to control the large reactivity insertions. Thus it is always used in conjunction with the control rod systems. In a reactor with both control systems:

- control rods are used to provide the reactivity control for fast shutdown and for compensating reactivity variance due to temperature change.
- chemical shim is used to keep the reactor critical during xenon transients and to compensate for the depletion of fuel and build-up of fission products during reactor lifetime.

The use of chemical shim reduces the number of control rods required in a reactor. Since control rod systems are expensive, any reduction in the number of control rods reduces the total cost of the reactor. Chemical shim is almost uniformly distributed in the core and thus perturbs power distribution less as the concentration of the boric acid is changed.

Chemical shim in thermal reactors primarily affects the thermal (fuel) utilization factor. Therefore, chemical shim worth can be computed from the following relation

$$\rho_w = \frac{\Sigma_a^{boron}}{\Sigma_a^{fuel} + \Sigma_a^{mod}} \tag{9-18}$$

By inserting Eq. (9-14) the reactivity worth reduces to

$$\rho_w = \left(1 - f_{out}\right)\frac{\Sigma_a^{boron}}{\Sigma_a^{mod}} \tag{9-19}$$

The boric acid concentration is usually specified in units of *ppm (parts per million)* of water.

The *ppm* represents 1 g of boron per 10^6 g of water. Therefore, if C represents the concentration in *ppm*, then the ratio of the mass of boron to the mass of water is

$$\frac{m_{boron}}{m_{H_2O}} = C \times 10^{-6} \tag{9-20}$$

giving

$$\frac{\Sigma_a^{boron}}{\Sigma_a^{mod}} = \frac{N^{boron}\sigma_a^{boron}}{N^{H_2O}\sigma_a^{H_2O}} = \frac{18\times759}{10.8\times0.66}\times C\times10^{-6} = 1.92\times C\times10^{-3} \qquad (9\text{-}21)$$

According to Eq. (9-20) the worth of chemical shim becomes

$$\rho_w = 1.92\times C\times10^{-3}\times\left(1-f_{out}\right) \qquad (9\text{-}22)$$

2.   FISSION PRODUCT POISONING

Fission products and their decay products absorb neutrons to some extent. The accumulation of the parasitic absorbers during the reactor operation tends to reduce the neutron multiplication factor.

Among all non-fission materials accumulated during the reactor operation, two are of the greatest importance for thermal reactors: ^{135}Xe and ^{149}Sm (with large thermal neutron absorption cross sections). Since the absorption cross section decreases rapidly with increasing neutron energy (see Chapter 2), the poisoning effect is of little importance in fast reactors. The change of neutron multiplication factor with the poison materials present in a thermal reactor are discussed as follows.

The neutron multiplication factor is written as (see Chapter 8)

$$k_{eff} = k_\infty P = \eta\varepsilon pfP \qquad (9\text{-}23)$$

where P stands for both thermal and fast neutron non-leakage probabilities. If a poison material (strong absorber) is added
- The non-leakage probability changes slightly because it is inversely related to $L^2 = 1/3\Sigma_{tr}\Sigma_a$.
- The fast fission factor remains unchanged, $\varepsilon =$ const.
- The reproduction factor does not change since it is only a function of fuel properties ($\eta = \nu\Sigma_f^{fuel}/\Sigma_a^{fuel}$).
- The resonance escape probability p may change depending on cross section of the poisoning material (Fig. 9-4).
- The fuel utilization factor is inversely related to absorption cross section and thus changes drastically

$$f = \frac{\Sigma_a^{fuel}}{\Sigma_a^{fuel} + \Sigma_a^{mod} + \Sigma_a^{poison} + \Sigma_a^{control}} \qquad (9\text{-}24)$$

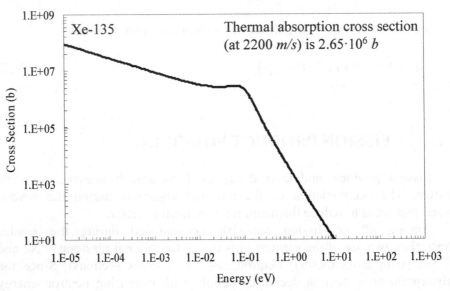

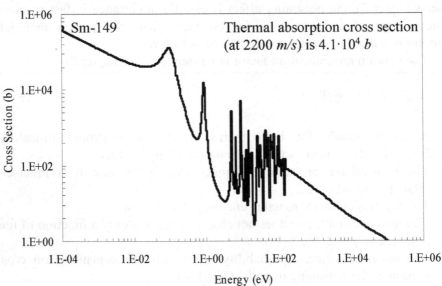

Figure 9-4. Radiative capture cross section for ^{135}Xe and ^{149}Sm

The effect of the poison material on reactivity change is

$$\Delta\rho = \rho' - \rho = \left(\frac{k'_{eff} - 1}{k'_{eff}}\right) - \left(\frac{k_{eff} - 1}{k_{eff}}\right) =$$

$$\frac{1}{k'_{eff}}\left(1 - \frac{k_{eff}}{k'_{eff}}\right) = \frac{1}{k_{eff}}\left(1 - \frac{k_\infty P}{k'_\infty P'}\right) \tag{9-25}$$

Because the non-leakage probability does not change significantly with the addition of poison material, $P/P' \sim 1$, the above equation reduces to

$$\Delta\rho = \frac{1}{k_{eff}}\left(1 - \frac{f}{f'}\right) \tag{9-26}$$

If the total absorption cross section is Σ_a it follows

$$f = \frac{\Sigma_a^{fuel}}{\Sigma_a} \tag{9-27}$$

$$f' = \frac{\Sigma_a^{fuel}}{\Sigma_a'} \tag{9-28}$$

where $\Sigma_a' - \Sigma_a = \Sigma_a^{poison}$ and $\Sigma_a = \Sigma_a^{fuel} + \Sigma_a^{mod} + \Sigma_a^{control}$. Finally

$$\Delta\rho = \frac{1}{k_{eff}}\left(1 - \frac{\Sigma_a'}{\Sigma_a}\right) = -\frac{1}{k_{eff}}\frac{\Sigma_a^{poison}}{\Sigma_a} \tag{9-29}$$

2.1   Xenon Poisoning

2.1.1   Production and Removal of ^{135}Xe During Reactor Operation

Xenon-135 (^{135}Xe) is the most important fission product poison and has a tremendous impact on the operation of a nuclear reactor. It is necessary to know its production and removal rate in order to predict how the reactor will respond to changes in power level. Xenon-135 is a non-$1/\upsilon$ absorber (Fig. 9-4) with a thermal neutron radiative capture (parasitic absorption) cross section of 2.6×10^6 b.

$$\text{Fission} \Rightarrow \; {}^{135}_{51}\text{Sb} \xrightarrow{\;\beta^-\;} {}^{135}_{52}\text{Te} \xrightarrow{\;\beta^-\;} {}^{135}_{53}\text{I} \xrightarrow{\;\beta^-\;} {}^{135}_{54}\text{Xe} \xrightarrow{\;\beta^-\;} {}^{135}_{55}\text{Cs} \xrightarrow{\;\beta^-\;} {}^{135}_{56}\text{Ba (stable)}$$

0.8sec ⇑ 19sec ⇑ 6.57h ⇑ 9.1h 2.3×10^6 yr

Fission   Fission   Fission

Figure 9-5. Production of ^{135}Xe in thermal reactor

Tellurium-135 (^{135}Te) decay chain is the primary production method of ^{135}Xe, however it can be directly produced from fission (see Fig. 9-5). The fission yield of ^{135}Xe is about 0.3 % and is about 6 % for ^{135}Te. ^{135}Xe is a product of the β decay of ^{135}I which is formed by fission and by the decay of ^{135}Te. Tellurium-135 is a fission product, but can also be formed from the β decay of ^{135}Sb (also a fission product). Nearly 95 % of all ^{135}Xe produced during reactor operation comes from the decay of ^{135}I.

Introducing y_i to represent the yield fraction for isotope i (the fraction of fission fragments that will be isotope i) and $PR = y_i \Sigma_f \Phi$ to be the production rate of isotope i, and following the decay scheme in Fig. 9-5 it follows:

- The decay times of the ^{135}Sb and ^{135}Te are very short. Thus, we may assume that all ^{135}Sb and ^{135}Te are ^{135}I by defining

$$y_I = y_{Sb} + y_{Te} + y_I \tag{9-30}$$

- The last nuclide in the decay chain has a very long half-life. Thus, the stable nuclide can be taken out of our analysis and we may simplify the decay chain as follows

$$^{135}\text{I} \rightarrow {}^{135}\text{Xe} \rightarrow {}^{135}\text{Cs} \tag{9-31}$$

- In the case of a homogeneous thermal reactor the iodine concentration can be determined as

$$\frac{dI}{dt} = \text{Production of iodine} - \text{Loss of iodine} \tag{9-32}$$

or

$$\frac{dI}{dt} = y_I \Sigma_f \Phi - \left(\lambda^I I + \sigma_a^I I \Phi \right) \tag{9-33}$$

where
I – concentration of ^{135}I

λ^I – radioactive decay constant of ^{135}I

σ_a^I – thermal neutron absorption cross section of ^{135}I

y_I – fission yield of ^{135}I (=0.061 for ^{235}U fuel)

Σ_f – macroscopic fission cross section of the fuel material in a reactor

Φ – thermal neutron flux.

- Under the same assumption, the xenon concentration change can be determined by:

$$\frac{dXe}{dt} = \text{Production of xenon} - \text{Loss of xenon} \tag{9-34}$$

or

$$\frac{dXe}{dt} = y_{Xe}\Sigma_f\Phi + \lambda^I I - \left(\lambda^{Xe} Xe + \sigma_a^{Xe} Xe\Phi\right) \tag{9-35}$$

where

Xe – concentration of ^{135}Xe

λ^{Xe} – radioactive decay constant of ^{135}Xe

σ_a^{Xe} – thermal neutron absorption cross section of ^{135}Xe

y_{Xe} – fission yield ^{135}Xe (=0.002 for ^{235}U fuel)

At steady state the rate change of concentration of both nuclides is constant (after the reactor has been operating for some time, the *equilibrium concentration* is attained), thus by setting Eqs. (9-34) and (9-35) equal to zero the equilibrium concentrations may be obtained.

- ^{135}I equilibrium concentration

$$I_0 = \frac{y_I\Sigma_f\Phi}{\lambda^I + \sigma_a^I\Phi} \approx \frac{y_I\Sigma_f\Phi}{\lambda^I} \tag{9-36}$$

The absorption cross section for ^{135}I is very small in the thermal energy region (see Fig. 9-6), so the above equation can be simplified by neglecting the absorption rate. The equilibrium concentration of ^{135}I is proportional to the fission reaction rate and power level.

- ^{135}Xe equilibrium concentration:

$$Xe_0 = \frac{y_{Xe}\Sigma_f\Phi + \lambda^I I_0}{\lambda^{Xe} + \sigma_a^{Xe}\Phi} \approx \frac{(y_{Xe} + y_I)\Sigma_f\Phi}{\lambda^{Xe} + \sigma_a^{Xe}\Phi} \tag{9-37}$$

The equilibrium concentration for ^{135}Xe increases with the power, because the numerator is proportional to the fission reaction rate. Since the thermal flux is also in the denominator, as it exceeds 10^{12} neutron/cm^2 s the term including the flux becomes dominant. Thus, at nearly 10^{15} neutron/cm^2 s the ^{135}Xe concentration approaches a limiting value.

The reactivity equivalent of the equilibrium xenon poisoning effect (by neglecting the presence of the control material) may be written in the following form

$$\Delta\rho_0 = \frac{1}{k_{eff}}\left(1 - \frac{\Sigma_a'}{\Sigma_a}\right) = -\frac{1}{k_{eff}}\frac{\Sigma_a^{poison}}{\Sigma_a} \tag{9-38}$$

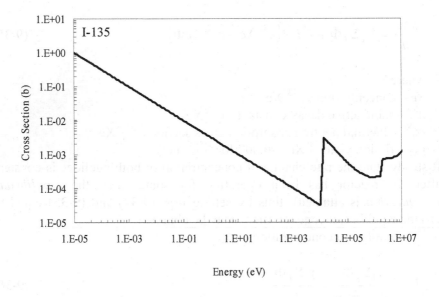

Figure 9-6. Radiative capture cross section for ^{135}I

where

$$\Sigma_a^{poison} = Xe\sigma_a^{Xe} \approx \frac{(y_{Xe} + y_I)\sigma_a^{Xe}\Sigma_f\Phi}{\lambda^{Xe} + \sigma_a^{Xe}\Phi} \tag{9-39}$$

To illustrate the reactivity change due to xenon accumulation, let's consider the thermal homogeneous reactor fueled with 2 % ^{235}U for which

$$\eta = 1.8 \quad \nu = 2.42 \quad \Sigma_f/\Sigma_a = 0.6 \quad y_I + y_{Xe} = 0.066$$
$$\sigma_a^{Xe} = 3\times10^6\,\text{b} \quad \lambda^{Xe} = 2.1\times10^{-5}\,\text{s}^{-1}$$

Using this data, and Eq (9-39), we may re-write Eq (9-38). in the following simplified form.

$$\Delta\rho_0 = -\frac{1.15\times10^{-23}\,\Phi}{2.1\times10^{-5}+3\times10^{-22}\,\Phi}$$

(9-40)

For a flux value of 10^{15} neutrons/cm^2 s the poisoning is negligible (-6×10^{-4}). For a flux which is 10 times higher, the poisoning is still low, - 0.005, i.e., 0.5% of all thermal neutrons are absorbed by the equilibrium amount of xenon. However, for a flux greater than 10^{16} neutrons/cm^2 s the poisoning increases rapidly, as shown in Fig. 9-7 and the limiting value is obtained for a flux of 10^{19} neutrons/cm^2 s. The equilibrium ^{135}I and ^{135}Xe concentrations as a function of neutron flux are illustrated in Fig. 9-8.

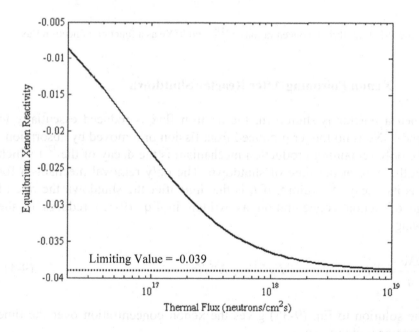

Figure 9-7. Reactivity equivalent of the equilibrium ^{135}Xe concentration for the example thermal reactor

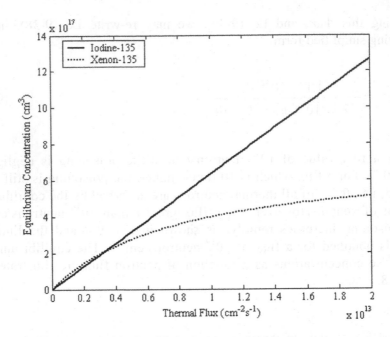

Figure 9-8. Equilibrium concentrations of ^{135}I and ^{135}Xe as a function of neutron flux

2.1.2   Xenon Poisoning After Reactor Shutdown

When a reactor is shutdown, the neutron flux is reduced essentially to zero and ^{135}Xe is no longer produced from fission or removed by absorption.

The only remaining production mechanism is the decay of the ^{135}I which was in the core at the time of shutdown. The only removal mechanism for ^{135}Xe is its decay. Therefore, if t_s is the time after the shutdown the rate of change of xenon concentration as written in Eq. (9-35) reduces to the following

$$\frac{dXe}{dt} = \lambda^{I} I - \lambda^{Xe} Xe = \lambda^{I} I_0 e^{-\lambda^{I} t_s} - \lambda^{Xe} Xe \qquad (9-41)$$

The solution to Eq. (9-35) gives the xenon concentration over the time after reactor is shutdown

$$Xe(t) = \frac{\lambda^{I} I_0}{\lambda^{Xe} - \lambda^{I}} \left(e^{-\lambda^{I} t_s} - e^{-\lambda^{Xe} t_s} \right) + Xe_0 e^{-\lambda^{Xe} t_s} \qquad (9-42)$$

The time at which the concentration is maximum may be attained by setting Eq. (9-42) equal to zero

$$t_{max} = \frac{1}{\lambda^{Xe} - \lambda^{I}} \ln \frac{\lambda^{Xe}}{\lambda^{I}} \left(1 - \frac{\lambda^{Xe} - \lambda^{I}}{\lambda^{I}} \frac{Xe_0}{I_0} \right) \qquad (9-43)$$

Because the decay rate of ^{135}I is faster than the decay rate of ^{135}Xe, the ^{135}Xe concentration peaks. The peak value is reached when $\lambda^{I}I = \lambda^{Xe}Xe$ which is in about 10 − 11 h for thermal reactors. The production of xenon from iodine decay is less than the removal of xenon by its own decay. This causes the concentration of ^{135}Xe to decrease. The concentration of ^{135}I at shutdown is greater for greater flux prior to shutdown which also influences the peak in ^{135}Xe concentration. Figure 9-9 illustrates the change in relative concentration of ^{135}Xe following reactor shutdown as a function of neutron flux and time after the shutdown. It can be seen that following the peak in ^{135}Xe concentration about 10 h after shutdown, the concentration will decrease at a rate controlled by the decay of ^{135}I and ^{135}Xe. A numerical example provided at the end of this chapter describes the accumulation of xenon after reactor shutdown and explains the Fig. 9-9.

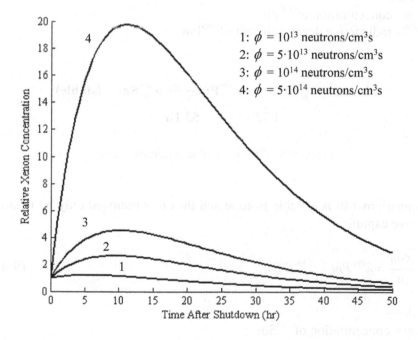

Figure 9-9. ^{135}Xe relative concentration (Xe/Xe_0) after reactor shutdown as a function of neutron flux

2.2   Samarium Poisoning

2.2.1   Production and Removal of ^{149}Sm During Reactor Operation

The fission product poison having the most significant effect on reactor operations, other than ^{135}Xe, is samarium-149 (^{149}Sm). Its effect is significantly different from that of ^{135}Xe. Samarium-149 has a thermal neutron radiative capture cross section of 4.1 × 10^4 b (see Fig. 9-10). It is produced from the decay of the ^{149}Nd which is itself a fission fragment as shown in Fig. 9-10. Since the ^{149}Nd decays fairly rapid in comparison to ^{149}Pm, it can be assumed that ^{149}Pm is produced directly from fission reactions with a yield of y_{Pm}.

The rate of change of its concentration is then determined by the following equation

$$\frac{dPm}{dt} = y_{Pm}\Sigma_f\Phi - \lambda^{Pm}Pm \tag{9-44}$$

where
Pm - concentration of ^{149}Pm
λ^{Pm} - radioactive decay constant of ^{149}Pm

$$\textbf{Fission} \Rightarrow {}^{149}_{60}\textbf{Nd} \xrightarrow{\beta^-} {}^{149}_{61}\textbf{Pm} \xrightarrow{\beta^-} {}^{149}_{62}\textbf{Sm} \quad \textbf{(stable)}$$
$$\textbf{1.72h} \qquad\qquad \textbf{53.1h}$$

Figure 9-10. ^{149}Sm production in thermal reactor

Samarium-149 is a stable isotope and thus it is removed only by neutron radiative capture

$$\frac{dSm}{dt} = \lambda^{Pm}Pm - \sigma_a^{Sm}\Phi Sm \tag{9-45}$$

where
Sm – concentration of ^{149}Sm
λ^{Sm} – radioactive decay constant of ^{149}Sm
$\sigma_a{}^{Sm}$ – thermal neutron absorption cross section of ^{149}Sm

Solving for the equilibrium yields the equilibrium concentrations of the two isotopes

$$Pm_0 = \frac{y_{Pm}\Sigma_f\phi}{\lambda^{Pm}} \tag{9-46}$$

and

$$Sm_0 = \frac{y_{Pm}\Sigma_f}{\sigma_a^{Sm}} \tag{9-47}$$

It can be seen from Eq. (9-47) that the equilibrium concentration of ^{149}Sm is independent of neutron flux and power level. With a change in power level, the equilibrium concentration of ^{149}Sm will go through a transient value and soon return to its original value.

2.2.2 Samarium Poisoning After Reactor Shutdown

After the reactor is shutdown, Eq. (9-45) for ^{149}Sm production reduces to

$$\frac{dSm}{dt} = \lambda^{Pm}Pm \tag{9-48}$$

Solving this simple differential equation gives the relation for Samarium concentration as a function of time after shutdown

$$Sm(t) = Sm_0 + Pm_0\left(1 - e^{-\lambda^{Pm}t_6}\right) \tag{9-49}$$

where Sm_0 and Pm_0 are concentrations at shut down. Because ^{149}Sm is a stable isotope, it cannot be removed by decay, which makes its behavior after reactor shutdown very different from that of ^{135}Xe, as illustrated in Fig. 9-11. The equilibrium is reached after approximately 20 days (500 h). The concentration of ^{149}Sm remains essentially constant during reactor operation (because it is not radioactive). When the reactor is shutdown, its concentration builds up from the decay of the accumulated ^{149}Pm. The build-up after shutdown depends on the power level before reactor shutdown. The concentration of ^{149}Sm does not peak as ^{135}Xe, but instead increases slowly to its maximum value of $Sm_0 + Pm_0$. After shutdown, if the reactor is again operated, ^{149}Sm is burned up and its concentration returns to the equilibrium value. Samarium poisoning is miniscule when compared to Xenon poisoning.

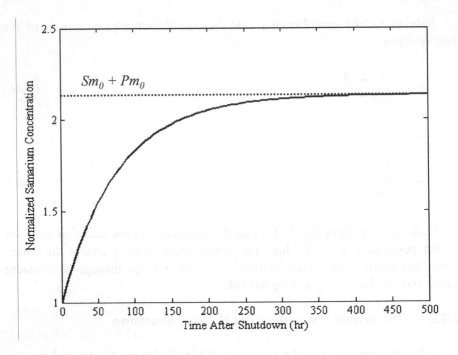

Figure 9-11. ^{149}Sm buildup as a function of time after shutdown

3.   TEMPERATURE EFFECTS ON REACTIVITY

3.1   Temperature Coefficients

The change in reactivity with temperature is described in terms of the temperature coefficient of reactivity. Different materials in a reactor are at different temperatures and produce various effects on reactivity. The temperature in a reactor does not change uniformly. An increase in reactor power would first cause an increase in fuel temperature (the region where power is generated). The coolant and moderator temperatures will change after the heat is transferred from the fuel. Thus, the two main temperature coefficients which are usually specified for thermal reactors are the fuel temperature coefficient and the moderator temperature coefficient. The general definition for the temperature coefficient of reactivity is

$$\alpha_T = \frac{d\rho}{dT} \tag{9-50}$$

or replacing the reactivity (see Chapter 8)

$$\rho = 1 - \frac{1}{k_{eff}} \rightarrow \alpha_T = \frac{1}{k_{eff}^2}\frac{dk_{eff}}{dT} \tag{9-51}$$

Since the multiplication factor is close to unity, Eq. (9-51) is simplified

$$\alpha_T \cong \frac{1}{k_{eff}}\frac{dk_{eff}}{dT} \left[\frac{1}{degree}\right] \tag{9-52}$$

The response of the reactor to a change in temperature depends on the algebraic sign of the temperature coefficient

1. $\alpha_T > 0$: since multiplication factor is always positive value, then dk_{eff}/dT is also positive. In other words, an increase in temperature leads to an increase in neutron population.
- Increase in temperature in a reactor thus increases the reactor power. This will, in turn, increase the temperature more and thus multiplication factor will be increased further which will increase power further and so on. Thus, when the temperature increases the power of a reactor increases and it can be stopped only by outside intervention.
- If temperature is decreased, the multiplication factor will decrease as well. The reactor power will decrease which will reduce temperature further and will reduce the neutron multiplication which will reduce reactor power and temperature and so on. Thus, in this case reactor will shutdown in the absence of external intervention.
2. $\alpha_T < 0$: since multiplication factor is always positive value, then dk_{eff}/dT is negative. In this case, an increase in temperature decreases the neutron multiplication factor.
- An increase in reactor temperature will cause reactor power to drop which will decrease the temperature. This temperature reduction will tend to return the reactor to its original state.
- A decrease in temperature will result in an increase in multiplication factor. Therefore, if temperature is reduced, the power of the reactor will increase and the reactor has a tendency toward its original operating conditions.

Thus, a reactor with a positive temperature coefficient is inherently unstable, while a reactor with a negative temperature coefficient is inherently stable.

3.2   Fuel Temperature Coefficient (Nuclear Doppler Effect)

The fuel reacts immediately to a change in a temperature. The fuel temperature reactivity coefficient is also called the prompt temperature coefficient or the nuclear Doppler coefficient. Fuel temperature promptly responds to a change in reactor power, a negative fuel temperature reactivity coefficient is more important than a negative moderator temperature coefficient. The time for heat generated in the fuel region to be transferred to the moderator is on the order of seconds. When a large positive reactivity insertion occurs, the negative moderator temperature coefficient cannot affect the power in that short time while the fuel temperature coefficient starts adding negative reactivity immediately. Two important nuclides which dominate the nuclear Doppler Effect are ^{238}U and ^{240}Pu.

In a typical light-water moderated low-enriched fuel thermal reactor the fuel temperature reactivity coefficient is negative as a result of the nuclear Doppler Effect (called Doppler broadening). Doppler broadening is caused by an apparent broadening of the resonances (see Chapter 7, Fig. 7-38) due to thermal motion of nuclei, explained as follows

- Stationary nuclei would absorb a neutron of energy E_0.
- If nucleus is moving away from a neutron the velocity and energy of the neutron must be greater than energy E_0 for it to undergo resonance absorption.
- If nucleus is moving toward the neutron, the required neutron energy would be less energy than E_0 in order to be captured by the resonance.
- Increased temperature of the fuel causes nuclei to vibrate more and thus broadening the neutron energy range where they are resonantly absorbed in the fuel region.

If the temperature is increased, the magnitude of the absorption cross section is decreased due to Doppler broadening effect which will increase neutron flux (analogous to the removal of a strong absorber from the core). The number of neutrons absorbed in the resonance region is proportional to the average neutron flux thus the number of resonance absorption increases with temperature. If the parasitic absorptions are increased, the multiplication factor will be reduced which accounts for the negative value of the prompt fuel temperature coefficient. The higher temperatures lead to larger widths of resonances and thus a broader energy region where neutrons can be absorbed.

The nuclear Doppler coefficient is obtained as follows

- By expressing the neutron multiplication as in Eq. (9-23), the resonance escape probability can be obtained in the following form

$$\ln k_{eff} = \ln(\eta \varepsilon f P) + \ln p_{esc} \tag{9-53}$$

- Differentiating with respect to temperature and assuming all parameters to be constant except the resonance escape probability results in a simple expression for the Doppler coefficient

$$\frac{d}{dT}(\ln k_{eff}) = \frac{1}{k_{eff}}\frac{dk_{eff}}{dT} = \frac{d}{dT}(\ln p_{esc}) = \frac{1}{p_{esc}}\frac{d\rho}{dT} \tag{9-54}$$

3.3   The Void Coefficient

The void coefficient of reactivity, α_v, is defined as a rate of change in the reactivity of a water-moderated reactor resulting from a formation of steam bubbles as the power level and temperature increase. The void fraction, x, is defined as the fraction of a given volume which is occupied by voids. If 30% of a volume is occupied by vapor with the rest being occupied by water then $x = 0.30$. The void coefficient of reactivity is defined as

$$\alpha_v = \frac{d\rho}{dx} \tag{9-55}$$

The response of the reactor to a change in void fraction depends on the algebraic sign of the void coefficient

1. $\alpha_v > 0$: an increase in void fraction will increase the reactivity. This will cause the reactor power to rise, which will increase the boiling and void formation. More voids will increase the reactivity and reactor power further which will increase the void fraction and so on. Without external action the reactor power will continue to increase until much of the liquid is boiled and reactor core melts down.
2. $\alpha_v < 0$: an increase in void fraction will reduce the reactivity and thus the reactor power. This condition tends to return the reactor to its initial state. Thus, a negative void coefficient is desirable.

The void coefficient is related to the moderator coefficient (see next Section) because the change in void fraction changes the density of the moderator, or coolant in thermal reactors. In water-cooled and water-moderated reactors the increase in void fraction decreases the reactivity and the void coefficient is negative. In fast reactors cooled with the liquid sodium the effect of void formation is the opposite. Namely, sodium slows down neutrons through inelastic scattering at high energies and absorbs neutrons at low energies. Thus, the removal of sodium causes reduced moderation and the neutron spectrum becomes harder which, in turn, increases the reactivity (the average number of fission neutrons released per neutron absorbed, η, increases with neutron energy for all fissile nuclides in fast reactors). Also, an increase in void formation increases neutron leakage because the density of coolant is reduced. This effect reduces the void coefficient and tends to make it negative. The sign of the void coefficient is determined by the value of these two factors. In large power fast reactors the void formation has a local effect. For example, if a void is formed in the central region of the core, the void coefficient will be positive since neutron leakage has little importance. The leakage becomes more important and reduces the void coefficient if void occurs toward the peripheral region of the core.

3.4   The Moderator Coefficient

3.4.1   Moderator Temperature Coefficient

The moderator temperature coefficient, α_{mod}, determines the rate of change of reactivity with moderator temperature. This coefficient determines the ultimate response of a reactor to fuel and coolant temperature change. It is desirable to have a negative moderator temperature coefficient because of its self-regulating effect. In thermal reactors when the moderator temperature is increased
1. the physical density of the moderator liquid is changed due to thermal expansions, and
2. thermal cross sections change.

The increased temperature of the moderator in water moderated reactors will cause the neutron flux to move toward higher neutron energies. This is an especially promoted effect when absorption cross section does not follow a $1/\upsilon$ dependence. Thus, the presence of, for example, ^{238}U at higher temperatures will increase parasitic absorptions and thus tend to keep the coefficient negative. The change in the neutron spectrum at increased moderator temperature has effect on reactivity which is more pronounced in the presence of poisons such as ^{135}Xe and ^{149}Sm because of their resonances placed at very low neutron energies (around 0.1 eV). The moderator expands at increased temperature which causes a reduction in the density of atoms present; therefore the efficiency of the moderator is reduced.

The magnitude and sign of the moderator temperature coefficient depends on the moderator-to-fuel ratio in such a manner that if

- reactor is under-moderated the coefficient will be negative
- reactor is over-moderated the coefficient will be positive.

3.4.2   Moderator Pressure Coefficient

The moderator pressure coefficient of reactivity is defined as the change in reactivity due to a change in system pressure. The reactivity is changed due to the effect of pressure on the moderator density. When the pressure is increased, the moderator density is increased which, in turn, increases the moderator-to-fuel ratio in the core. In the case of an under-moderated core, the increase in moderator-to-fuel ratio will result in a positive reactivity insertion. In water moderated reactors, this coefficient is much smaller than the temperature coefficient of reactivity.

NUMERICAL EXAMPLE

Xenon and Iodine concentration after shutdown

For the data listed in Table 9-1 calculate the xenon and iodine concentrations as a function of time after shutdown of a ^{235}U thermal reactor which operated at a flux of 10^{15} neutrons/cm^2 s.

The solution was obtained using MATLAB and is shown in Fig. 9-12.

Table 9-1. Data for ^{235}U thermal reactor

Uranium density	19.1 g/cm^3
Xenon-135 fission yield	0.00237
Iodine-135 fission yield	0.0639
Xenon-135 decay constant	$2.09 \cdot 10^{-5}$ s^{-1}
Iodine-135 decay constant	$2.87 \cdot 10^{-5}$ s^{-1}
Xenon-135 absorption cross section	$2.65 \cdot 10^6$ b
Uranium-235 fission cross section	582.2 b

Solution in MATLAB:

```
clear all
lambdaXe = 2.09*10^−5; % s^−1
lambdaI = 2.87*10^−5; % s^−1
gammaI = 0.0639; % I-135 fission yeild
gammaXe = 0.00237; % Xe-135 fission yield
sigmaf = 19.1*6.022e23*(582.2*10^−24)/235; % U-235 fission cm^−1
sigmaaXe = (2.65e6)*10^−24; % Xe-135 absorption cm^2
flux = 10^15; % cm^−2 * s^−1
t = linspace(0,180000);
figure
hold on
% Equilibrium Concentrations
```

```
I0 = gammaI*sigmaf*flux/lambdaI;
Xe0 = (lambdaI*I0 + gammaXe*sigmaf*flux)/(lambdaXe + sigmaaXe*flux);
for i = 1:100
% Build-up After Shutdown
    I(i) = I0*exp(-lambdaI*t(i));
    Xe(i)=Xe0*exp(-lambdaXe*t(i))+(lambdaI*I0/(lambdaI-
        lambdaXe))*(exp(-lambdaXe*t(i)) - exp(-lambdaI*t(i)));
end
plot(t/3600,I,'k')
hold on
plot(t/3600,Xe,'k:')
xlabel('Time After Shutdown (hr)')
ylabel('Concentration (cm^-^3)')
legend('Iodine-135','Xenon-135')
```

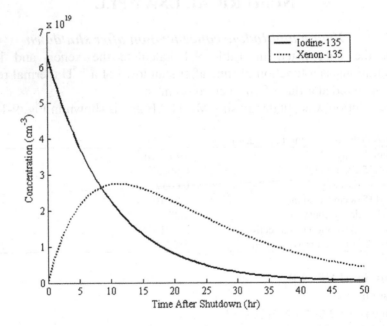

Figure 9-12. ^{135}Xe and ^{135}I concentrations after shutdown of a ^{235}U thermal reactor

PROBLEMS

9.1 Plot the differential and integral control rod worth curves if the differential rod worth data are given as follows:

Fractional distance from the bottom of the core	Inserted reactivity
0 – 0.125	0.1
0.125 – 0.25	0.2
0.25 – 0.375	0.4
0.375 – 0.5	0.6
0.5 – 0.625	0.6
0.625 – 0.75	0.4
0.75 – 0.875	0.2
0.875 – 1	0.1

9.2 Explain the role of soluble poisons (chemical shim) in thermal reactors.

9.3 In order to control and minimize the corrosion in the reactor coolant system the pH of the coolant is monitored. In nuclear reactors that do not use chemical shim pH is maintained at values high as 10. In reactor systems that use chemical shim (boric acid) how would the pH limit change?

9.4 Chemical shim in thermal reactors primarily affects the thermal (fuel) utilization factor. Derive the relation for the chemical shim worth.

9.5 A common pair of fragments from ^{235}U fission is xenon and strontium. Write the reaction and calculate energy released per this fission event; investigate the decay scheme of xenon and strontium; discuss how xenon is removed from the reactor; discuss the equilibrium level of xenon.

9.6 Describe the effect of the poison material on reactivity change.

9.7 Sketch the behavior of xenon poisoning.

9.8 Discuss the loss and production of xenon on reactor start-up and on power decrease from steady state to full power.

9.9 Discuss the production and removal of samarium.

9.10 Discuss the samarium response to reactor shutdown.

9.11 Solve the differential equation that describes the xenon concentration change after reactor shutdown.

9.12 Derive the relation to obtain the time needed to achieve the maximum concentration of xenon after the reactor shutdown.

9.13 Define the temperature coefficients.

9.14 For the moderator coefficient of −15 pcm/K calculate the reactivity change that results from a temperature decrease 3.5 K.

9.15 How will macroscopic cross section of a moderator change if with increased temperature its density decreases? How will thermal utilization factor change?

9.16 Discuss the fuel temperature coefficient and why it is negative?

Appendix 1: World Wide Web Sources on Atomic and Nuclear Data

Periodic table of the elements
- First ionization potential:
  http://web.mit.edu/3.091/www/pt/pert9.html
- Atomic and chemical characteristics of elements:
  http://pearl1.lanl.gov/periodic/default.htm
- Comprehensive set of data:
  http://www.chemistrycoach.com/periodic_tables.htm

Table of nuclides
- Cross section plots and fundamental characteristics of nuclides:
  http://atom.kaeri.re.kr/
- Nuclear physics data: http://physics.nist.gov/PhysRefData/
- Ionization potentials:
  http://environmentalchemistry.com/yogi/periodic/1stionization.html

Electron and photon attenuation data: http://atom.kaeri.re.kr/ex.html

Physical constants: http://physics.nist.gov/PhysRefData/

Atomic and molecular spectroscopic data:
http://physics.nist.gov/PhysRefData/

X-ray and γ ray data: http://physics.nist.gov/PhysRefData/

Stopping power and range tables for electrons, protons and helium ions:

http://physics.nist.gov/PhysRefData/Star/Text/contents.html

National nuclear data center: http://www.nndc.bnl.gov/index.jsp
- Nuclear structure and decay database:
  http://www.nndc.bnl.gov/databases/databases.html#structuredecay
- Nuclear reactions databases:
  http://www.nndc.bnl.gov/databases/databases.html#reaction

Appendix 2: Atomic and Nuclear Constants

Fundamental Constants

Quantity	Symbol	Value	Unit
Atomic mass unit	amu or u	1.660538×10^{-27}	kg
		931.481	MeV
Avogadro's number	N_A	6.02217×10^{23}	mole^{-1}
Boltzmann's constant	k	1.38062×10^{-23}	J/K
Electron rest mass	m_e	9.109382×10^{-31}	kg
		5.48593×10^{-4}	amu
		0.510998	MeV
Elementary charge	e	$1.60217653 \times 10^{-19}$	C
Neutron rest mass	m_n	1.674927×10^{-27}	kg
		1.008665	amu
		939.565	MeV
Newtonian gravitational constant	G	6.6742×10^{-11}	$\text{m}^3/\text{kg s}^2$
Planck's constant	h	6.626069×10^{-34}	Js
		4.135667×10^{-15}	eV s
Proton rest mass	m_p	1.672621×10^{-27}	kg
		1.007276	amu
		938.272	MeV
Speed of light	c	$2.99792458 \times 10^{-10}$	cm/s
Stefan–Boltzmann constant	σ	5.670400×10^{-8}	$\text{W m}^{-2}\,\text{K}^{-4}$

Atomic and Nuclear Constants

Quantity	Symbol	Value	Unit
Bohr radius	$a_0 = \hbar^2 / ke^2 m$	0.5291771	nm
Classical electron radius	r_e	2.817940×10^{-15}	m
Compton wavelength	λ_C	2.426310×10^{-12}	m
Rydberg constant	$R = ke^2 /(2a_0)(hc)$	10973731.568	m^{-1}
Rydberg energy	$R = hcR$	2.179872×10^{-18}	J
		13.6	eV

Appendix 3: Prefixes

Factor	Prefix	Symbol
10^{18}	exa	E
10^{15}	peta	P
10^{12}	tera	T
10^{9}	giga	G
10^{6}	mega	M
10^{3}	kilo	k
10^{2}	hecto	h
10^{1}	deka	da
10^{-1}	deci	d
10^{-2}	centi	c
10^{-3}	milli	m
10^{-6}	micro	μ
10^{-9}	nano	n
10^{-12}	pico	p
10^{-15}	femto	f
10^{-18}	atto	a

Appendix 4: Units and Conversion Factors

Angle

Unit	Symbol	Value
Radian	rad	0.01745
Degree	°	**1**
Minute	'	60
Second	"	3,600

Energy

Unit	Symbol	Value
Joule	**J**	**1**
Erg	erg	10^7
Watt second	Ws	1
Kilowatt hour	kWh	2.7778×10^{-7}
Mega electron volt	MeV	6.242×10^{12}
British thermal unit	Btu	9.478×10^{-4}

Length/Distance

Unit	Symbol	Value
Angstrom	A	10^{10}
Nanometer	nm	10^9
Micrometer	μm	10^6
Millimeter	mm	10^3
Centimeter	cm	10^2
Meter	**m**	**1**
Kilometer	km	10^{-3}
Inch	in (")	39.37008
Foot	ft (')	3.28084
Yard	yd	1.09361
mile	mi	6.2137×10^{-4}

Mass

Unit	Symbol	Value
Milligram	mg	10^6
Gram	g	10^3
Kilogram	**kg**	**1**
Ounce	oz	35.274
Pound	lb	2.2046
Tonne (metric)	t	10^{-3}

Temperature

Unit	Symbol	Value
Fahrenheit	F	$C \times (9/5) + 32$
Celsius	**C**	**C**
Kelvin	K	$C + 273.15$

Time

Unit	Symbol	Value
Second	s or sec	3.1536×10^7
Minute	m or min	5.256×10^5
Hour	h or hr	8,760
Day	da	365
Week	wk	52.14286
Month	mo	11.99203
Year	**yr or a**	**1**

Appendix 5: Neutron Angular Momentum and Spin

Angular momentum is a product of mass, size and speed of rotation

Angular momentum = (mass) × (radius) × (rotation speed)

For example, the orbital angular momentum (Fig. A.5-1) describes the motion of the Earth around the Sun, while the intrinsic angular momentum of the Earth describes its rotation about the axis.

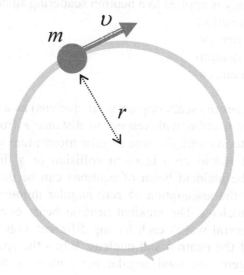

Figure A.5-1. Angular momentum

The experiments show that most subatomic particles have an eternal, built-in (or intrinsic) angular momentum, which is not due to their motion through space. Neutron and proton possess such an intrinsic angular momentum (also called the spin) of the magnitude equal to $\frac{1}{2}\hbar$. An electron standing at rest has an angular momentum of $\frac{1}{2}\hbar$. The quantum mechanical property of the spin of such a magnitude is that its orientation has only two states defined as *parallel* or *anti-parallel* in respect to the direction of reference. Therefore, the components of the spin along given direction are $+\frac{1}{2}\hbar$ or $-\frac{1}{2}\hbar$.

The combined intrinsic spins of neutrons and protons define the angular momentum of a nucleus, l, in addition to nucleus orbital angular momentum. For even-A nuclei the total angular momentum of a nucleus is an integer multiple of $\hbar$. For odd-A nuclei the total angular momentum is an odd multiple of $\frac{1}{2}\hbar$.

The angular momentum of the excited states of a nucleus can be different from the angular momentum of the ground state of that nucleus. The angular momentum is quantized such that there are in total $(2l + 1)$ possible orientations in space with respect to a given axis. The magnitude of the angular momentum components in any of these states is a multiple of Planck constant, $m\hbar$, where m is the magnetic quantum number taking values $-l$, $-l+1$, ..., l. Even-A nuclei do not have angular momentum, i.e., $l = 0$. This lowest energy state with angular momentum quantum number $l = 0$ is called the *s*-state; $l = 1$ corresponds to *p*-state, $l = 2$ to *d*-state, $l = 3$ to *f*-state, etc. This is explained in Chapter 2.

Same terminology is applied to a neutron scattering such that

$l = 0$ is *s*-wave neutron

$l = 1$ is *p*-wave neutron

$l = 2$ is *d*-wave neutron

$l = 3$ is *f*-wave neutron

etc.

Since neutron sees the scattering center (a nucleus) as a point, the angular momentum of that neutron with respect to distance r from the nucleus is $m\upsilon r$. Neutron scattering with the zero angular momentum would imply $r = 0$ and therefore that would be a head-on collision or a direct hit [Ott and Bezella, 1989]. The incident beam of neutrons can be described as simple plane wave under the assumption of zero angular momentum for both the neutron and the nucleus. The incident neutron beam consists always of a combination of several waves each having different angular momenta, i.e., spin orientation of the neutron and nucleus. When the spins of the neutron and nucleus are zero, the total angular momentum is determined by the orbital angular momentum of neutron.

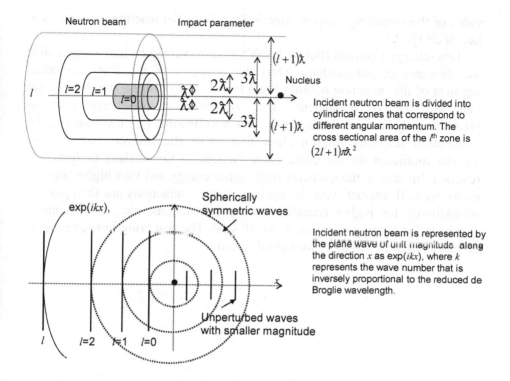

*Figure A.5-2. s-*wave neutron scattering

The plane wave neutron beam interaction with the point nucleus is illustrated in Fig. A.5-2. The incident neutron beam is divided into cylindrical zones. The inner cylindrical zone contains neutrons of zero angular momentum and has the impact parameter less than neutron wavelength, λ. The next cylindrical zones contain neutrons of higher angular momentum with impact parameter between λ and 2λ. The neutron angular momentum in the lth cylindrical zone is between $lm\upsilon\lambda$ and $(l+1)m\upsilon\lambda$. Since the angular momentum is quantized, the range becomes $l\hbar$ to $(l+1)\hbar$. The cross sectional area of the last cylindrical zone (segment) is $(2l+1)\pi\lambda^2$. This is at the same time the upper limit for the reaction (*absorption*) cross section for the incoming channel. Note that whenever the compound nucleus is formed from the incoming channel point of view the incoming neutron is absorbed, regardless of the outgoing channel that can define the interaction as scattering or absorption. Any absorption of the plane wave results in its weakening. Scattering of a neutron without absorption therefore takes place only if the outgoing plane wave is not weakened in its magnitude but is shifted in phase. The incoming and outgoing waves are coherent in case of elastic scattering. The maximum value of the scattering cross section is four times the reaction cross section, i.e., $4(2l+1)\pi \lambda^2$.

Low-energy neutrons (below 1 keV) have wavelengths much larger than the diameter of the nucleus. Thus the diameter of the first cylindrical segment of the incoming neutron beam (Fig. A.5-2) with $l = 0$ will be larger than the size of a nucleus. That is why the low-energy neutrons predominantly interact through the **s-wave collisions**, i.e., the zero angular momentum neutrons enter nucleus first while the neutrons with higher angular momenta do not come close enough to the nucleus to initiate a reaction. In general the neutrons with higher energy and thus higher angular momenta will interact with the nucleus. Such interactions are then *p*-wave interactions; for higher energies *d*-wave interactions, etc., are becoming dominant for neutron energies above 10 keV. The scattering in such cases is no more isotropic but more forward directed.

Appendix 6: Gradient

By definition, the gradient (Fig. A.6-1 and Fig. A.6-2) of a straight line is an indication of how steep that line is and may be calculated as follows:

Gradient = change in Y/change in X

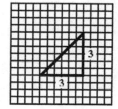 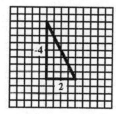 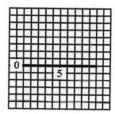

Figure A.6-1. Examples of a gradient: starting from the left end of the line, going to the right the gradient is *positive*, up is positive, and down is negative, across to the left is also negative

From Fig. A.6-2, it follows that the gradient can be defined as

$$\text{Gradient (slope)} = \frac{\Delta f}{\Delta x}$$

Δf and Δx represent finite quantities; for infinitesimal quantities, it follows that the gradient is represented as

$$\lim_{\Delta \to 0} \frac{\Delta f}{\Delta x} = \frac{df}{dx}$$

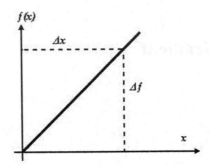

Figure A.6-2. Definition of a gradient

Appendix 7: k_{eff} in Two-Group Diffusion Theory

Following Fig. 8-14 we shall derive k_{eff} using the two-group diffusion approach and discuss the coefficients that appear in the six-factor formula. Starting from

$$k_{eff} = \frac{v_1 \Sigma_{f1} \Phi_1(\vec{r}) + v_2 \Sigma_{f2} \Phi_2(\vec{r})}{-D_1 \nabla^2 \Phi_1(\vec{r}) - D_2 \nabla^2 \Phi_2(\vec{r}) + \Sigma_{a1} \Phi_1(\vec{r}) + \Sigma_{a2} \Phi_2(\vec{r})} \qquad (8\text{-}140)$$

it can be re-written as follows

$$-D_1 \nabla^2 \Phi_1(\vec{r}) - D_2 \nabla^2 \Phi_2(\vec{r}) + \Sigma_{a1} \Phi_1(\vec{r}) + \Sigma_{a2} \Phi_2(\vec{r}) =$$
$$\frac{1}{k_{eff}} \left(v_1 \Sigma_{f1} \Phi_1(\vec{r}) + v_2 \Sigma_{f2} \Phi_2(\vec{r}) \right) \qquad (A.7\text{-}1)$$

Neglecting the in-group scattering, the equation describing the thermal neutrons becomes

$$-D_2 \nabla^2 \Phi_2(\vec{r}) + \Sigma_{a2} \Phi_2(\vec{r}) = \Sigma_{s1 \to 2} \Phi_1(\vec{r}) \qquad (A.7\text{-}2)$$

By subtracting these two equations and defining that the neutrons are removed from a group through fission and capture and scattering out of that energy group, so that $\Sigma_{r1} = \Sigma_{a1} + \Sigma_{s1 \to 2} = \Sigma_{f1} + \Sigma_{\gamma 1} + \Sigma_{s1 \to 2}$

$$- D_1 \nabla^2 \Phi_1 \left(\vec{r} \right) + \left(\Sigma_{a1} + \Sigma_{s1 \to 2} \right) \Phi_1 \left(\vec{r} \right) =$$
$$\frac{1}{k_{eff}} \left(\nu_1 \Sigma_{f1} \Phi_1 \left(\vec{r} \right) + \nu_2 \Sigma_{f2} \Phi_2 \left(\vec{r} \right) \right) \tag{A.7-3}$$

For the critical bare reactor the following relations are valid

$$\nabla^2 \Phi_1 \left(\vec{r} \right) + B^2 \Phi_1 \left(\vec{r} \right) = 0 \quad \nabla^2 \Phi_2 \left(\vec{r} \right) + B^2 \Phi_2 \left(\vec{r} \right) = 0 \quad \Rightarrow$$
$$\nabla^2 \Phi_1 \left(\vec{r} \right) = -B^2 \Phi_1 \left(\vec{r} \right) \quad \nabla^2 \Phi_2 \left(\vec{r} \right) = -B^2 \Phi_2 \left(\vec{r} \right) \tag{A.7-4}$$

Combining last equations

$$D_1 B^2 \Phi_1 \left(\vec{r} \right) + \Sigma_{r1} \Phi_1 \left(\vec{r} \right) - \frac{1}{k_{eff}} \left(\nu_1 \Sigma_{f1} \Phi_1 \left(\vec{r} \right) + \nu_2 \Sigma_{f2} \Phi_2 \left(\vec{r} \right) \right) = 0$$
$$D_2 B^2 \Phi_2 \left(\vec{r} \right) + \Sigma_{a2} \Phi_2 \left(\vec{r} \right) - \Sigma_{s1 \to 2} \Phi_1 \left(\vec{r} \right) = 0 \tag{A.7-5}$$

The solution of these two equations is obtained as follows

det = 0

$$\begin{vmatrix} D_1 B^2 + \Sigma_{r1} - \dfrac{1}{k_{eff}} \left(\nu_1 \Sigma_{f1} \right) & -\dfrac{1}{k_{eff}} \left(\nu_2 \Sigma_{f2} \right) \\ -\Sigma_{s1 \to 2} & D_2 B^2 + \Sigma_{a2} \end{vmatrix} = 0 \tag{A.7-6}$$

When solved for k_{eff}

$$k_{eff} = \frac{\nu_1 \Sigma_{f1}}{D_1 B^2 + \Sigma_{r1}} + \frac{\Sigma_{s1 \to 2}}{D_1 B^2 + \Sigma_{r1}} \frac{\nu_2 \Sigma_{f2}}{D_2 B^2 + \Sigma_{a2}} \tag{A.7-7}$$

Equation (A.7-7) gives k_{eff}

$$k_{eff} = k_{eff1} + k_{eff2} \tag{A.7-8}$$
[fast + thermal neutron contribution]

Knowing that the fast fission factor is defined as

$$\varepsilon = \frac{\text{number of fast} + \text{number of thermal fission neutrons}}{\text{number of thermal fission neutrons}} \tag{A.7-9}$$

this equation can be also written as

$$\varepsilon = \frac{k_{eff1} + k_{eff2}}{k_{eff2}} = 1 + \frac{k_{eff1}}{k_{eff2}} \qquad (A.7\text{-}10)$$

The second term in Eq. (A.7-10) explains the increase of fast fission neutrons. Also, Eq. (A.7-10) gives the following

$$\varepsilon k_{eff2} = k_{eff1} + k_{eff2} \equiv k_{eff} \qquad (A.7\text{-}11)$$

From Eq. (A.7-7)

$$
\begin{aligned}
k_{eff2} &= \frac{\Sigma_{s1 \rightarrow 2}}{D_1 B^2 + \Sigma_{r1}} \frac{\nu_2 \Sigma_{f2}}{D_2 B^2 + \Sigma_{a2}} = \\
&\frac{\Sigma_{s1 \rightarrow 2}}{\Sigma_{r1}} \frac{\Sigma_{r1}}{D_1 B^2 + \Sigma_{r1}} \frac{\nu_2 \Sigma_{f2}}{\Sigma_a^{fuel}} \frac{\Sigma_a^{fuel}}{\Sigma_{a2}} \frac{\Sigma_{a2}}{D_2 B^2 + \Sigma_{a2}} = \\
&p \cdot P_{fast} \cdot \eta \cdot f \cdot P_{thermal}
\end{aligned}
\qquad (A.7\text{-}12)
$$

Finally, we obtain the six-factor formula

$$
\begin{aligned}
k_{eff2} &= p \cdot P_{fast} \cdot \eta \cdot f \cdot P_{thermal} \\
\varepsilon k_{eff2} &= k_{eff1} + k_{eff2} \equiv k_{eff} \\
k_{eff} &= \varepsilon \cdot p \cdot P_{fast} \cdot \eta \cdot f \cdot P_{thermal}
\end{aligned}
\qquad (A.7\text{-}13)
$$

Bibliography

Allen BJ, McGregor BJ, and Martin RF. (1989). Neutron capture therapy with gadolinium-157. Strahlenther Onkol 165, 156–158.

Born M. (1969). Atomic Physics. Dover Publications Inc, New York.

Clark M, Jr., Hansen RJ. (1964). Numerical Methods in Reactor Analysis. Academic Press, New York.

Cohen RE, Lide DR, Trigg GL. (2003). AIP Physics desk Reference. 3rd ed., Springer, AIP Press, New York.

Connolly TJ. (1978). Foundations of Nuclear Engineering, John Wiley & Sons, New York.

Cottingham WN, Greenwood DA. (2001). An Introduction to Nuclear Physics. Cambridge University Press, Cambridge.

Culbertson CN, Jevremovic T. (2003). Computational assessment of improved cell-kill by gadolinium-supplemented BNCT. Phys Med Biol 48, 3943–3959.

Duderstadt JJ, Hamilton LJ. (1976). Nuclear Reactor Analysis. John Wiley & Sons Inc. New York/London/Sydney/Toronto.

Emfietzoglou D, et al. (2003) Monte Carlo simulation of the energy loss of low-energy electrons in liquid water. Phys Med Biol 48, 2355–2371.

Feyman R. (1963). Six easy Pieces. Helix Books, Massachusetts

Feyman R. (1965). The Character of Physical Law. The MIT Press, Cambridge.

Foster AR, Wright RL. (1968). Basic Nuclear Engineering. Allyn and Bacon Inc, Massachusetts.

Gabel D, Foster S, Fairchild R. (1987). The Monte Carlo Simulation of the Biological Effects of the 10B(n,a)7Li Reaction in Cells and Tissue and Its Implications for Boron Neutron Capture Therapy. Radiat Res 111, 14–25.

Glasstone S, Sesonske A. (1994). Nuclear Reactor Engineering: Reactor Design Basics. 4th ed., vol. 1. Chapman & Hall, New York.

Griffiths DJ. (1995). Introduction to Quantum Mechanics. Prentice Hall, New Jersey.

Hatanaka H, Sano K, Yasukocki H. (1992). Clinical results of Boron Neutron capture therapy, Progress in Neutron Capture Therapy for Cancer. Plenum Press, New York, 561–568.

Heisenberg W. (1953). Nuclear Physics. Philosophical Library, New York.

Henley EM, Garcia A. (2007). Subatomic Physics. Third Edition. World Scientific.

ICRP, International Commission on Radiological Protection. (1977). Recommendations of the International Commission on Radiological Protection, Publication No. 26. Pergamon Press, Oxford and New York.

ICRP, International Commission on Radiological Protection. (1982). Cost-Benefit Analysis in the Optimization of Radiation Protection, Publication No. 37. Pergamon Press, Oxford and New York.

Jevremovic T, et al. (April 25–29, 2004). AGENT Code Open – Architecture Analysis and Configuration of Research Reactors – Neutron Transport Modeling with Numerical Examples, PHYSOR 2004 – The Physics of Fuel Cycles and Advanced Nuclear Systems: Global Developments. Chicago, Illinois.

Joneja OP, et al. (Nov 2003). Comparison of Monte Carlo simulations of photon/electron dosimetry in microscale applications. Australas Phys Eng Sci Med 26(2): 63–69.

Kagehira K, Sakurai Y, Kobayashi T, Kanda K, Akine Y. (1994). Physical dose evaluation on gadolinium neutron capture therapy. Annu Rep Res Reactor Inst Kyoto Jpn Kyoto Univ 27, 42–56.

Krane KS. (1996). Modern Physics. 2nd ed. John Wiley & Sons, New York.

Lamarsh JR. (1983). Introduction to Nuclear Engineering. 2nd ed. Addison-Wesley, Massachusetts.

Lapp RE, Andrews HL. (1954). Nuclear Radiation Physics. 2nd ed. Prentice-Hall, New York.

Maeda T, Sakurai Y, Kobayashi T, Kanda K, Akine Y. (1995). Calculation of the electron energy spectrum for Gadolinium capture therapy. Annu Rep Res Reactor Inst Kyoto Univ 28, 39–43.

Marcel V, Bartelink H. (2000). Radiation-induced Apoptosis. Cell Tissue Res 33, 133–142.

Masunaga S, Ono K, Suzuki M, Sakurai Y, Takagaki M, Kobayashhi T, Kinashi Y, Akaboshi M. (1999). Repair of potentially lethal damage by total and quiescent cells in solid tumors following a neutron capture reaction. J Cancer Res Clin Oncol 609(125), 614.

Mayo RM. (1998). Introduction to Nuclear Concepts for Engineers. American Nuclear Society, Illinois.

Miller G, Hertel N, Wehring B, Horton J. (1993). Gadolinium neutron capture therapy. Nucl Technol 103, 320–330.

Miller JH, et al. (2000). Monte Carlo simulation of single-cell irradiation by an electron microbeam. Radiat Environ Biophys 39, 173–177.

Mundy D, Harb W, Jevremovic T. (2006). Radiation binary targeted therapy for HER-2 positive breast cancers: Assumptions, theoretical assessment and future directions. Phys Med Biol 51, 1377–1391 [Top 10 paper in 2006 selected by the Editor]

Murray RL. (1954). Introduction to Nuclear Engineering. Prentice-Hall, New York.

Murray RL. (1957). Nuclear Reactor Physics. Prentice-Hall, New Jersey.

Nakanishi A, et al. (1999). Toward a cancer therapy with Boron-Rich oligomeric phosphate diesters that target the cell nucleus. Proc Natl Acad Sci USA 96(1), 238–241

National Council on Radiation Protection. (1989). Guidance on Radiation Received in Space Activities, NCRP Report #98, July 31, 1989.

Ono K, Masunaga SI, Kinashi Y, Takagaki M, Akaboshi M, Kobayashi T, Akuta K. (1996) Radiobiological evidence suggesting hetergeneous microdistribution of boron compounds in tumors: Its relation to quiescent cell population and tumor cure in neutron capture therapy. Int J Radiat Oncol Biol Phys 34(5) 1081–1086.

Ono K, Masunaga SI, Suzuki M, Kinashi Y, Takagaki M, Akabosahi M. (1999). The combined effect of boronophenylalanine and borocapate in boron neutron capture therapy for SCVII tumors in mice. Int J Radiat Oncol Biol Phys 43(2), 431–436.

Ott KO, Bezella WA. (1989). Introductory Nuclear Reactor Statics. Rev. ed. American Nuclear Society, Illinois.

Ott KO, Neuhold RJ. (1985). Introductory Nuclear Reactor Dynamics. American Nuclear Society, Illinois.

Phillips AC, (2003). Introduction to Quantum Mechanics. John Wiley & Sons Ltd.

Rydin RA. (1977). Nuclear Reactor Theory and Design. Physical Biological Sciences Ltd., Virginia.

Shults KJ, Faw RE. (2002). Fundamentals of Nuclear Science and Engineering. Marcel Dekker Inc., New York.

Silver BL. (1998). The Ascent of Science. Oxford University Press, New York.

Singleterry RC, Jr., Thibeault SA. (2000). Materials for Low-Energy Neutron Radiation Shielding, NASA/TP-2000-210281

Taylor JR, Zafiratos CD, Dubson MA. (2004). Modern Physics for Scientists and Engineers. 2nd ed., Prentice Hall.

Tokuyye K, Tokita N, Akine Y, Nakayama H, Sakurai Y, Kobayashi T, Kanda K. (2000). Comparison of radiation effects of Gadolinium and boron neutron capture reactions, Strahlenther Onkol 176, 81–3.

Trefil J. (2003). The Nature of Science. Houghton Mifflin, Massachusetts.

Wheeler F, Nigg D, Capala J, Watkins P, Vroegindeweij C, Auterinen I, Seppala T. (1999). Boron neutron capture therapy (BNCT): Implications of neutron beam and boron compound characteristics, Med. Phys. 27(7), p.1237

Wilson JW, et al. (1997). Shielding Strategies for Human Space Exploration. NASA Johnson Space Center, Houston, Texas.

Wilson WE, et al. (2001). Microdosimetry of a 25keV electron microbeam, Radiat Res 155, 89–94.

Index